Greening Urban Spaces: A Healthy Community Design

Greening Urban Spaces: A Healthy Community Design

Editors

Hongxiao Liu
Tong Wu
Yuan Li

MDPI • Basel • Beijing • Wuhan • Barcelona • Belgrade • Manchester • Tokyo • Cluj • Tianjin

Editors
Hongxiao Liu
South China Botanical Garden
Chinese Academy of Sciences
Guangzhou
China

Tong Wu
The Natural Capital Project
Stanford University
Stanford
United States

Yuan Li
Northwest Land and Resource Research Center
Shaanxi Normal University
Xi'an
China

Editorial Office
MDPI
St. Alban-Anlage 66
4052 Basel, Switzerland

This is a reprint of articles from the Special Issue published online in the open access journal *International Journal of Environmental Research and Public Health* (ISSN 1660-4601) (available at: www.mdpi.com/journal/ijerph/special_issues/healthy_community_design).

For citation purposes, cite each article independently as indicated on the article page online and as indicated below:

LastName, A.A.; LastName, B.B.; LastName, C.C. Article Title. *Journal Name* **Year**, *Volume Number*, Page Range.

ISBN 978-3-0365-8351-8 (Hbk)
ISBN 978-3-0365-8350-1 (PDF)

Contents

About the Editors

Hongxiao Liu

Liu Hongxiao is an associated researcher in South China Botanical Garden, Chinese Academy of Sciences. Her research focuses on urban ecosystem service assessment as well as greenspace and human mental health. She is leading projects on urban cultural ecosystem service assessment and mental health service modelling, integrating such information in green infrastructure planning. Currently, she also works on urban biodiversity restoration and ecosystem service improvement project. She obtained her PhD at the eco-environmental research center, Chinese Academy of Sciences.

Tong Wu

Tong Wu is a Senior Scientist for the Natural Capital Project at Stanford University. His research is centered on the challenges and prospects of sustainable development in China and the Global South. Topics that he has worked on include the mainstreaming of ecosystem services in management and planning, the role of nature in improving public health outcomes, and strategies for climate change mitigation and adaptation. Currently, a major research focus is the development of gross ecosystem product (GEP) as a policy instrument for aligning environmental and socio-economic imperatives. Tong received his undergraduate degree from the University of California, Berkeley and his doctorate from Arizona State University.

Yuan Li

Yuan Li is an associate researcher at the Northwest Land and Resources Research Center of Shaanxi Normal University. He received his PhD degree in Water and Soil Conservation and Desertification Control in 2016. His research interests mainly focus on the harmonious coexistence between humans and nature and sustainable development, including rural revitalization, urban green space, ecological restoration, and efficient utilization of agricultural resources. The National Natural Science Foundation of China (41807041), Shaanxi Key R&D Program (2022NY-191), and other funding agencies have supported his research. In addition, he has published many works in high-impact journals and published books and book chapters with internationally renowned publishing houses. He is also a young scholar involved in the ecological protection of the Qinling Mountains in Shaanxi Province and a science and technology expert in Xi'an City.

Preface to "Greening Urban Spaces: A Healthy Community Design"

The focus of this reprint is to draw attention to Greening Urban Spaces. Cities play a central role in advancing economic and social developments and addressing the challenges that humanity face today. Although cities provide access to better health services, urban life is also associated with factors that are deleterious to human health, such as increased stress, mental fatigue, pollution, as well as sedentary lifestyles and a disconnection to the natural environment. There is a growing consensus across many academic fields and health promotion policy areas that the provision of accessible and high-quality greenspace is a vital element in the effort against the adverse health effects of urbanization. However, there are still research gaps that need to be addressed before the health benefits of greenspaces can be fully integrated into practice.

Hongxiao Liu, Tong Wu, and Yuan Li
Editors

International Journal of
*Environmental Research
and Public Health*

MDPI

Article

The Impact of the Implementation of International Law on Marine Environmental Protection on International Public Health Driven by Multi-Source Network Comment Mining

Anqi Yang * and Shudong Yang

School of Law, Chongqing University, Chongqing 400044, China
* Correspondence: 20172301006@cqu.edu.cn

Abstract: With the increase of people's living space, global warming caused by the decrease of greening urban spaces and the serious decline of greenspace quality has led to extreme weather events and coastal erosion, which has become the biggest threat to the ocean and has also led to the occurrence of international public safety incidents. Therefore, it is of great practical significance to explore the tense relationship between the current marine environmental protection and global public safety for the development of an international healthy community. Firstly, this paper discusses the influence of implementing the international law of marine environmental protection on global public health after the reduction of green urban space and the decline of green space quality. Secondly, K-means and discrete particle swarm optimization algorithms are introduced and the particle swarm optimization-K-means clustering (PSO-K-means) algorithm is designed to screen and deal with the mapping relationship between latent variables and word sets about the impact of implementing the international marine ecological protection law on the international public health community in network data information. Moreover, the influencing factors are clustered and the scenarios are evaluated. The results show that the clustering analysis of the marine environment can promote the clustering of marine characteristic words. Meanwhile, the PSO-K-means algorithm can effectively cluster vulnerability data information. When the threshold is 0.45, the estimated recall rate of the corresponding model is 88.75%. Therefore, the following measures have been formulated, that is, increasing greening urban spaces and enhancing the quality of green space to enhance the protection of marine environment, which has practical reference value for realizing the protection of marine environment and the sustainable development of marine water resources and land resources.

Keywords: international public safety; international law of marine environmental protection; greening urban spaces; green space quality; healthy community; PSO-K-means

Citation: Yang, A.; Yang, S. The Impact of the Implementation of International Law on Marine Environmental Protection on International Public Health Driven by Multi-Source Network Comment Mining. *Int. J. Environ. Res. Public Health* **2023**, *20*, 5130. https://doi.org/10.3390/ijerph20065130

Academic Editors: Hongxiao Liu, Tong Wu and Yuan Li

Received: 4 January 2023
Revised: 10 March 2023
Accepted: 10 March 2023
Published: 14 March 2023

1. Introduction

In recent years, under the trend of the rapid expansion of people's activity space, the green area of urban space has been seriously reduced and the quality of green space has decreased significantly. The resulting global warming has even led to a dramatic deterioration of the marine environment. In the era of big data, online comments profoundly affect individuals, groups, societies, and countries [1,2]. Public opinion formed through online comments has also led people to pay attention to the impact of marine pollution on human survival and development. The public constantly reflects on the relationship between human beings and the environment through the news about the situation to gradually form a substantial public opinion pressure and supervision role [3]. People expect to find the reasons for the increase in environmental pollution and the frequent occurrence of pollution incidents to analyze how to prevent environmental pollution [4]. In addition, the local division of data can be realized through the clustering algorithm of unsupervised learning. After clustering the data set, the clusters are analyzed separately to obtain more detailed results. Cluster analysis can be used in the data preprocessing stage to focus on complex data structures with specific algorithms to standardize complex data structures. Therefore, applying intelligent analysis methods to the analysis of global warming and marine environmental degradation, timely and accurately grasping relevant online commentary, and formulating and implementing environmental protection laws and policies have become the focus of many scholars in related fields.

Agglomerative hierarchical clustering (AHC) has been applied to opinion analysis several times in recent years. The AHC method is generally a small to large method. It starts by dividing each piece of data into a complete cluster. Then, according to the similarity of the data, the nearest clusters are gradually merged into a cluster until all clusters have merged into one cluster or some established conditions have been reached [5]. The K-means algorithm, density-based spatial clustering of applications with noise algorithm, and ordering points to identify the clustering structure algorithm of the AHC algorithm are preferred algorithms in the analysis methods [6,7]. Among them, density-based clustering has good effect and fast clustering speed. For each attribute divided, a single scan determines the number of mesh cells and grid cells per object. The K-means clustering algorithm is the most commonly used of all clustering algorithms. After artificially setting the initial K value, K objects are randomly selected to represent the center of K clusters. Next, the distance of other objects from the K centers is calculated and assigned to the nearest cluster. Then, the K-means algorithm iteratively improves the intra-cluster variation, calculates a new mean for each cluster, and takes the updated mean as the new cluster center point. All objects are reassigned to continue to iterate until they stabilize. The clustering algorithm has attracted much attention from experts and scholars because of the simplicity of the algorithm presented in the application, its easy implementation, and the good results of the application [8]. Currently, cluster analysis methods are widely used in medical diagnoses, image processing, information retrieval, statistics, biology, and other fields [9–12].

Therefore, the classical K-means algorithm in the AHC analysis method is selected to combine with the particle swarm optimization (PSO) algorithm. The purpose of this paper is to explore the relationship between the current situation of marine environmental protection and international public health safety by digging up online comments on marine environmental protection. Its novelty lies in the design of the particle swarm optimization-K-means clustering (PSO-K-means) algorithm. Through the PSO-K-means algorithm, the mapping relationship between potential variables and word sets of marine protection status is clustered and the influencing factors are evaluated. In this way, the impact on global public health is explored by taking measures such as greening urban spaces, enhancing the quality of green spaces, and implementing the international marine environmental protection law. In addition, the development trend of online public opinion is predicted, which provides a reference for formulating and implementing marine ecological protection policies.

2. Literature Review

At present, there are studies on global warming and the governance of the marine environment. Benedetti et al. (2021) found that zooplankton richness was expected to decline slightly in the tropics but to increase significantly in temperate to subpolar regions. Meanwhile, climate change threatened the contribution of plankton communities to plankton-mediated ecosystem services such as biocarbon sequestration [13]. Bache and Reynolds (2022) reviewed ocean goals as part of the sustainable development goals and considered the process of linkage thinking in more detail, particularly in relation to ocean-climate linkages. It was found that relying on SDG interaction analysis was risky because the results were inaccurate or could not adapt to rapid transformation or knowledge acquisition in some areas. They also recognized that planetary boundary tipping points would help bring the oceans into climate consideration [14]. Therefore, the strategies of marine environmental protection and climate management are analyzed through the literature collation of environmental protection and public health. It promotes the improvement of algorithm operation efficiency while saving storage space resources, which has practical reference significance for establishing international public security model.

As one of the data mining techniques, clustering algorithms can extract useful information from massive data. Numerous scholars have studied it. Bu et al. (2019) applied the classical K-means technique to model a realistic social media network as a discrete-time dynamic system. They performed social media information analysis regarding opinion matrix and community structure interactions [15]. Tang (2022) used the K-means algorithm for mining and analyzing the public opinion hotspots of Chinese microblogs, which could grasp the public opinion hotspots easily and quickly after experiments [16]. Rahim Taleqani et al. (2019) performed data analysis using cluster analysis on 32,000 comments on Twitter on the issue of bicycle sharing, which precisely derived from people's concerns about bicycle sharing [17]. K-means clustering analysis and data mining techniques based on K-means clustering analysis can also be used in finance. Li et al. (2020) used K-means clustering analysis to analyze the investment efficiency of smart investments and had good results [18]. The losses of major enterprises during the new coronavirus can also be explored in the data. Zheng et al. (2022) analyzed the deviation from the state support policy for enterprises in southern China, thus providing ideas for their continued development [19]. The results of the literature review data are summarized in Table 1.

Table 1. Systematic summary of the literature review data.

Literature Number	Author	Year	Research Content
[15]	Bu et al.	2019	K-means clustering analysis based on leader identification and dynamic game
[16]	Tang	2022	Chinese detection and research of microblog public opinion analysis system
[17]	Rahim Taleqani et al.	2019	Public perception of no-dock bike sharing
[18]	Li et al.	2020	Machine learning intelligent model based on K-means algorithm
[19]	Zheng et al.	2022	Evaluation of small and medium-sized enterprise support policies based on K-means clustering

Through the research of the above scholars, it is found that, under the global warming trend, people have paid attention to urban space greening and marine environmental governance. Moreover, cluster analysis methods have become indispensable in the era of big data and cluster analysis methods will greatly save human and material resources. However, applying clustering algorithms to marine environmental analysis is extremely rare. Therefore, this paper improves the cluster analysis algorithm and applies it to the analysis of the marine environment and the formulation of countermeasures, which can provide a reference for the subsequent green and sustainable development of the marine environment.

3. Methods and Models

3.1. K-Means Clustering Algorithm

K-means is a partitional clustering algorithm called the K-means algorithm; it has unique advantages in extensive data analysis and information mining [20]. Figure 1 shows the workflow of the K-means clustering algorithm.

In Figure 1, K-means starts from K randomly selected centroids to define the prototype of the cluster structure. Each data point is assigned to the cluster to which the nearest centroid belongs according to the distance. Then, the centroid position of each cluster is updated iteratively (the new centroid is usually the mean of the data points within the cluster and it hardly corresponds to the actual data points) until the positions of all the particles no longer change. The K-means algorithm is efficient and scalable when dealing with datasets. The optimal state is where the algorithm's K division squared error values tend to be the smallest. The effect is good when the distribution characteristics of the clustering results are apparent. The disadvantage of the K-means algorithm is that it can only be used if the cluster's mean is pre-defined. Therefore, choosing a suitable initial centroid is a critical step in the K-means algorithm.

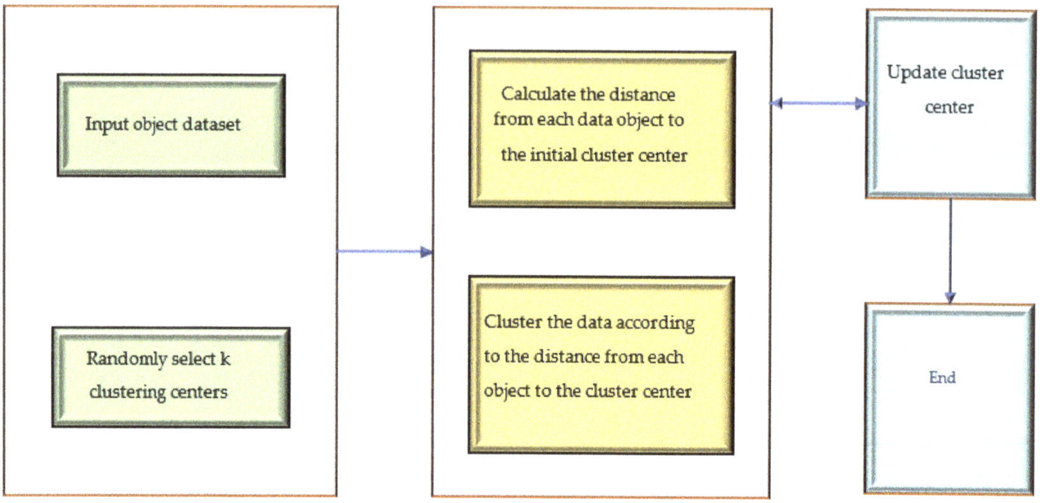

Figure 1. K-means algorithm flow.

3.2. Discrete Particle Swarm Optimization (DPSO) Algorithm

The PSO algorithm is proposed for the continuous function optimization problems. The algorithm is built on a constant domain to optimize the solution [21]. A PSO algorithm suitable for the discrete binary version is proposed after modifying the basic PSO to solve the discrete optimization problem. In the DPSO algorithm, the value of each dimension of the particle's position vector t is limited to integer zero or one. There is no limit to the particle's velocity, but velocity is usually used to characterize the probability that each dimension in the position vector takes an integer zero or one. If the velocity value v of a particular dimension j of particle i is larger, the probability that the corresponding position of the particle takes an integer one is higher. Otherwise, the probability of taking zero is greater. The sigmoid function conforms to this characteristic, so the sigmoid function is used in the PSO algorithm to convert the range of the speed v. The function conversion method is as follows.

$$sigmoid\left(v_i^j(t+1)\right) = \frac{1}{1+e^{-v_i^j(t+1)}} \tag{1}$$

The sigmoid function increases monotonically on the interval (+∞,−∞). It has a value of 0.5 at coordinate $r = 0$. The equation of the discrete binary PSO algorithm based on the above sigmoid function is as follows. In addition to the particle's moving speed needing to be further converted, its basic speed equation and the standard particle swarm algorithm's speed equation are as follows.

$$v_i^j(t+1) = v_i^j(t) + \left[c_1 r_1\left(t_i^j - x_i^j t\right)\right] + \left[c_1 r_2\left(t_i^j - x_i^j t\right)\right] \tag{2}$$

$$v_i^j(t+1) = sigmoid\left(v_i^j(t+1)\right) \tag{3}$$

$$v_i^j(t+1) = \begin{Bmatrix} 0, r_i^j(t) > v_i^j(t+1) \\ 1, r_i^j(t) \leq v_i^j(t+1) \end{Bmatrix} \quad (4)$$

In Equations (2)–(4), $r_i^j(t) \sim U(0,1)$ represents a random value that follows a normal distribution. It is mainly used to limit the probability of the speed value. c_1 is the initial center vector and c_2 is the largest center vector. In addition, the maximum speed limit is also preserved in the PSO algorithm. The greater the value of $\left|v_i^j(t)\right| < V_{MAX}$, the greater the corresponding probability value and the greater the variation probability of the particle dimension value. From the curve of the sigmoid function $f(x) = \frac{1}{1+e^{-v_i^j(t+1)}}$, the value of the sigmoid function tends to zero and $x_i^j = 0$ when the jth dimension velocity of particle i is $v_i^j < 10$. Similarly, the value of the sigmoid function tends to one and $x_i^j = 1$ when the particle's velocity $v_i^j > 10$. Therefore, it is necessary to limit the maximum speed range to prevent the algorithm from being stagnant in the search in the process of binary encoding.

3.3. PSO-K-Means Algorithm

The K-means clustering algorithm is considered one of the most influential and popular data mining algorithms among the AHC algorithms. Despite its popularity, this algorithm has certain limitations, including issues related to the random initialization of the centroids. This can lead to unexpected convergence. Therefore, this paper chooses to use the PSO algorithm to determine the initial centroids of the K-means clustering algorithm. In this way, the K-means clustering algorithm can select the optimal K value in the marine environmental protection comment analysis. The optimal analysis results can also be obtained. Here, the combination of the two algorithms is called the PSO-K-means algorithm. From Figure 2, the number of clusters K is obtained based on the optimal solution obtained by the PSO algorithm. K-means finds the cluster structure represented by K centroids.

Applying the PSO algorithm to the K-means algorithm can quickly search and accurately find the initial cluster center of the K-means algorithm. The function of evaluating the clustering effect is taken as the fitness function of the particle swarm. The frequent word space set $K_{sst} = \left(k_s^1, k_s^2, k_s^3 \cdots k_s^m\right)^T$ is divided into K categories. Then, the fitness function is:

$$g(x) = \sqrt{\sum_{i=1}^{m} \sum_{j=1}^{k} \left(k_s^i - c_j\right)^2} \quad (5)$$

In Equation (5), m is the total number of particles in the PSO algorithm and i is the particle number. k is the initial number of particles in the K-means algorithm. s is the total number of texts. c is the cluster. j is the number of clusters.

The fitness variance can reflect the convergence degree of the particle swarm and the group fitness variance can be obtained according to Equation (6).

$$\delta^2 = \frac{1}{n} \sum_{i=1}^{n} \left[g(x) - \overline{g}\right] \quad (6)$$

In Equation (6), g is the population's average fitness. When δ^2 is less than the set threshold, the fitness value fluctuates less. The particle swarm is in a convergent state. At this point, terminating the PSO algorithm and executing the K-means algorithm can make the later convergence speed fast.

PSO-K-means is applied to multi-source network comment mining to explore the impact of the implementation of international law on marine environmental protection on global public health. From Figure 3, the review information text is collected first. The PSO-K-means algorithm is used for cluster analysis and topic extraction to obtain the analysis after data preprocessing.

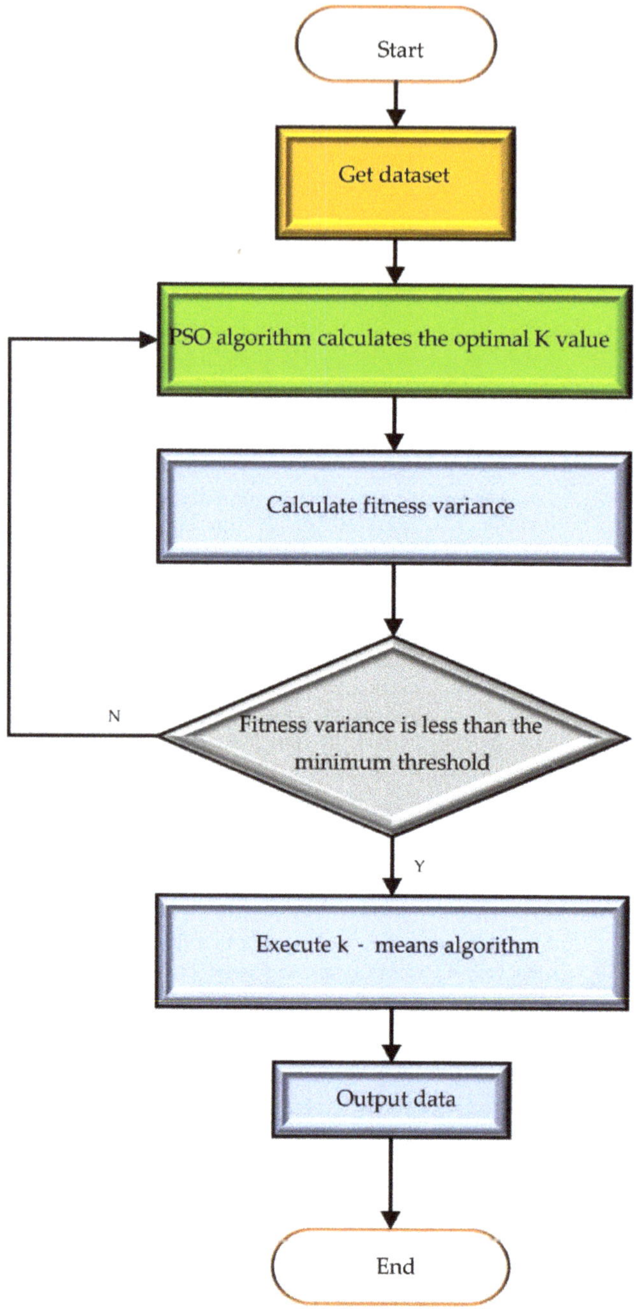

Figure 2. PSO-K-means algorithm flow.

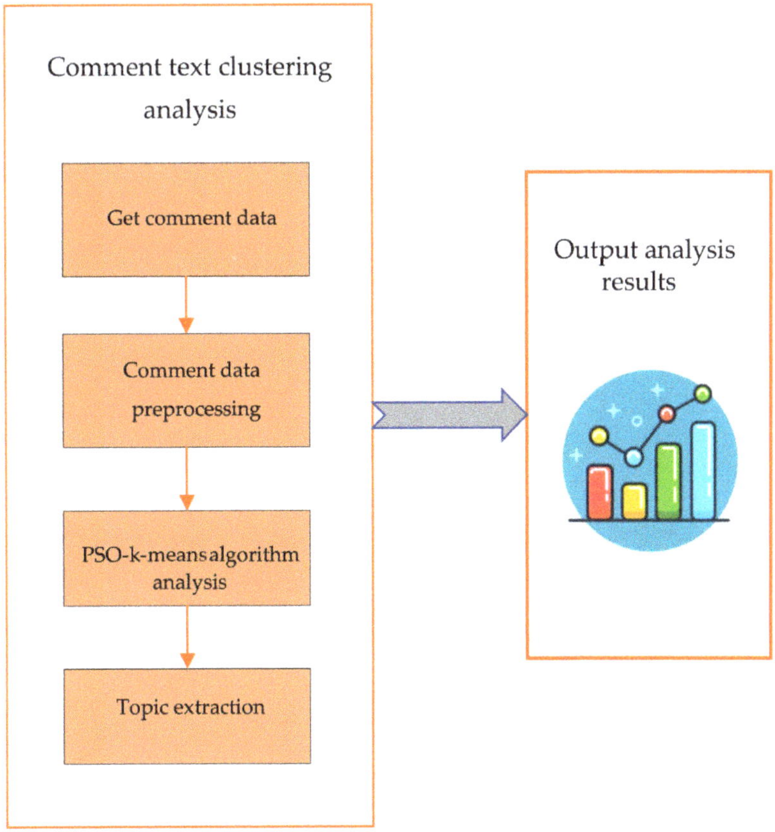

Figure 3. Application of PSO-K-means algorithm.

3.4. Preprocessing of Online Comment Data

From Figure 4, comment information mining needs to go through four processes: data collection, preprocessing, text clustering, and result in analysis. First, the data source of this analysis is the information related to marine environmental protection obtained from major social networking and news websites through crawler programs. After preliminary screening, a total of 5931 data texts are obtained and recorded as the original marine environmental protection review data set. Secondly, data preprocessing is performed to extract the information on the features of the online reviews and put them into the text. Finally, the text is preprocessed to obtain the feature vector, generating compelling information. In the process of text clustering, the PSO algorithm is used to find the cluster center intelligently. Then, clustering is implemented using the K-means algorithm. During the analysis of the clustering results, the marine environmental protection information after topic extraction is obtained by comparing the relevant high-frequency comments. The accuracy of the PSO-K-means clustering algorithm in clustering comment information is evaluated. Then, the algorithm's performance is analyzed and compared with the pure K-means algorithm for accuracy convergence. Finally, the application of the algorithm in practice is evaluated.

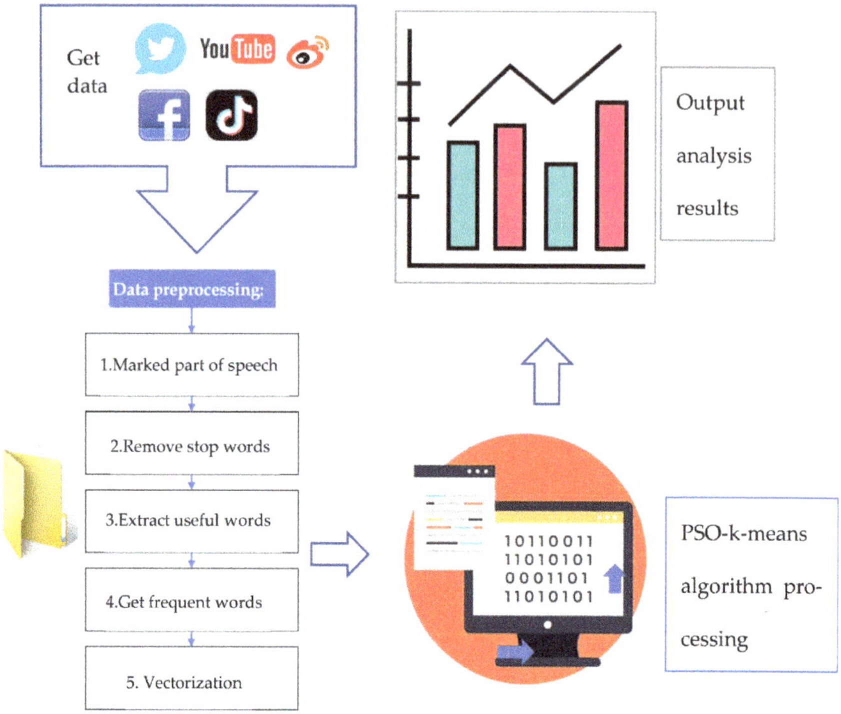

Figure 4. Analysis of the comments data of the marine environmental protection network.

The detailed steps of data preprocessing are as follows.

1. The first is to mark the part of speech. A tokenizer is used to tag high-frequency words in each comment.
2. The second is to remove stop words. Stop words are meaningless words that need to be removed. Remove meaningless words (such as am, is, are, and for).
3. The third is to extract valuable words. Proper nouns, verbs, and adjectives are extracted to ensure that the extracted words have actual meanings and can accurately express the characteristic information of comments.
4. The fourth is to obtain frequent words. All recorded frequencies are counted. High-frequency words have a representative role in this description field. Words that are representative and exceed the word frequency threshold are selected as representative elements combined with comment elements and high-frequency vocabulary.

A set of documents $D(d1, d2, d3 \cdots dn)$ and a set of words $W(w1, w2, w3 \cdots wn)$ are given. d represents a document and w represents a high-frequency word. Assuming that the order and position of each word in the text are ignored, a d-w matrix can be formed, as shown in Equation (7).

$$A = [n(d_i, w_i)]_{|D| \times |W|} \tag{7}$$

In Equation (7), $n(d_i, w_i)$ represents the word frequency of the word w_i in the document d_i. In the latent variable set $Z(z_1 z_2 \cdots z_n)$, z represents the latent variable that has not been observed. Therefore, d and w are independent of each other and k is based on experience. The correspondence between the three-level variables of "text-implicit-word" is shown in Figure 5.

5. The fifth is the vectorization of representative comment information. Through preprocessing, the dataset is segmented and stop words removed to turn it into a set of words. Each comment in the comment dataset is converted into a vector format of $(t_1, w_{i1}; t_2, w_{i2} \cdots t_n, w_{in})$ to convert textual data into format vector data. t_n refers to a word in a comment. The next is the weight value w_{in} that the word has. All articles are integrated into vector format as input to the PSO-K-means algorithm.

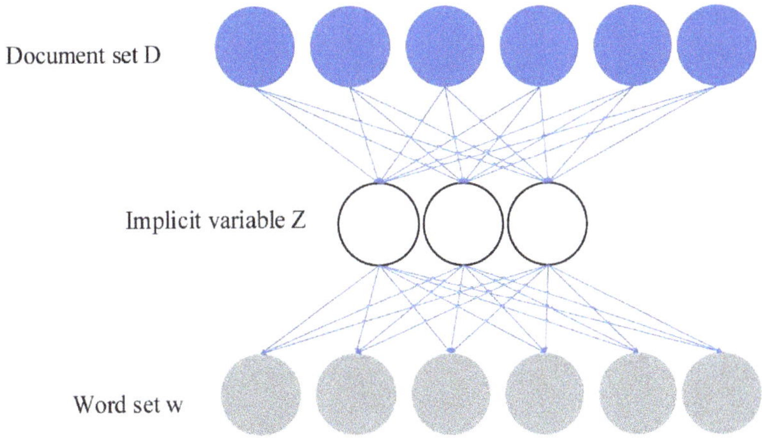

Figure 5. Mapping relationship among document set, latent variable, and word set.

Additionally, in the performance evaluation of model data mining, word frequency refers to the frequency of feature words in the text dataset. Recall refers to the proportion of data in the algorithm that is true positive and judged to be positive. The number of network comments for feature words can be judged by word frequency and the optimization performance of different algorithms can be evaluated through recall. Word frequency and recall can be obtained according to Equations (8) and (9).

$$\text{Word frequency} = \frac{F1}{F2} \quad (8)$$

$$\text{Recall} = \frac{TP}{TP + FN} \quad (9)$$

In Equations (8) and (9), $F1$ represents the number of occurrences of a particular phrase in the text. $F2$ refers to the total number of independent phrases that divide the dataset text into pieces. TP represents the total number of samples that are actually positive and predicted to be positive. FN represents the total number of samples that are actually positive and predicted negative.

3.5. Clustering of Related Online Comment Texts

After processing the original marine environmental protection review data set obtained above, the relevant online review texts are initialized by the PSO algorithm, as shown in Figure 6. The remaining objects are divided into the nearest classes according to their distances from each cluster center and iterate continuously until the function converges. The original complex comment data set can be normalized and the comment hotspots from different aspects of the commenting object can be extracted from it. Then, the feature information extraction of online comments is realized to accurately grasp the hotspots and key points of the online comments.

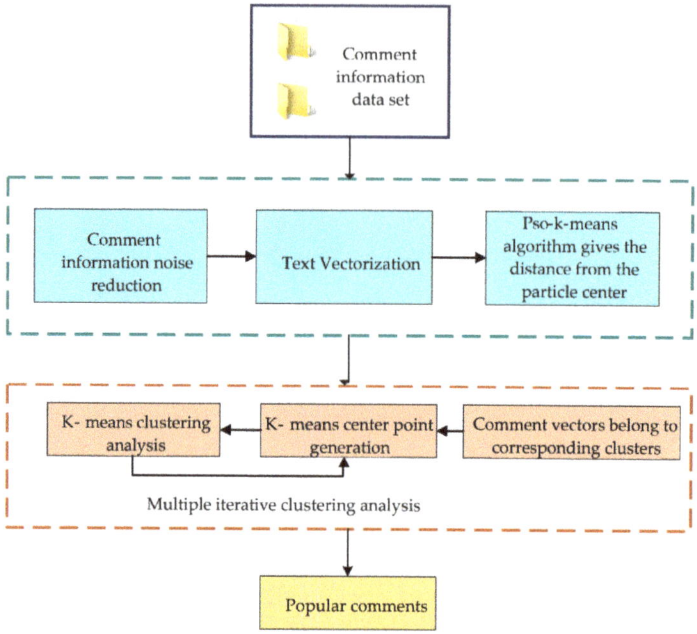

Figure 6. Clustering of online comment texts.

The PSO-K-means clustering algorithm is used to obtain the distance of each vectorized information from the particle after the noise reduction and vectorization of the comment information. Then, cluster according to distance. Firstly, the comment vector with the largest local density is selected from the set as the initial center vector C1 of the maximum and minimum distance algorithm. The maximum and minimum distance algorithm can find the center point accurately combined with the reduced candidate vector set. The distance between other vectors in the dataset and C1 is calculated and the vector C2 with the largest distance is selected as the second center vector. The distance from each comment vector to each center vector in the dataset is calculated and denoted as D_{ij}. The minimum values in D_{ij} are selected to form the minimum distance from the sample to the center vector set. The largest distance D_w in this set is chosen. In addition, the vector C_w corresponding to D_w as a new center vector is added to the center vector set to continue to search for the next center vector until the maximum D_{ij} is satisfied. Meanwhile, D_{ij} is assigned to $T1$. The purpose is to select the center of each comment vector as large as possible to avoid falling into the optimal local solution. Through the cluster analysis stage, multiple clusters can be obtained. Each cluster represents a hot spot in this review dataset. According to the corresponding comment vector in each hotspot, the word vector combination with the largest weight is extracted, representing the hotspot's main text content. Sorting out the hotspot information in each cluster realizes the extraction of all hotspot information in this review dataset.

4. Results and Discussion

4.1. Cluster Analysis of Marine Environmental Reviews

Only 4830 of the original 5931 are left after denoising, deduplication, word segmentation, stop word filtering, and other text preprocessing operations on the original marine environmental protection review dataset. When performing text preprocessing operations, the evaluation command is first issued to the background by any user and the background system selects the appropriate data range through the functional logic layer. After the initial population data query is carried out through the data layer, the clustering calculation of each item is performed on the model range data. The data of the clustering model is retained before the judgment operation of the clustering model is performed. After the original text set is converted into structured data, term clustering and document clustering are performed on the data set in the cluster analysis process. For the document clustering process, the representative output term is set to three at a time. K-means cluster analysis is performed on structured data. The number of initial clusters is set to four, five, and six by the PSO algorithm. Different K values will have different effects on the clustering results. The clustering results using PSO-K-means are shown in Table 2.

Table 2. Clustering results of marine environmental protection review data.

Number of Clusters	Cluster	Feature Word	Number
K = 6	1	Pollution, marine, and environment	1956
	2	Water quality, resources, and supervision	712
	3	Plastic, leaking, and creature	695
	4	Life, ship, and development	679
	5	Food chain, health, and transportation	476
	6	Legislation, ocean currents, and red tides	312
K = 5	1	Pollution, marine, and environment	2214
	2	Water quality, resources, and supervision	806
	3	Plastic, leaking, and creature	714
	4	Life, ship, and development	695
	5	Food chain, health, and transportation	401
K = 4	1	Pollution, marine, and environment	2367
	2	Water quality, resources, and supervision	984
	3	Plastic, leaking, and creature	758
	4	Life, ship, and development	721

From Table 1, after clustering analysis of 4830 valid comments, different K values will output different clustering results. The less the K value is set, the fewer the classification results of the representative entries and the easier it is to obtain the analysis results. The larger the K value is set, the more detailed the classification will be and the more complex the analysis results will be. Furthermore, the clustering time and the number of iterations are also affected. After preprocessing the comment text and text representation, the feature items are counted. Figure 7 displays the cluster analysis graph.

For the form of feature item visualization, the word cloud diagram is shown in Figure 8. "Word cloud" is to visually highlight "keywords" that frequently appear in web texts by forming a "keyword cloud layer" or "keyword rendering". There are more than 4000 feature items in the review set, so the feature items with a word frequency higher than 300 are used when drawing the word cloud and those with a word frequency lower than 300 are eliminated. When drawing the word cloud map, a sparsity of 0.98 is selected and 18 high-frequency feature items are retained to display the relationship between feature items.

From Figure 8, feature items with high frequency are clustered in the middle area of the word cloud; however, high frequency does not mean important. Although some feature items are not in the middle area of the word cloud, their importance to comments is more important than the feature items in the middle area. From the word cloud network constructed from 18 feature items, the words near the central region are large. This indicates that its importance is high and the connection between the feature terms is strong.

4.2. PSO-K-Means Algorithm Experimental Evaluation

The selection of the threshold value in the research algorithm directly affects the clustering effect. Therefore, choosing an appropriate threshold is significant for improving the clustering effect. Here, the review corpus is used as the training set. The recall is adjusted to observe the clustering effect. When the recall value reaches the maximum value, the clustering effect is the best and the value at this time is also used as the threshold. First, product features for Chinese online products are extracted, filtered, and optimized. The similarity between features is calculated using the point mutual information of the equation. Then, the clustering algorithm is performed. The algorithm selects the maximum recall value as the final selection threshold. The selection threshold is revealed in Figure 9. From the figure, when the threshold value is 0.45, the corresponding recall rate is the highest, which is 88.75%. At this time, the corresponding clustering effect is also the best.

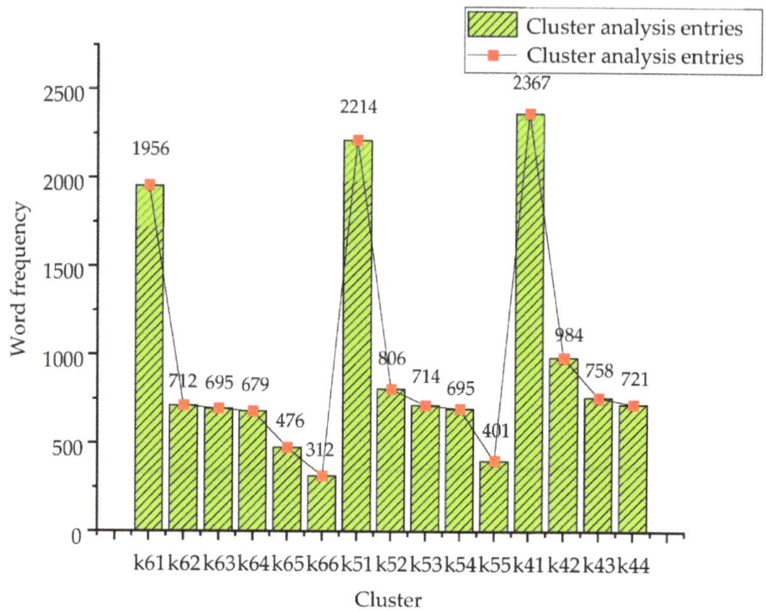

Figure 7. Cluster analysis diagram of marine environmental protection comments.

Figure 8. Schematic diagram of the feature word cloud.

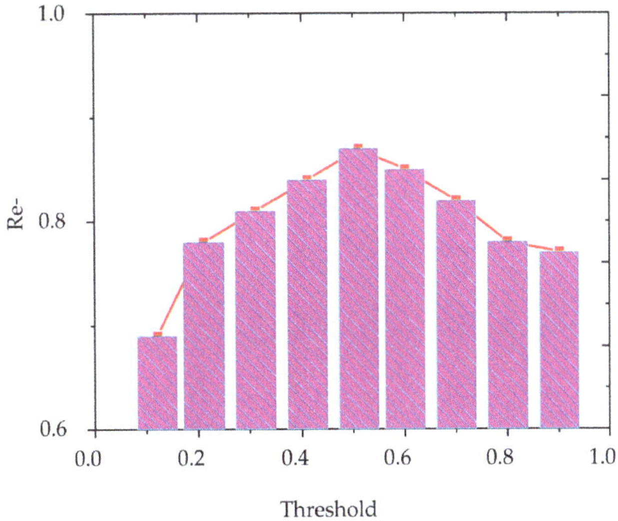

Figure 9. The effect of different thresholds on the regression rate.

Multi-source web reviews are mined through experimental analysis. Table 3 shows the performance comparison results of the PSO-K-means algorithm proposed here and the K-means method only.

Table 3. Comparison of recall rates of two algorithms.

Threshold	Recall	
	PSO-K-Means	**K-Means**
0.1	0.68	0.68
0.2	0.77	0.76
0.3	0.80	0.79
0.4	0.82	0.81
0.5	0.85	0.83
0.6	0.83	0.83
0.7	0.82	0.81
0.8	0.77	0.77
0.9	0.77	0.75

Table 3 indicates that using the PSO-K-means algorithm has a higher recall rate than simply using the K-means algorithm. The recall performance of the model represents the proportion of data in the algorithm that is true positive and judged positive. It can be seen that the PSO-K-means algorithm has a promising application in the cluster analysis of related reviews. The main function of PSO is to find the optimal value. The programming implementation of genetic algorithm is relatively complicated and the utilization rate of network feedback information is not high. Therefore, the improved optimization algorithm of K-means + PSO is used here, which has a reference effect on the model experiment of the big data analysis algorithm.

The accuracy values of the PSO-K-means algorithm and the K-means algorithm under different thresholds are further analyzed, as shown in Table 4.

Table 4 compares the accuracy values of the PSO-K-means algorithm and the K-means algorithm. It is found that using the PSO-K-means algorithm has a higher accuracy value than simply using the K-means algorithm and it reaches 90% with a threshold of 0.5. Therefore, the PSO-K-means algorithm has application prospects in the cluster analysis of related reviews. It also has a reference effect on the research of big data analysis algorithms.

Table 4. Comparative analysis table of accuracy values of the two algorithms.

Accuracy	Threshold								
	0.1	**0.2**	**0.3**	**0.4**	**0.5**	**0.6**	**0.7**	**0.8**	**0.9**
PSO-K-means	0.72	0.81	0.85	0.86	0.90	0.88	0.86	0.81	0.81
K-means	0.72	0.80	0.83	0.85	0.87	0.86	0.86	0.81	0.80

Finally, the time required for clustering the PSO-K-means algorithm and the K-means algorithm under different thresholds is analyzed, as shown in Table 5.

Table 5. Comparative analysis table of the time required for clustering of the two algorithms.

Time (ms)	Threshold								
	0.1	**0.2**	**0.3**	**0.4**	**0.5**	**0.6**	**0.7**	**0.8**	**0.9**
PSO-K-means	88.72	84.11	82.03	78.44	77.99	77.72	76.99	76.32	77.62
K-means	89.12	85.01	83.55	81.81	80.13	80.14	78.96	79.02	81.84

Table 5 compares the time required for the PSO-K-means algorithm and the K-means algorithm at different thresholds. It is found that with the threshold increase, the clustering completion time of the PSO-K-means algorithm and the K-means algorithm showed a trend of first decreasing and then rising. Using the PSO-K-means algorithm takes less time than simply using the K-means algorithm and the time required is only 76.32ms at a threshold of 0.8. Therefore, the PSO-K-means algorithm can complete the cluster analysis in a shorter time and the clustering accuracy is better for providing a reference for the research of big data analysis algorithms.

4.3. Research on the Impact of the Implementation of International Marine Environment Protection Law on Global Public Health

It is necessary to explore the impact of the implementation of international law on marine environmental protection on global public health. Years of implementation of international law on marine environmental protection increase. The frequency of representative entries related to public health in relevant comments is counted. Among the original 5931 reviews, good reviews containing public health-related terms are preprocessed into 2867. The PSO-K-means algorithm is set to perform year statistics on different time windows; the results are shown in Table 6.

Table 6. Statistics of comment information in different time windows.

Feature Word	**Years**	**Word Frequency**
Health	1982–1992	3
Pollution	1992–2000	16
Environment	2000–2010	491
Legislation	2010–2021	2357
Feature word	Years	Word frequency

From Table 6, the frequency of public health-related entries in online reviews has increased yearly with the implementation of the international law on marine environmental protection. Especially after 2000, there has been an explosion of commentary on marine environmental protection and public health. This is closely related to the improvement of international marine protection law after 2000. Therefore, implementing marine environmental protection laws is vital in promoting global public health.

4.4. Discussion

Gao et al. (2020) studied the application of the PSO algorithm in the K-means algorithm [22]. The PSO-K-means algorithm has good clustering performance and is obviously superior to the existing classical or most advanced clustering algorithms, which is consistent with the research results of Gao et al. (2020) [22]. In summary, the PSO-K-means algorithm can efficiently cluster vulnerability data information. It has the characteristics of high accuracy of the AHC algorithm when applied to the clustering analysis of multi-source online reviews. The development trend of online public opinion information can be judged by adjusting the time window of the algorithm to predict the implementation of the international marine environmental protection law and the global public health image. The results show that the clustering analysis of marine environment by the PSO-K-means algorithm can promote the clustering of marine characteristic words, which has a practical reference value for realizing marine environmental protection and sustainable development of marine resources. Furthermore, through the analysis of the international marine environment, it is found that global warming is one of the possibilities for the deterioration of the international marine environment. As a result, global warming will be mitigated by taking measures such as greening urban spaces and improving the quality of green spaces. This is of great practical significance for establishing healthy communities and can provide concrete measures for the governance of the international marine environment.

5. Conclusions

In this paper, the PSO-K-means algorithm is designed and, based on this algorithm, the original marine environmental protection review data set obtained in the network is mined and analyzed in detail. The results show that the designed algorithm has good clustering performance and can promote the clustering of marine feature words, thus promoting the protection of marine environment and the development of global public health communities. Therefore, mitigating global warming by greening urban spaces and improving the quality of green spaces is of great significance to establishing healthy communities and the governance of the international marine environment. The PSO-K-means algorithm applied to the analysis of online review data is more robust than traditional methods, which provides ideas for online review data mining. Through this algorithm, the reactions and opinions of network users can be quickly understood, which affects the judgment of public social opinion. The combined use of the PSO-K-means algorithm is better than K-means alone. After the recall rate test, the recall rate of the PSO-K-means algorithm can reach 88.75%. Therefore, the PSO-K-means algorithm has a good application prospect. However, there are still many details in the process of information processing that are not considered carefully. The solutions to many problems still need to be improved. The effect of data preprocessing to eliminate invalid comments needs to be further strengthened. The visualization of analysis results still needs improvement. This will also be an important research direction for future related work.

Author Contributions: Conceptualization, methodology, validation, formal analysis, investigation: A.Y., S.Y.; resources, A.Y.; writing—original draft preparation, A.Y.; writing—review and editing, A.Y., S.Y.; visualization, A.Y.; supervision, S.Y.; funding acquisition, S.Y. All authors have read and agreed to the published version of the manuscript.

Funding: This research received no external funding.

Institutional Review Board Statement: Not Applicable.

Informed Consent Statement: Not applicable.

Data Availability Statement: Not applicable.

Conflicts of Interest: The authors declare no conflict of interest.

References

1. Zhang, Y.X.; Feng, Y.X.; Yang, R.Q. Network public opinion propagation model based on the influence of media and interpersonal communication. *Int. J. Mod. Phys. B* **2019**, *33*, 1950393. [CrossRef]
2. Rim, H.; Lee, Y.; Yoo, S. Polarized public opinion responding to corporate social advocacy: Social network analysis of boycotters and advocators. *Public Relat. Rev.* **2020**, *46*, 101869. [CrossRef]
3. Kandziora, J.H.; Van Toulon, N.; Sobral, P.; Taylor, H.L.; Ribbink, A.J.; Jambeck, J.R.; Werner, S. The important role of marine debris networks to prevent and reduce ocean plastic pollution. *Mar. Pollut. Bull.* **2019**, *141*, 657–662. [CrossRef] [PubMed]
4. Chaudhuri, S.; Kumar, A. Urban greenery for air pollution control: A meta-analysis of current practice, progress, and challenges. *Environ. Monit. Assess.* **2022**, *194*, 235. [CrossRef]
5. Wu, S.; Lin, J.; Zhang, Z.; Yang, Y. Hesitant Fuzzy Linguistic Agglomerative Hierarchical Clustering Algorithm and Its Application in Judicial Practice. *Mathematics* **2021**, *9*, 370. [CrossRef]
6. Ahmed, M.; Seraj, R.; Islam, S.M.S. The k-means algorithm: A comprehensive survey and performance evaluation. *Electronics* **2020**, *9*, 1295. [CrossRef]
7. Jo, T. Semantic string operation for specializing AHC algorithm for text clustering. *Ann. Math. Artif. Intell.* **2020**, *88*, 1083–1100. [CrossRef]
8. Peterson St-Laurent, G.; Hagerman, S.; Kozak, R. What risks matter? Public views about assisted migration and other climate-adaptive reforestation strategies. *Clim. Change* **2018**, *151*, 573–587. [CrossRef]
9. Lin, M.; Huang, C.; Chen, R.; Fujita, H.; Wang, X. Directional correlation coefficient measures for Pythagorean fuzzy sets: Their applications to medical diagnosis and cluster analysis. *Complex Intell. Syst.* **2021**, *7*, 1025–1043. [CrossRef]
10. Behura, A. The cluster analysis and feature selection: Perspective of machine learning and image processing. *Data Anal. Bioinform. Mach. Learn. Perspect.* **2021**, *67*, 249–280.
11. Kuo, R.J.; Lin, J.Y.; Nguyen, T.P.Q. An application of sine cosine algorithm-based fuzzy possibilistic c-ordered means algorithm to cluster analysis. *Soft Comput.* **2021**, *25*, 3469–3484. [CrossRef]
12. Jasiński, M.; Sikorski, T.; Kaczorowska, D.; Rezmer, J.; Suresh, V.; Leonowicz, Z.; Kostyła, P.; Szymańda, J.; Janik, P.; Bieńkowski, J.; et al. A case study on data mining application in a virtual power plant: Cluster analysis of power quality measurements. *Energies* **2021**, *14*, 974. [CrossRef]
13. Benedetti, F.; Vogt, M.; Elizondo, U.H.; Righetti, D.; Zimmermann, N.E.; Gruber, N. Major restructuring of marine plankton assemblages under global warming. *Nat. Commun.* **2021**, *12*, 5226. [CrossRef] [PubMed]
14. Bache, S.J.; Reynolds, A. Systems Thinking in a Fluid Environment: SDG 14 and the Ocean-Climate Nexus. *Environ. Sci. Proc.* **2022**, *15*, 17.
15. Bu, Z.; Li, H.J.; Zhang, C.; Cao, J.; Li, A.; Shi, Y. Graph K-means based on leader identification, dynamic game, and opinion dynamics. *IEEE Trans. Knowl. Data Eng.* **2019**, *32*, 1348–1361. [CrossRef]
16. Tang, J.; Xu, X. Research on Detection of Chinese Microblog Public Opinion Analysis System. In *Proceedings of the 11th International Conference on Computer Engineering and Networks*; Springer: Singapore, 2022; Volume 67, pp. 766–773.
17. Rahim Taleqani, A.; Hough, J.; Nygard, K.E. Public opinion on dockless bike sharing: A machine learning approach. *Transp. Res. Rec.* **2019**, *2673*, 195–204. [CrossRef]
18. Li, L.; Wang, J.; Li, X. Efficiency analysis of machine learning intelligent investment based on K-means algorithm. *IEEE Access* **2020**, *8*, 147463–147470. [CrossRef]
19. Zheng, Z.; Qianqian, H.; Nannan, T. Evaluating SMEs-Supporting Policies During COVID-19 Pandemic with K-Means Clustering. *Data Anal. Knowl. Discov.* **2022**, *5*, 148–157.
20. Abernathy, A.; Celebi, M.E. The incremental online k-means clustering algorithm and its application to color quantization. *Expert Syst. Appl.* **2022**, *3*, 117927. [CrossRef]
21. Khooshechin, S.; Mansourzadeh, F.; Imani, M.; Safdari, J.; Mallah, M.H. Optimization of flexible square cascade for high separation of stable isotopes using enhanced PSO algorithm. *Prog. Nucl. Energy* **2021**, *140*, 103922. [CrossRef]
22. Gao, H.; Li, Y.; Kabalyants, P.; Xu, H.; Martinez-Bejar, R. A novel hybrid PSO-K-means clustering algorithm using Gaussian estimation of distribution method and Lévy flight. *IEEE Access* **2020**, *8*, 122848–122863. [CrossRef]

International Journal of
Environmental Research and Public Health

MDPI

Article

Design and Realization of Rural Environment Art Construction of Cultural Image and Visual Communication

Fulong Liu [1,*], Baogang Lin [1] and Kun Meng [2]

[1] College of Art, Xi'an University of Architecture and Technology, Xi'an 710055, China
[2] College of Art and Design, Shaanxi University of Science and Technology, Xi'an 710021, China
* Correspondence: liufulong@xauat.edu.cn

Abstract: The practice of rural construction has been exploring and trying to adapt to the needs of rural development in various periods. In recent years, under the attention and promotion of the central policy, various social forces have joined the ranks of rural construction, and art intervention in rural construction has begun as a new method. Entering the public eye, it deeply intervenes in the construction and development of the countryside in a more gentle way, from the key point of interaction between the social and cultural orientation and the material needs of the countryside. However, most of the art interventions in rural construction practice only unilaterally use artistic techniques to beautify local areas or display works, without realizing the hidden artistic and cultural value of the village and ignoring the participation and role of the villagers in the whole process. After the construction is completed, once the foreign construction forces are withdrawn, the development of the village will stagnate. Therefore, mobilizing the main body of rural construction (original villagers) to participate in the joint construction of the village is an important link to solve the current problems of art intervening in the construction of rural settlements.

Keywords: cultural image; rural environment; art construction; applied research

Citation: Liu, F.; Lin, B.; Meng, K. Design and Realization of Rural Environment Art Construction of Cultural Image and Visual Communication. *Int. J. Environ. Res. Public Health* **2023**, *20*, 4001. https://doi.org/10.3390/ijerph20054001

Academic Editors: Hongxiao Liu, Tong Wu, Yuan Li and Paul B. Tchounwou

Received: 28 December 2022
Revised: 17 February 2023
Accepted: 18 February 2023
Published: 23 February 2023

1. Introduction

At the beginning of the new journey of building a well-off society in my country, the rural foundation was weak. At the Fifth Plenary Session of the 16th Central Committee of the Communist Party of China, the goal of building a new countryside was put forward—production development, affluent living, civilized rural customs, clean and tidy village appearance, and democratic management. Entering a new era of socialism, in 2020, when the first centenary goal is about to be achieved, the main social contradictions have changed. Compared with the general requirements for the construction of new countryside at the Fifth Plenary Session of the 16th Central Committee of the Communist Party of China, in addition to "township and civilization", the changes in the deepening of connotation are reflected for the first time. Rural development has risen to a strategic height. In recent years, research on rural revitalization has shown a spurt of growth [1]. Traditional villages have paid a large price in the transformation and development of modernization. The 19th National Congress of the Communist Party of China provided an important action guide for rural development in the new era, pointing out that rural development should implement the rural revitalization strategy. Sociologists, artists, local sages, and other experts from all walks of life have also invested in rural construction projects. Various projects related to art rural construction are also in full swing. With the advancement of new urbanization, the state pays great attention to rural issues, rural economy, and rural construction [2]. Cognition and thinking should be a special study that combines rural art practice with the specific concept of rural landscape sites [3].In the context of the rapid development of rural revitalization in China, the homogeneity of its cultural and environmental perception products is relatively serious. The article analyzes the cultural image and the professional method of visual communication in the rural environment, integrates the construction of the rural environment of the plastic arts intervention institution, and will ultimately bring you to experience the spiritual meaning of local culture in the environmental perception, leading you to have the aesthetic feeling of the scenery.

2. Relevant Theoretical Basis

2.1. Rural Revitalization

A village refers to a village or village settlement, a regional complex with natural, social, and economic characteristics, and has multiple functions such as production, life, ecology, and culture. The rural revitalization strategy is the strategy put forward by President Xi Jinping in the report of the 19th National Congress of the Communist Party of China. Efforts will be made to solve outstanding problems such as insufficient and comprehensive rural development. At the same time, the rural revitalization strategy is proposed on the basis of a profound understanding of China's urban-rural relations, evolution trends, and laws.

2.2. Art Creation

The modern Chinese dictionary explains the word "intervention" in this way: intervention means going deep into it and intervening [4]. The positive experience of real life endows the real significance of art intervention, which is mainly reflected in two aspects as shown in Table 1: on the one hand, art that can integrate its own life experience and care for real life is an art activity that can be inherited; "Art intervention" has the intervention of independent personality, the spirit of criticism, the courage to participate in society, promote communication, and enhance the sense of identity [5].

Table 1. Two aspects of artistic involvement.

Dimensions	Intervention Features
Dimensions in terms of art	Gives art a greater possibility of inheritance
Life aspects	Make life more artistic

When art intervenes, the creator does not produce an "art product", but forms a larger social event through the effectiveness of design art. They have an effective influence, causing people to think about related topics perception, so as to effectively intervene and reshape current social events. At the end of the 20th century, a community building movement emerged in Taiwan [6]. This community movement not only transforms the space environment, but also pays attention to the participation of community residents in public affairs, the happiness index in life, and the aesthetic perception of the community environment. The concept coincides with the concept of art intervention in rural construction under the concept of co-construction in this article. The community building of Tugou Village lets art connect residents' lives and production spaces, so that villagers can feel the beauty brought by art to the community and actively participate in the public activities of building the community. The process of art entering into rural environment construction is shown in Figure 1:

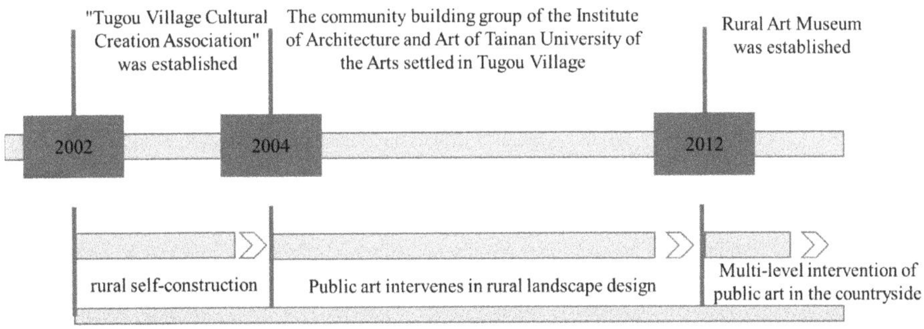

Figure 1. The process of rural environment construction.

2.3. Rural Landscape Environment

Rural landscape can be defined as a landscape space with human settlements and related behaviors in addition to urban landscapes. The core of rural landscape is rural settlement landscape. The landscape environment complex involves the individual architectural features, house structure, settlement pattern, etc., of rural settlements. It involves the environmental characteristics of the external space of the settlement, including the connection system with the external landscape environment [7]. In this study, the rural landscape environment with artistic intervention is a sustainable rural landscape environment that has an efficient artificial ecosystem, maintains the integrity and diversity of natural landscapes, and can inherit rural regional culture and a good rural living environment. In the face of the traditional architecture and landscape pattern of traditional villages being eroded by modern styles, traditional culture has been broken by modern life, and handmade products have lost their status as necessities of life and have been replaced by industrialized and machined products. Along with all these factors, the inherent living soil possessed by the countryside is losing its inheritance. It is in the embarrassing situation of being assimilated by urban culture.

2.4. Jointly Create Ideas

Based on the "people-oriented" living environment construction theory, the Ministry of Housing and Urban-Rural Development creatively put forward the concept of joint construction in response to community governance issues. The Ministry of Housing and Urban-Rural Development officially clarified the jointly created theory and implementation model in relevant documents and promoted the transformation of environmental construction from government-led to multi-party participation. In 2014, the concept of co-creation was applied to practice for the first time in the reconstruction and construction of old urban communities in Xiamen and achieved remarkable results.

Participation is the essence of co-creation. The innovation of this concept lies in the formation of a new governance model with villagers as the main body, decision-making, development, construction, construction, effect evaluation, and achievement sharing to solve the stubborn problems that hinder rural development. In this whole process, the status and role of the government and designers have changed compared with the past. They are no longer decision-makers, but the main participants, in guiding villagers to build beautiful homes. The application of the concept of co-creation in rural construction is conducive to accelerating the process of rural revitalization. The most important thing is to arouse the villagers' subject consciousness, which is the key to solving the problem of rural development.

3. Research on the Application of Cultural Intention and Visual Communication Design in Rural Environment

3.1. Systematic "Five Blessings" Cultural Image

The application of cultural intention and visual communication design in rural environment is very extensive. Taking Zhanzishi Village in southeast Hubei as an example, the cultural intention of "Five Blessings" has been applied in various fields such as architecture and tourism. In the process of the construction of Chinese rural culture and local natural environment, the "auspicious" psychological state is widespread. Integrating the interpretation of the regional cultural markers of Zhanzishi Village, its regional culture has an inherent systematization. The local natural environment culture shows the comprehensive traditional Chinese "five blessings" culture. This value concept was handed down from Chongqing's historical time and is not a reflection of the local living subject's behavior norms. It forms a cultural artistic conception of "Shun", "De", "Gui", "Fu", "Shou", and one by one. For example, the sweet persimmons match everything well and the number one bridge and the number one Lang pavilion match the spirit essence of the ten years of hard study. The capital bridge represents the pursuit of perfection of capital, the good official bridge metaphor the official life, the immortal bridge metaphor refers to the cultivation of immortality, which has extremely rich and very practical content.

3.2. Organization of Cultural System: Construct the Sequence of Virtual and Real Space

The comprehensive "Five Blessings" cultural artistic conception of Zhanzishi Village is integrated and the bridge section organization strategy of its garden landscape spatial sequence is carried out in the following manner: 1. Refine a clear cultural theme style and make it a clue to the structure of institutional environment art space; 2. The design scheme has an ingenious rhythmic sequence. This is similar to the traditional garden design principle of "layout planning".

The aesthetic sequence rhythm of natural environment space is regularly changed and reflected by the cultural elements or cultural modules in each space according to the designer's orientation. The connection point of environmental art and the base price of the art site are set up according to the "life situation" and the traditional "five blessings" culture in the local culture. In the space language described by the regional culture, the Xianren River in Zhanzishi Village is the boundary of Wu Haiying. The south bank is the noisy vulgar, and the south bank is the spiritual spectacle. The four roads ("virtue", "expensive", "rich", and "smooth") all take the four "bridges" in the Xianren River as the starting point to open the different life realm of life. When we are accompanied by the winding rise of the countryside, the natural scenery along the line implies the warning of life, and there will be some fatigue and embarrassment along the way. The design of tourist attractions along these four roads focuses on the movement track of life. When reaching the end point, you can only choose one road ("longevity") to go out of the mountain. It also implies that it is inevitable: however, it is not a pity, and life is not happy.

4. Problems in the Creation of Rural Environmental Art

4.1. The Main Needs and Artistic Input Are Not Equal

Differences in growth environment, knowledge structure, artistic accomplishment, etc., lead to different attitudes of artists and villagers towards the countryside. Artists often consider how to activate the rural history and culture through their own creations, so as to reflect the cultural value of their works. It is easy to forget that the villagers are the creators of local culture, ignoring the subjective status of the villagers [8]. Therefore, it is difficult for artists to create art based on the actual needs of the villagers, which to a certain extent, discourages villagers from having positive attitudes towards art intervention. As the cultural level and aesthetic level are not generally high, the art creations performed by artists in the villages are only an "unknown way" to increase income in the eyes of the villagers. It is difficult for the villagers to participate in the project and create together with the artists. Art intervenes in the countryside to activate culture. The ideal is always the "wishful thinking" of the artist. In recent years, the Chinese government has vigorously advocated the protection and planning of traditional historical villages. More and more villages have been included in the list of protection and planning, as shown in Table 2:

Table 2. Preparatory stage.

Preliminary Preparation Steps	Project Approval	Event Planning	In-Depth Research	Information Collection	Data Analysis
Participating subject	Government, designers, villagers	Government, designers, villagers	Government, designers, villagers	Government, designers, villagers	Analyst
Main content	In view of the goal of village construction, all parties discussed together	Before entering the village to carry out work, carry out a series of publicity activities that can bring the relationship between the villagers closer together	Comprehensive and in-depth understanding of various elements in the countryside, including environment, architecture, villagers, production, and living organization	Organize and record research data	Objective and comprehensive analysis of survey data
Work method	Hold a meeting	Public welfare activities, volunteers in the village to explain	Interviews, questionnaires, surveys, participatory observation, developing detailed survey checklists	Drone shooting, GIS geographic information system, on-site photo shooting	SWOT analysis, chart analysis

4.2. *Insufficient Participation of Villagers*

The lack of attention to the subjectivity of the villagers makes the reality of rural development and artistic input unequal, which in turn, leads to a very low participation of villagers in the process of artistic intervention [9]. In the final analysis, art intervention in rural construction should take villagers as the main body and give full play to the wisdom and creativity of villagers. Participation methods also need to be improved. The current organizational form is shown in Table 3 below:

Table 3. Comparison of the process of rural construction with traditional art and rural construction under the concept of co-construction.

Compare Content	Traditional Art Intervenes in Rural Environment Construction	Artistic Intervention in Rural Construction under the Concept of Joint Construction
Participating subject	Government, artist, designer	Villagers, government, artists, social institutions, village organizations
Way of participation	Top down	Bottom-up, diverse collaboration
Participation time	Not stationed in the village	Resident in the village, participating in the whole process of construction

5. Research Results and Suggestions

5.1. *Research Results and Reason Analysis*

5.1.1. "Source"—Extraction of Original Rural Site Information

The countryside is the birthplace of traditional culture. After the intervention of art in the construction, to a certain extent, cultural customs have been restored and continued, and the sustenance of rural culture and emotions has been preserved. It is also a way to show rural history and culture to the new generation of people. The rural environment has attracted urban residents to visit because of the intervention of art and the charm of the countryside. Tourists see the brilliance of traditional farming culture in the new era. Not only did the countryside win attention, but the rural culture was also valued. Villagers also felt different spaces and concepts due to the intervention of art [10]. Art is the best glue between urban and rural areas. It allows urban and rural cultures to communicate with each other, allowing art to penetrate deeply into the countryside and make it regain its brilliance.

Where does the "source" come from? It can be understood as the "raw material" of the countryside. When art intervenes in the countryside, artists have already appeared here, and the scene needs to exist. The scene is equivalent to the identity of "raw material" to a large extent. The artist needs to obtain information from the "source", which is the artist's creative method and inspiration source. In the process of the artist's creation, there will be two "source" creation paths. One is that the artist abandons the extraction and refinement of the "source" and insists that his own creation has nothing to do with the original venue and audience; here, the artist does not care about the original source. The venue is dynamic. On the other hand, the artist's creative path reflects that some artists insist that, in their artistic creation, they must understand the local situation, have a thorough understanding of the "source", and extract the corresponding cultural element symbols from it. Then, the direction of artistic creation can be determined: see Figure 2.

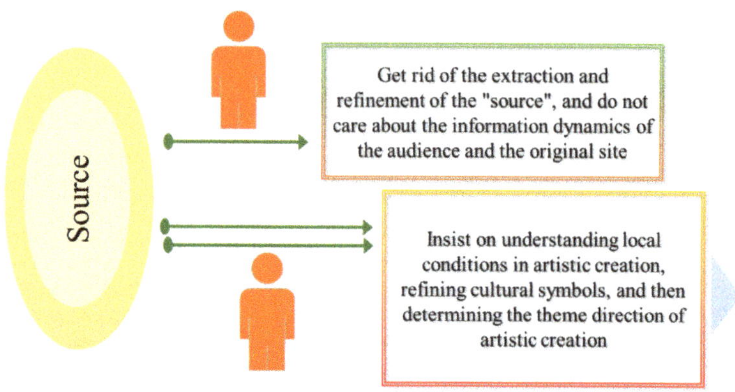

Figure 2. "Source" authoring path.

5.1.2. "Sink"—The Output of Rural Site Information

In the process of rural construction, traditional rural residential buildings are often considered to be "shabby houses" and are demolished. The effect of the construction no longer has the taste of the countryside and loses the traditional cultural characteristics of the countryside. Under the concept of co-construction, art intervention in the rural environment is not a cultural invasion, but a benign guidance [11]. Artists and designers should gradually change the villagers' perception of rural things through the intervention of art, have a certain understanding of the rural aesthetic standards, and learn how to express their own aesthetic ideas. When the overall aesthetic level of the village is improved, the traditional culture of the village can be continued well.

The output of "Hui" about rural site information also means that we need to think about what kind of products we can provide when designing and creating. Taking the land landscape as an example, the vast rural land is increasingly favored by artists to create land art. This is a transformation of the content of the rural space after the intervention of art. The vast land has been freed from agricultural production. The environment is bound, and the remodeling into a work of art through design art should be promoted. In the work of art, the environment itself must be purified. The landscape should also be purified to slightly weaken the things that are too artificial.

5.1.3. "Field"—To Create the Presence of Rural Space

The place represents the viewpoint and vision of the rural users. The place shows memory through time and changes, and time will leave a memory in the spiritual space of the place [12–15]. When a work of art is involved in rural practice, the external material and internal spiritual connotation depend, to a large extent, on the particularity of the realized spatial landscape environment. The spatial scale, spatial texture, and local materials in the rural landscape environment all play an important role. The symbolic value in the rural place reflects the accumulation of spiritual connotation value and represents the rural spiritual place. The spirit of the place in the spiritual place carries the sense of belonging and identity in the rural space and place. The type of field is shown in Figure 3.

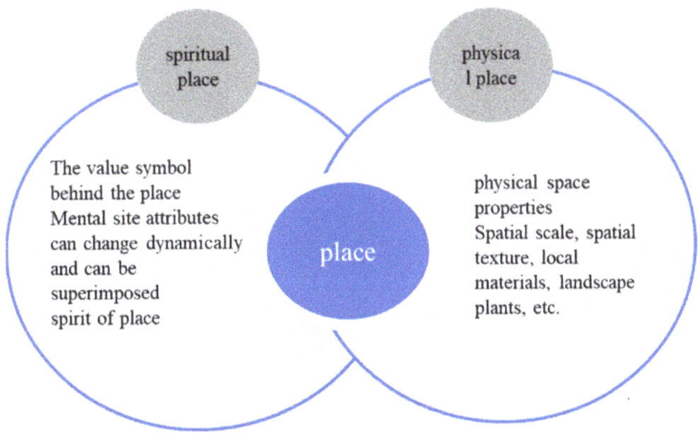

Figure 3. Types of "fields".

Farming is the most traditional way of production in rural areas, and agriculture has been a pillar industry in Chinese rural areas since ancient times. However, after the social development and progress of today, farmers are no longer willing to go to the fields; relying on growing crops cannot bring better economic benefits. The traditional rural farming culture is in crisis. Should art intervene in the countryside, and should it change the rural status quo of the industrial structure dominated by agriculture? The answer is negative. When art intervenes in rural construction, we should first protect the traditional rural farming culture. Then, under the guidance of the concept of joint construction, we should combine the secondary and tertiary industries to realize the linkage of multiple industries, coordinated development, and improve the endogenous development momentum of rural areas. This should be done to create their own industries and brands and integrate art into the rural environment: see Figure 4.

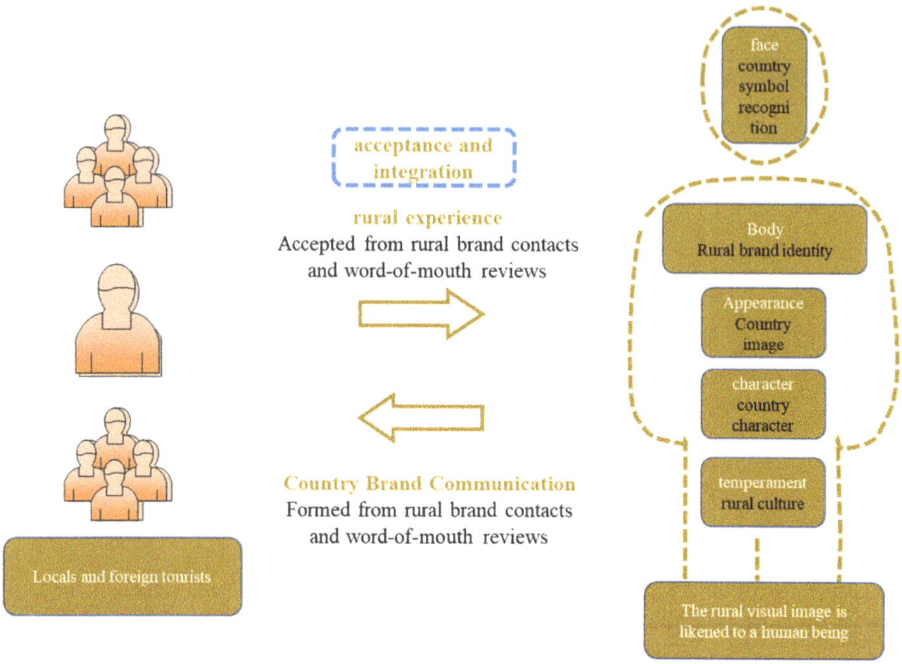

Figure 4. Country brands.

5.2. The Design Method of Art Intervening in Rural Environment

5.2.1. Co-Participation in the Design Process—Bottom-Up, Multi-Collaboration

In the past, when art intervened in rural environment construction, artists or designers led the construction plan. However, farmers hardly participated in the entire construction process [16]. Although this "all-in-one" intervention method can quickly improve the appearance of the village, it cannot stimulate the villagers' sense of belonging and responsibility to their homes. In the later maintenance stage after the construction is completed, it is difficult for the villagers to spontaneously maintain the construction results and buildings. The effect is far less effective than expected. Under the concept of co-construction, art intervention in rural environmental construction must change the conventional construction method, guide villagers to actively participate in the construction of their homes, allow designers to participate in design guidance and services throughout the process, and form a bottom-up design process model with the government. More attention should be paid to the cultivation of the participation process, so that the villagers can truly participate in the construction of their homes.

The government, designers, and villagers participate in the whole process of the construction process. First, they can understand the advantages and disadvantages of the village and the actual needs of the villagers in more detail. Then, they can have an impact on the direction of the village's development. The second is to better coordinate the relationship between the government, designers, villagers, social forces, and other parties. Third, it can not only complete the transformation of environmental space, but also promotes industrial upgrading, enhances the self-governance ability of the village, and finally realizes sustainable development of the village.

5.2.2. Common Progress of Design Concepts—Artistic Vision of the Countryside

The difficulty of rural construction lies in "people", and the difficulty of art rural construction is more prominent in "people", because there have always been different opinions in terms of aesthetic awareness. The aesthetic concepts of villagers and artists are often completely different, and their ideas are inconsistent. How can they move towards the same goal? The first step in the design of rural environment under the concept of joint construction is to change the concept of villagers, look at the countryside from an artistic perspective, re-understand the value of the countryside, and re-understand the wisdom of the countryside [17]. The success of Japanese art village construction lies in the fundamental change in the concept of farmers.

Rural revitalization is essentially rooted in the cultural customs, the way of life, and the villagers' inheritance of traditional handicraft skills. The seemingly random and irregular construction in the countryside contains the traditional construction techniques of local dwellings and conveys the cultural attributes and regionality of the village itself. Craftsmen with traditional craftsmanship in the countryside use their own craftsmanship and wisdom to build village houses, manufacture living utensils, and connect rural life. Before the reconstruction of the village, it is necessary for artists, designers, and villagers to regain the wisdom of the functional layout of the rural space and the traditional rural craftsmanship to awaken the rural memory. In my country's traditional villages, the spatial texture of the village looks natural, but it is also the most reasonable. The location of each building, the location of the village entrance, and the location of the ancestral temple have their own rules of development. From any angle, the wisdom of the countryside is revealed [18].

5.2.3. The Beginning of the Change of Design Concept—Starting from the Needs of the Villagers

In the previous construction, designers and artists simply took "beautiful and livable" as the goal of village construction. They think that as long as the living space environment of the villagers is improved, the living standards of the villagers can be improved, and the village can develop. However, only focusing on the "appearance" of rural construction cannot make the road of rural construction go very far, and some villages have even stopped developing. To realize the transformation from "beautiful and livable" to sustainable development of rural construction, designers and artists need to start from the actual problems of the village, change their design thinking, and explore more effective construction methods for the countryside. Starting from the foundation of rural industries, creating opportunities for industrial development created by multiple functions is a necessary condition to stimulate the enthusiasm of village development. According to the actual conditions of different villages, the feasibility of rural industry cultivation should be investigated, existing industries should be continued, and new industries that can be cultivated should be added; this will stimulate the endogenous power of village revitalization.

Rural environment designers, under the guidance of the concept of co-construction, should abandon the ideal thinking of rural construction, listen to the real voice of the village, face the actual demands of the villagers, and let the designers truly serve the village. In this way, the enthusiasm of the villagers will be mobilized and devoted to home construction, as shown in Figure 5.

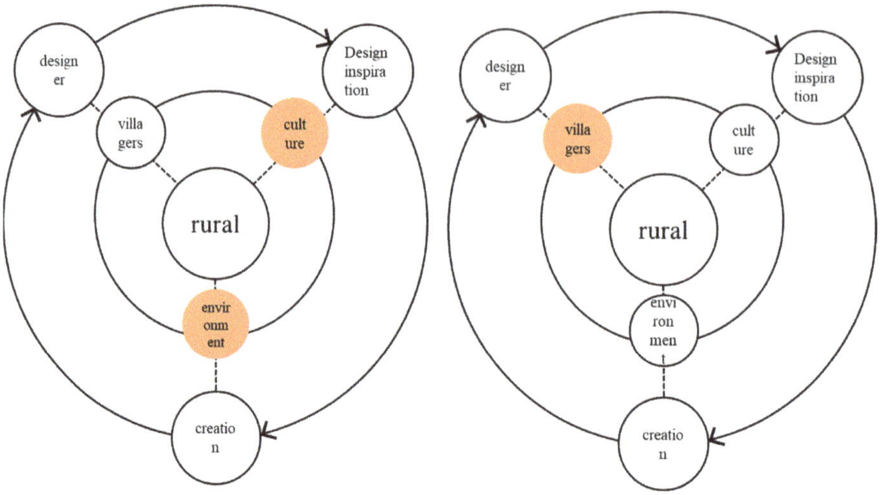

Figure 5. Changes in designers' rural construction ideas.

5.2.4. Mutual Transfer of Design Ideas—An Easy-to-Communicate Way of Expression

In the past, in cases of art intervention in rural environment construction, participation of villagers in the design stage was very low. One of the important reasons was that the ideas of designers, artists, and villagers were not well communicated to each other. Designers and artists cannot stand in the position of villagers, and the villagers cannot understand and accept the ideas of designers and artists. Therefore, it is important for art to intervene in the rural environment under the concept of joint construction to make it easier for both parties to obtain each other's ideas and information [19].

The common way of expressing the environmental design scheme is to use a computer to construct a site model and to render the design. This method is suitable for use in the city, but it does not seem to be the best method for the countryside. The more acceptable way of expressing drawings for villagers is the complete integration of the design scheme and the actual site. The designer draws the design ideas directly from the photos of the original site and compares them before and after the renovation. This reflects the design effect more intuitively [20].

6. Conclusions

Taking Zhanzishi Village in southeastern Hubei Province as an example, this paper analyzes the purpose of cultural intention and visual communication design of the rural environment. On the one hand, the village environment is beautiful, and the villagers' happiness index is improved; On the other hand, we should develop rural tourism and improve the economic situation of the villagers. In the new era, the purpose of rural environmental art design is to improve the quality of rural residential environment. With the strong development trend of Chinese cities, rural areas are slowly trapped in green ecological dilemma and aesthetic misconceptions. In many places, the problem of simplification is more serious in the process of development, ignoring the establishment of the existing rural aesthetic situation and the overall level of rural environmental art design in the new countryside. As the level of global integration and economic development globalization continues to increase, the traditional farming culture and technology in China have been greatly affected. The inadequate and uneven development of regions and cities has seriously affected the rural living environment, and traditional rural aesthetic culture and art have been severely tested. "Beautiful rural construction" should be realized from time to time to inject new vitality into the new rural infrastructure, generally learn the experience and lessons of urban infrastructure, and use visual communication art introduction to highly combine aesthetics. It can improve the quality of rural plastic arts, the natural environment, and technology on the premise of meeting the growing material civilization and cultural needs of farmers. Therefore, the new rural environmental art design and visual communication must actively refer to the experience and lessons of the city. It should use the improvement of local comprehensiveness as a guidance, reflect the progressive atmosphere of the times, and promote the effective development of the new rural environmental art design and visual communication work. This should be done on the premise of meeting the chemical life vision of farmers.

Author Contributions: Conceptualization, methodology, F.L.; software, validation, B.L.; formal analysis, investigation, resources, data curation, K.M.; writing—original draft preparation, F.L.; writing—review and editing, F.L.; visualization, B.L.; supervision, K.M.; project administration, F.L.; funding acquisition, B.L. All authors have read and agreed to the published version of the manuscript.

Funding: The research is supported by: Annual Project of Shaanxi Social Science Fund "Research on Public Art Intervention Path Based on Rural Spatial Cultural Identity in Guanzhong Area" (Project No.: 2021J047). Special scientific Research Project of Education Department of Shaanxi Province: Systematic design of cultural and creative products of red Culture tourism in northern Shaanxi project (Project No.: 20JK0076). Project of Shaanxi Province to improve public Science Quality (Project No.: 2021PSL131).

Institutional Review Board Statement: Not applicable.

Informed Consent Statement: Not applicable.

Data Availability Statement: The labeled data set used to support the findings of this study is available from the corresponding author upon request.

Conflicts of Interest: The authors declare no conflict of interest.

References

1. Shams, S.; Asl, M.R.S. An Analysis of Earthquake Preparedness in Tehran with Emphasis on the Resilience Theory (Case Study of Tehran District 4). *J. Hous. Rural. Environ. JHRE* **2020**, *39*, 129–144.
2. Ionescu, C.; Malița, M.A.; Perieanu, V.Ș.; Burlibașa, M.; Dina, M.N.; Costea, R.; Perieanu, M.V.; Costea, R.; Babiuc, I.; Beuran, I.A.; et al. Treatments' Complexity in Dental Care Assistance, Urban Versus Rural Environment—Preliminary Study. *Acta Med. Transilv.* **2021**, *26*, 75–78. [CrossRef]
3. Ansari, A.H.M.; Umar, R.; Us Saba, N.; Sarah, S. Assessment of Current and Future Groundwater Stress through Varied Scenario Projections in Urban and Rural Environment in Parts of Meerut District, Uttar Pradesh in Ganges Sub-Basin. *J. Geol. Soc. India* **2021**, *97*, 927–934. [CrossRef]
4. Yin, Y.; Yin, H.P. Study on Impacts of Mining Development in Wanjiang City Belt on Urban and Rural Environment. *IOP Conf. Ser. Earth Environ. Sci.* **2021**, *831*, 012053. [CrossRef]
5. Nipen, M.; Vogt, R.D.; Bohlin-Nizzetto, P.; Borgå, K.; Mwakalapa, E.B.; Borgen, A.R.; Jørgensen, S.J.; Ntapanta, S.M.; Mmochi, A.J.; Schlabach, M.; et al. Spatial trends of chlorinated paraffins and dechloranes in air and soil in a tropical urban, suburban, and rural environment. *Environ. Pollut.* **2021**, *292*, 99–105. [CrossRef] [PubMed]
6. Wang, H. Filling the Distance between Village and Art with Footsteps—Taiwan Professor Wu Lixing's Rural Revitalization Road. *Taisheng* **2021**, *16*, 76–77.
7. Reed, K.; Reed, B.; Bailey, J.; Beattie, K.; Lynch, E.; Thompson, J.; Vines, R.; Wong, K.C.; McCrossin, T.; Wilson, R. Interprofessional education in the rural environment to enhance multidisciplinary care in future practice: Breaking down silos in tertiary health education. *Aust. J. Rural. Health* **2021**, *29*, 127–136. [CrossRef] [PubMed]
8. Moraitis, N.; Tsipi, L.; Vouyioukas, D.; Gkioni, A.; Louvros, S. Performance evaluation of machine learning methods for path loss prediction in rural environment at 3.7 GHz. *Wirel. Netw.* **2021**, *27*, 4169–4188. [CrossRef]
9. Liu, P.; Qi, S.; Li, D.; Ravenscroft, N. Promoting agricultural innovation as a means of improving China's rural environment. *J. Environ. Manag.* **2020**, *280*, 111675. [CrossRef] [PubMed]
10. Liu, Y.; Liang, X. Application of Chinese Regional Traditional Cultural Elements in New Rural Landscape Design. *IOP Conf. Ser. Earth Environ. Sci.* **2020**, *598*, 012043. [CrossRef]
11. Moncho, M.; Palomar-Abril, V.; Soria-Comes, T. Palliative Care Unit at Home: Impact on Quality of Life in Cancer Patients at the End of Life in a Rural Environment. *Am. J. Hosp. Palliat. Med.* **2022**, *39*, 523–529. [CrossRef] [PubMed]
12. Al-Gohary, A.M.; Attia, E.A.; El-Shennawy, L.H. Training Needs for Female Agricultural Extension Agents in Planning Extension Programs to Preserve the Rural Environment in Gharbia Governorate. *Alex. Sci. Exch. J.* **2021**, *42*, 1127–1141.

13. Omoze, E.L.; Edeko, F.O. Statistical tuning of cost 231 Hata model in deployed 1800mhz GSM networks for a rural environment. *Niger. J. Technol.* **2021**, *39*, 1216–1222. [CrossRef]
14. Yang, Y. Exploration on the Path of Intangible Cultural Heritage Endowed Beautiful Rural Construction From the Perspective of Art Education—The case of Yunfu Stone Art. In Proceedings of the 2nd International Conference on Literature, Art and Human Development (ICLAHD 2020), Xiamen, China, 23–25 October 2020; Advances in Social Science, Education and Humanities Research. Atlantis Press: Amsterdam, The Netherlands, 2020. [CrossRef]
15. Wan, A.; Qi, X.; Yue, W.; Tu, R. Construction and case verification of rural environmental value-added evaluation system. *Environ. Dev. Sustain.* **2022**, *24*, 116–123.
16. Ao, Y.; Martek, I. Public Satisfaction-Based Performance Appraisal of Rural Infrastructure Construction. In *Rural Built Environment of Sichuan Province, China*; Springer: Singapore, 2020. [CrossRef]
17. Zhi, L.I.; Mao, H.Y. Discussion on Residents' Satisfaction with Rural Cultural Construction in Poverty-stricken Areas and Its Promotion Countermeasures: A Survey Based on Chongqing. *J. Chongqing Technol. Bus. Univ. (Soc. Sci. Ed.)* **2020**, *55*, 16–22.
18. Zhu, H.; Wang, Z. Construction and Research of Formative Evaluation System in Fine Art Classroom Teaching in Rural Primary Schools. *Asian Agric. Res.* **2020**, *12*, 77–84.
19. Brycht, N. Construction Waste Management in Rural Areas of the Czestochowa District in the Aspect of Environmental Safety. *Conf. Qual. Prod. Improv. CQPI* **2020**, *2*, 60–68.
20. Amerani, E.; Partalidou, M. The construction of the rural through social media: An imaged-based methodology to understand perceptions for the Greek countryside. *Int. J. Sustain. Agric. Manag. Inform.* **2020**, *6*, 250–260. [CrossRef]

International Journal of
Environmental Research and Public Health

MDPI

Article

Healthy City Community Space-Oriented Structural Planning and Management Optimization under COVID-19

Ya Yang [1,†], Zhengyu Jiang [2], Yawei Hou [3], Huaxing Wang [4,5,*] and Zeyu Wang [6,†]

[1] School of Architecture & Urban Planning, Anhui Jianzhu University, Hefei 230009, China
[2] School of Economics and Finance, Xi'an Jiaotong University, Xi'an 710061, China
[3] School of Mechanics and Civil Engineering, China University of Mining and Technology, Xuzhou 221116, China
[4] Law School, Hangzhou City University, Hangzhou 310000, China
[5] School of Economics, Zhejiang University, Hangzhou 310000, China
[6] School of Public Administration, Guangzhou University, Guangzhou 510006, China
* Correspondence: 18603848020@163.com
† These authors contributed equally to this work.

Abstract: This work studies ways of Healthy City Construction (HCC) and Urban Governance Optimization (UGO) during the COVID-19 pandemic. The specific urban community space planning structure is proposed following a literature review on the healthy city's theoretical basis and historical development. Then, the proposed HCC-oriented community space structure is tested by surveying residents' physical and mental health and infectious risk using a questionnaire survey and Particle Swarm Optimization (PSO). Specifically, the particle fitness is calculated according to the original data conditions, and the community space with the highest fitness is determined. Based on the calculation, the community space's neighbors are investigated from different aspects through a questionnaire survey on patients' daily activities and community health security coverage. The results showed that: (1) The score of daily activities of community patients with respiratory diseases was 2312 before the implementation of the proposed community structure and 2715 after the implementation. Therefore, the service quality of residents increases after implementation. (2) The proposed HCC-oriented community space structure improves the physical self-control ability of chronic patients and helps them reduce their pain. This work aims to create a people-oriented healthy city community space, improve the city's "immune system," and regenerate the energy and environmental sustainability of the urban living environment.

Keywords: healthy city; health security; Community spatial structure planning; community space

Citation: Yang, Y.; Jiang, Z.; Hou, Y.; Wang, H.; Wang, Z. Healthy City Community Space-Oriented Structural Planning and Management Optimization under COVID-19. *Int. J. Environ. Res. Public Health* **2023**, *20*, 3863. https://doi.org/10.3390/ijerph20053863

Academic Editors: Hongxiao Liu, Tong Wu and Yuan Li

Received: 5 January 2023
Revised: 19 February 2023
Accepted: 20 February 2023
Published: 21 February 2023

1. Introduction

Since early 2020, countless human, financial, and material resources have been invested in disease control and prevention with the outbreak of the Corona Virus Disease 2019 (COVID-19) pandemic. The pandemic has aroused panic among the population and has also rung an alarming bell against humans to protect their health better against diseases while pursuing economic development [1]. This is especially true for urban residents who live in extremely crowded community spaces with relatively fewer health facilities and inadequate medical services. Thus, the voice for a Healthy City Construction (HCC) is a rising concern for public health [2]. The HCC involves community space planning and intelligent management on a legal basis, among others. Its ultimate goal is to safeguard the resident's health and safety and improve their living space and subjective well-being [3].

Urban management has always tried to improve residents' health and has achieved some success and academic achievements. Over the years, scholars have found that the sustainable development philosophy is the key to HCC. Urban transport planning, public health facilities, and supportive policies are all essential for the HCC [4]. Yang et al. evaluated several Chinese cities using the Healthy City Construction Index (HCCI). The team explained that different urban districts and streets' HCCI could be positively correlated, negatively correlated, or not correlated [5]. In China, HCC is just taking its maiden flight. Other studies examined residents' mental states during the COVID-19 pandemic. They observed that the national lockdown policies had harmed the residents' health, causing social distancing, loneliness, and eating disorders [6]. Another research group found the HCC played a major role in relieving residents' mental stress during the COVID-19 pandemic. The research highlighted the potential and key aspects of urban space regeneration. The research results enabled people to live new social lives, helping people to heal disease with a better mindset and get through hard times [7]. There is also evidence that social relationships, physical vitality, life attitudes, diet, and weight are important health indexes. Creating an environment that supports healthy behaviors helps engage citizens and builds their confidence [8]. This further demonstrates the importance of HCC. In urban community spaces, structure planning and the residents' physical and mental health are correlated [9]. In order to encourage residents' physical exercise, the community spatial structure must consider both the individual household's life habits and overall demographic distribution as much as possible. On the other hand, environment protection and public resource conservation can strengthen social cohesion and promote a green and open community space.

Therefore, this study will study the specific requirements of HCC-oriented community space planning and management and proposes an urban community space planning structure. The paper then discusses the residents' physical and mental health under the proposed structure. Based on this, it summarizes the HCC development requirements and standards. In order to do so, based on the smart city residents' health assessment index, a questionnaire survey is designed. Ultimately, the intelligent management measures of the smart and healthy city are explored, highlighting the importance of healthy city-oriented Urban Governance Optimization (UGO).

2. Methods

2.1. Theoretical Sources of Healthy and Smart Cities

The smart city is an information-based digital city development model by emerging network technologies such as the Internet, Internet of Things (IoT), and edge computing. Its main features are the integration of smart technologies with urban facilities and urban governance to generate a low-carbon and human-friendly living environment [10]. Based on the smart city model, the "healthy city" is a complex system of engineering. Thus, HCC intersects with the field of "smart city" construction [11]. An HCC-oriented community space planning focuses on people orientation. The healthy city is a new and future sustainable urban development paradigm in the Internet era. In urban governance, people orientation must be reflected in all stages, including urban planning, development, and construction, to achieve sustainable growth. A healthy city is expected to be a panacea for improving public social services and urban residents' health and security [12]. Table 1 shows the difficulties of constructing a healthy city.

Table 1. Difficulties in creating healthy cities.

City Disease	Influence Situation
Polluted environment	Residents have low health conditions and poor living conditions.
Mass infection	
Traffic congestion	Residents' simple but basic needs are not always met, such as traveling and some living materials.
Stressful life	
Frequent disasters	Job opportunities are rare, and defaults on wage payments occur occasionally.
Prominent social problems	

HCC should consider the macro, meso, and micro factors. Then, in terms of residents' health, HCC will involve socialized health, urbanized health, and population-based health. The specific structure is displayed in Figure 1.

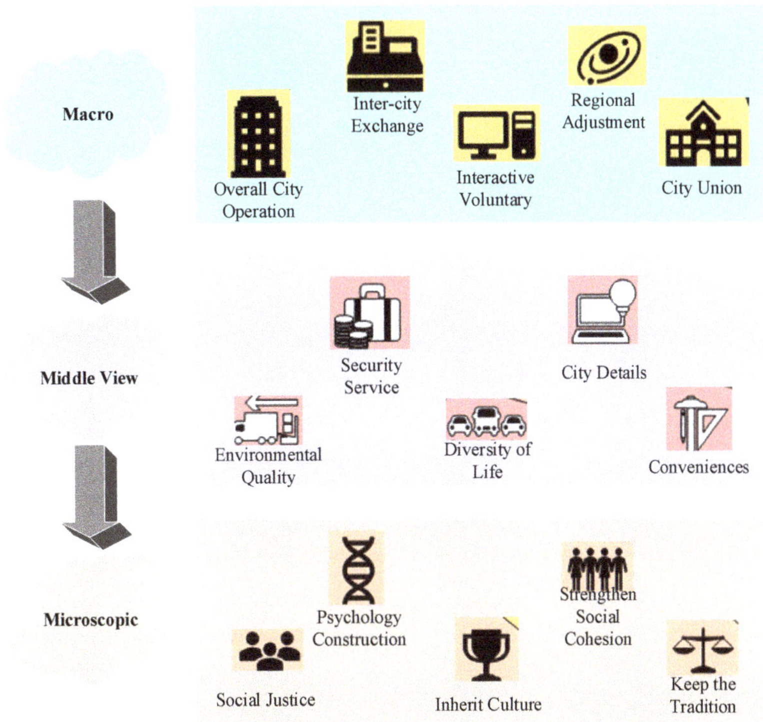

Figure 1. Three important structures of healthy cities.

According to Figure 1, theoretically, the HCC must meet the following requirements. The community space is the basic unit of a city and the main place for residents' daily activities. Accessing health services and facilities is a community residents' common demand [13]. A healthy city integrates, connects, and supplies various health service resources in the community and strengthens residents' health-management awareness and ability. It also improves the community health services' efficiency and effect [14,15]. HCC-oriented community spatial structure planning, management, and optimization provide a basis for evaluating the current situation of community health service security in China. They will help analyze the factors affecting the efficiency of community health service centers to formulate appropriate and effective measures. This has practical significance for solving the residents' medical treatment difficulties and realizing China's people-oriented medical reform goal.

2.2. Using Particle Swarm Optimization (PSO) to Evaluate Residents' Health Status and Infectious Risk in a Healthy City

COVID-19 is a highly contagious respiratory infectious disease. The virus can spread directly or indirectly through the host. Three primary factors increase the resident's infectious risk, as explained in Figure 2:

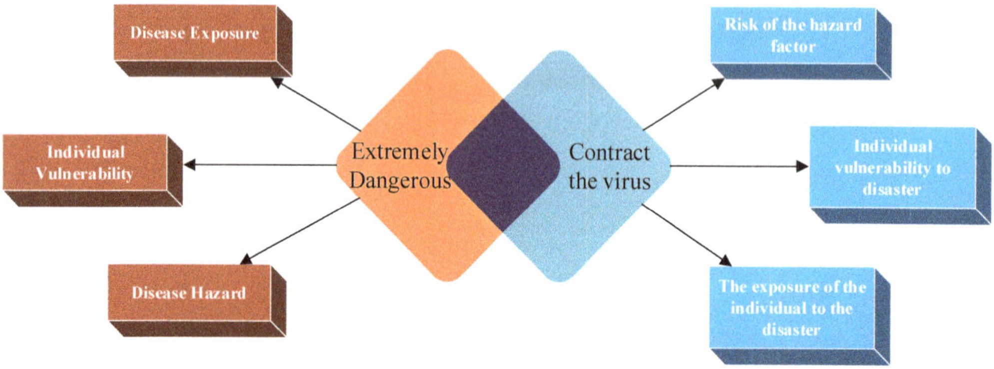

Figure 2. Residents' primary infectious factors.

The disease's infectious risk will be analyzed and calculated as follows: Define the risk value = $R(A, T, V)$. R is calculated by Equation (1).

$$R = H \times E \times V \tag{1}$$

In Equation (1), R means the resident infectious risk index for respiratory diseases, and H refers to the risk index of disasters caused by respiratory infectious diseases among urban residents. E represents the exposure risk index of residents to the disease, and V expresses the vulnerability of residents. Therefore, the specific evaluation value of the risk is calculated by Equation (2) based on the vulnerability, exposure, and upper respiratory tract infectious risk:

$$R_i = E_i W_e \times V_i W_v \times H_i W_h \tag{2}$$

In Equation (2), the E, V, and H are calculated by Equations (3)–(5):

$$E_i = \sum_{i=1}^{n} K_{ei} W_{ei} \tag{3}$$

$$V_i = \sum_{i=1}^{n} K_{vi} W_{vi} \tag{4}$$

$$H_i = \sum_{i=1}^{n} K_{hi} W_{hi} \tag{5}$$

R_i is the residents' comprehensive infectious risk against respiratory diseases in the i community. The larger the value, the greater the community residents' infectious risk is. In addition, E_i, V_i, and H_i, respectively, represent the residents' exposure, vulnerability, and upper respiratory tract infectious risk in the i community; K_{ei}, K_{vi}, and K_{hi} demonstrate the quantities of the three indicators, respectively; W_{ei}, W_{vi}, and W_{hi} express the weights of the three indicators, respectively.

Further, original data with different degrees and quantification will be standardized to eliminate the quantitative relationship between different variables and make comparison easier. To this end, Equations (6) and (7) are used to screen all data indicators according to the positive and negative indicators.

$$x_i = \frac{x_i}{\max(x_{i'})} \tag{6}$$

$$x_i = \frac{\min(x_{i'})}{x_{i'}} \tag{7}$$

Equation (6) is a positive index operation, and Equation (7) is a negative index operation, where x_i and $x_{i'}$ represent the standard value and the original value of the ith index. $\max(x_{i'})$ and $\min(x_{i'})$ denote the maximum and minimum value of the ith index, respectively.

Subsequently, PSO will use the original high-level data characteristics to calculate the index score. The main steps are: 1. the population with different conditions is divided; 2. the most suitable position is found according to the conditions; 3. the particle swarm is initialized; 4. the fitness of the current particle position is calculated; 5. the current position of the particle is set as Pbest; 6. the position Gbest with the highest fitness is taken; 7. a questionnaire survey is issued to the community with the highest fitness, and 8. the best risk assessment is obtained by summarizing. The quickest way for a bird swarm to search for food randomly is to infer the individual closest to the food source if there is only one food source in this specific area. In this "bird flock", individuals communicate to acquire new knowledge and change their behaviors according to the environmental situation [16]. The evolution of the entire bird population includes the birth of new organisms, the process of group classification, and the process of searching for food.

There are a total of n particles in the space of dimension d, and each particle has a potential solution obtained by the objective function. The fitness is the function value of the particle position, which is used to judge the quality of the particle position. Equations (8) and (9) display the particle's position and update speed:

$$v_{id}(t+1) = W * v_{id}(t) + c_1 r_1 (P_{id} - X_{id}(t)) + c_2 r_2 (P_{gt} - X_{id}(t)) \quad (8)$$

$$X_{id}(t+1) = X_{id}(t) + v_{id}(t+1) \quad (9)$$

$v_{id}(t+1)$ is the velocity of particle i after t + 1 iteration; W is the weight of particle velocity; $v_{id}(t)$ is the velocity of i in dimension d at iteration t; $X_{id}(t)$ is the position of i in dimension d of the tth iteration; c_1 and c_2 indicate learning factors and are positive real numbers, and r_1 and r_2 are random distribution functions on the interval [0, 1].

2.3. Community Space Structure Planning for Healthy Cities

A healthy city focuses on people's health in all aspects, from urban planning, construction, and management to ensure the healthy life and work of the general public. The health city is an organic integration of healthy people, environment, and society necessary for the development of human society [17,18]. A healthy city's space planning is to regenerate urban resources and redistribute urban materials. Rational construction of land planning can help adjust the overall urban spatial design to promote sustainable city growth. It is a predetermined procedure for controlling land use and planning for fair welfare. The purpose of urban planning is to realize residents' high-quality life through urban construction and other strategies and initiatives. From another perspective, urban planning is to determine the size, nature, and development direction of a city. Planning the urban physical space achieves the economic and social development goals over a certain period of time and promotes the quality of life of the urban population [19,20]. Figure 3 shows the healthy city' community spatial planning structure.

The specific planning scheme of the healthy city community space is presented in Figure 4.

The proposed community space planning scheme plans not only material resources and visualizes functions through smart technologies but also covers various cultural factors. The proposed community spatial structures center around resents' health and offer a novel community space planning.

The following factors influence the planning and management of the community spatial structure:

1. Rational organization of urban road traffic. In particular, the division of arterial roads is often the main factor determining the scale of residential land.
2. The radius of public facilities for residents living services. Reasonable facility scale (i.e., the economy of operation and management) and service radius (i.e., the walking

distance of residents using public buildings) affect the convenience and safety of residents' lives.

3. A unified urban administrative management system. As the basic cell of society, the family is bound to be linked with the administrative organization of society.
4. Terrain. The organizational form of residential areas is affected and restricted by topography to a certain extent. It is generally difficult to compactly arrange residential land on undulating terrain [21].

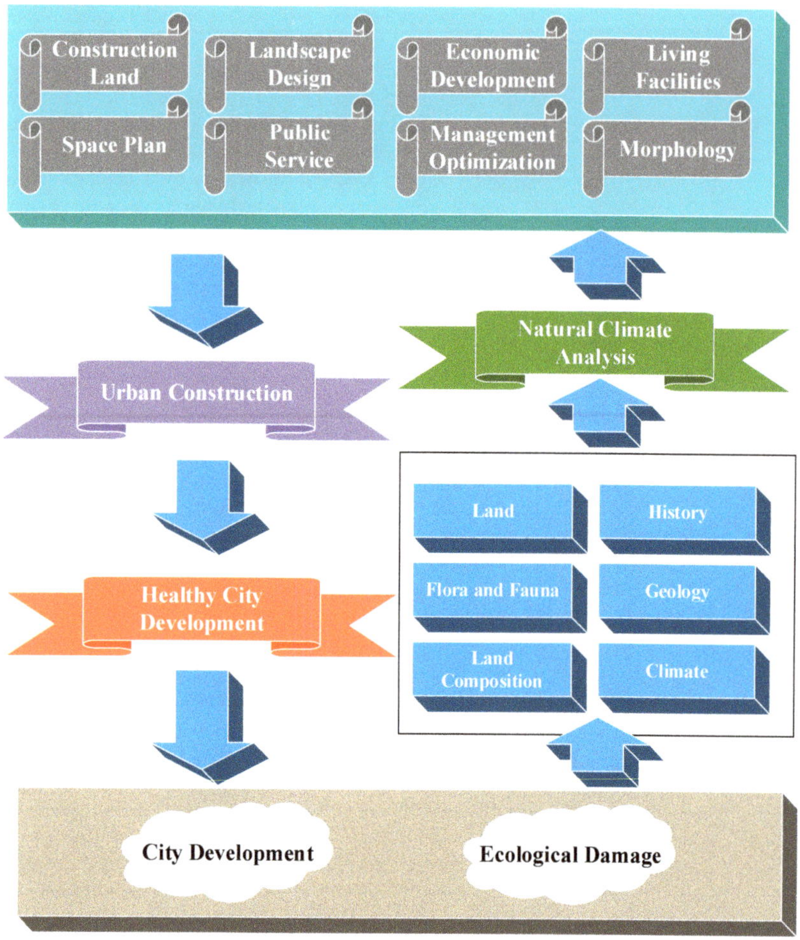

Figure 3. Healthy city's community spatial structure planning.

According to the concept of "three areas and two channels", healthy urban planning divides the community space into three levels: block, residential area, and group, with public roads running through them. The layout of the spatial form features an overall openness and specific enclosedness. Under such a layout, when respiratory diseases spread, the community space can be rapidly transformed into a three-level prevention and control system [22], as shown in Figure 5.

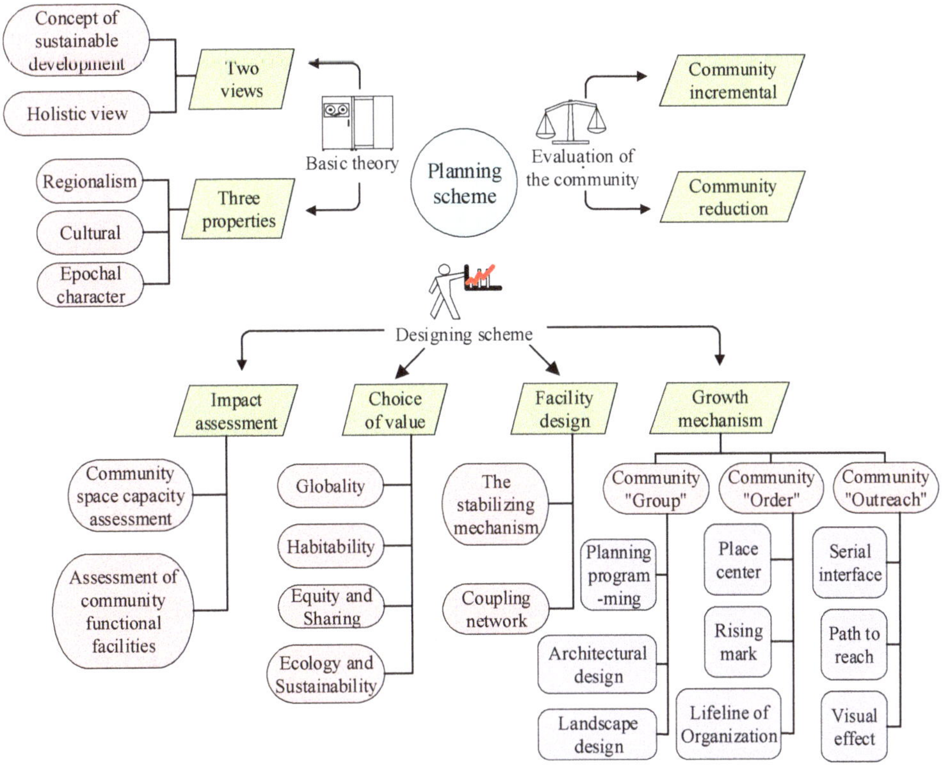

Figure 4. The healthy city community space' specific planning process and schemes.

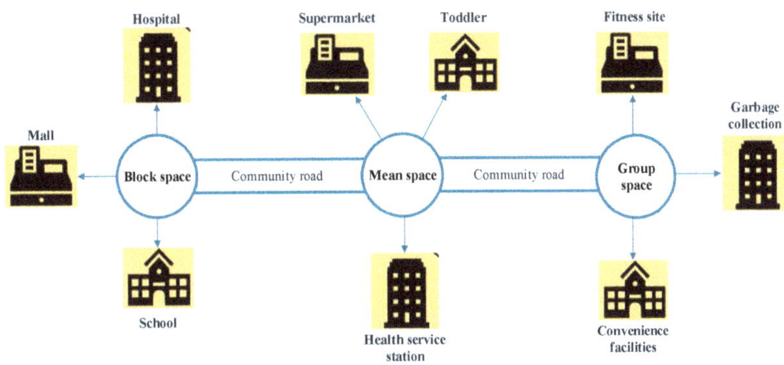

Figure 5. Community three-level prevention and control system.

In Figure 5, the block space is mainly an open mode, providing the public with municipal, commercial, transportation facilities, and other services; it serves within the block and also serves its neighbors, realizing the fully open and shared 15 min living circle. The residential space is mainly semi-open to create good lighting, ventilation, and a green environment; it sets up adequate outdoor fitness places for vulnerable groups and susceptible people to carry out basic fitness activities downstairs. The group space is mainly shared in private mode, and the convenience facilities, public elevators, corridors, garbage collection facilities, and other public service supporting facilities within are group-specific services, not shared externally [23].

Different from the traditional layout of the residential area, the community's three-level spatial layout can further divide the closed level of the residential area according to the road grade. "Sina" is a new community pilot project. "Sina", an overall openness and specific enclosedness pattern, increases the space for neighborhood communication and can be relatively enclosed to form isolated units under abnormal conditions. Meanwhile, the medical concept of "three areas and two channels" is combined in the plane layout. It follows the design principles of the healthy city in detail design and relies on the intelligent system in management. The layout equips the community with the resilience to handle sudden crises and realizes the original intention of "managing and controlling" an open community [24].

2.4. Reliability and Validity Analysis

Reliability means the consistency or homogeneity of the measurement results. According to the literature, reliability can be divided into internal reliability and external reliability. Internal reliability refers to whether a group of questions measures the same concept, that is, how consistent is the group of questions. Common analysis methods include Cronbach's and split-half reliability. External reliability refers to the degree of consistency of repeated measurements on the same research object for the same questionnaire at different times. The commonly used analysis method is retesting by SPSS. Validity is to test the validity of the questionnaire; generally speaking, it is to determine whether the designed items are reasonable and whether they can effectively reflect the researcher's goals. A questionnaire survey usually contains single-choice, multiple-choice, fill-in-the-blank questions, and other forms of questions. Validity analysis is only for scale data, and non-scale items such as multiple-choice, single-choice gender, and other items cannot be used for validity analysis [25].

2.5. Questionnaire Design of Investigating Healthy City Resident's Health Status and Public Services

As a developing country, China still struggles to transform all community health services into public institutions in the short term. Since 2008, Hefei City has greatly improved the environment and facilities of community health service institutions through innovation and increasing investment in community health services. In November 2011, 127 community health service institutions were established in the urban area of Hefei, and the coverage rate of community health service for urban residents was over 95% [26]. This experiment conducted an online questionnaire survey on the residents of the smart community in Hefei, and the distribution objects were divided into the elderly, the middle-aged, the adolescent, the sick (including various physical diseases), and the healthy group. The questionnaire survey investigated residents' health, mainly from respiratory diseases and community health services. Respiratory diseases are based on the residents' basic knowledge of pneumonia, community epidemic prevention work, self-vaccination, and physical conditions. Community service involves the community environment, community staff, work content, and physical and mental health status. Table 2 exhibits the content of the questionnaire.

The validity of the questionnaire needs to be judged by a specific validity test. The validity of the questionnaire is tested by statistical software. The relevance of the survey topic, the correctness of the question, and the appropriateness of the survey time are tested based on the obtained initial data. Through the evaluation of statistical methods and relevant experts, the obtained results of the validity test are expressed in Table 3.

Table 2. The questionnaire.

	Basic Information					
Age	Do you have a chronic disease?					
Gender	Have you been infected with COVID-19?					
No.	Content of the questionnaire	Options: A. Strongly agree B. Relatively agree C. Neutral D. Disagree E. Extremely disagree				
	Respiratory diseases (mainly COVID-19)					
1	COVID-19 is extremely contagious.	A	B	C	D	E
2	Maintaining personal and environmental hygiene is an effective measure to prevent COVID-19.	A	B	C	D	E
3	In a crowded environment, I always wear a mask.	A	B	C	D	E
4	I am actively vaccinated against COVID-19.	A	B	C	D	E
5	You are satisfied with your community's COVID-19 prevention efforts.	A	B	C	D	E
6	Have you recently contracted other respiratory diseases?	A	B	C	D	E
7	Community workers will often visit your home to check on your health.	A	B	C	D	E
	Status of community safety and health services					
1	I think my living environment is clean and up to standard regarding hygiene.	A	B	C	D	E
2	I can achieve my goals in my living environment in a healthy state.	A	B	C	D	E
3	My life and activities will not be threatened by environmental sanitation.	A	B	C	D	E
4	Community staff will conduct regular health and hygiene training.	A	B	C	D	E
5	You are satisfied with the health service attitude of the community workers.	A	B	C	D	E
6	Community workers can achieve health protection purposes in a variety of ways.	A	B	C	D	E
7	You are satisfied with the work content of health care in the community.	A	B	C	D	E
8	You think the spread of healthy communities is very important.	A	B	C	D	E
9	The appearance of community workers is decent.	A	B	C	D	E
10	The community environment is quiet, clean, and human-focused.	A	B	C	D	E
11	Community service workers are very concerned about the residents, and they can help answer any concerns in a timely manner.	A	B	C	D	E
12	Community health service centers are generally trustworthy.	A	B	C	D	E
Other supplements:						

Table 3. Results of the validity test of the questionnaire.

Performance \ Methods	Statistics	Expert Assess
Initial data	Good	Good
Relevance	Good	Good
The correctness of the question	Good	Better
The appropriateness of the survey time	Good	Good

According to Table 3, the overall effect and performance of the questionnaire design were good, and the validity was up to standard. Follow-up data collection can be continued. The questionnaire was distributed twice: before the implementation of the proposed HCC-oriented community spatial planning structure and after the implementation. Overall, 200 copies of questionnaires were distributed to residents in the community in a healthy city in a certain place, and 186 valid questionnaires were collected, with an effective rate of 93%.

3. Results

3.1. Analysis of the Daily Living Ability of Patients with Respiratory Diseases

This section uses a questionnaire survey to analyze before and after the proposed HCC-oriented community spatial planning structure. Community services are investigated from the aspects of a community environment, community staff, community service work content, and physical and mental health status. A scoring system is adopted for statistical analysis, and each item is scored from 0 to 30 points. The results will reflect the patients' physical mobility and their ability to perform activities of daily living. The questionnaire data needs to be sorted and analyzed before processing. In this experiment, the questionnaire is divided into two aspects for investigation. The first is a survey on the disease status of community patients with respiratory diseases (mainly COVID-19). In the experiment, the effectiveness of the proposed HCC-oriented community spatial planning structure was identified by comparing the scores of patients' ability to act by themselves and scores of their daily life impact before and after the implementation of the proposed community structure. The relevant data in the questionnaire is drawn as a statistical graph for analysis of the results. The data results are illustrated in Table 4.

Table 4. Comparison of scores of daily activities of patients with community respiratory diseases before and after the implementation of the proposed HCC-oriented community spatial planning structure.

Time of Evaluation \ Judgment Standard	Patient's Physical Mobility	Daily Activities
Before the implementation of the proposed community structure	17.43 ± 7.28	16.57 ± 7.24
After the implementation of the proposed community structure	20.18 ± 5.42	19.75 ± 5.45

The total score of the statistical questionnaire is 2312 before the implementation of the proposed community structure and 2715 after the implementation. Apparently, the service quality of residents after the implementation has improved. The analysis of the two modules: epidemic prevention and community health services, shows that both before and after implementation, epidemic prevention work has higher scores than daily work. Thus, the community is relatively secure. However, it also manifests that while the service workers are increasing their efforts to overcome the epidemic, they are negligent in daily health services. Therefore, these public servants are expected to pay more attention to managing community health, coordinate their work, and take corresponding incentive measures. The data results are expressed in Figure 6.

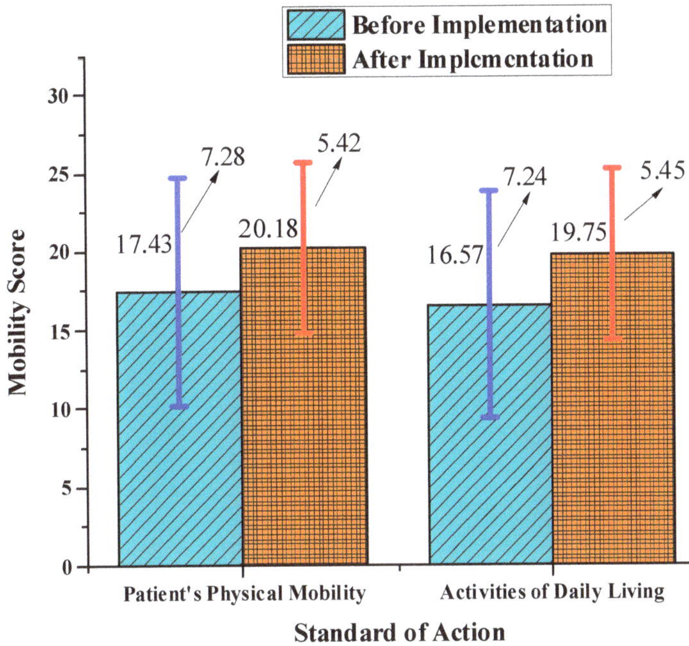

Figure 6. Comparison of daily activity scores of community patients with respiratory diseases before and after the implementation of the proposed community structure.

Figure 6 manifests that before the implementation of the proposed community structure, the average score of physical mobility of patients with respiratory diseases in the community was 17.43, and the score of positive and negative errors was 7.28. After the implementation, the score of mobility increased to 20.18, and the score of positive and negative errors has also been reduced to 5.42. Thus, the HCC-oriented community spatial planning has improved the physical self-control ability of patients with respiratory diseases and helped patients reduce the pain of the disease. As for the daily activities of patients, the average score before the implementation of the healthy city was 16.57, and the error in the positive and negative directions was 7.24. After the implementation of the proposed community structure, the score of activities increased to an average of 19.75, and the score of error in the positive and negative directions was reduced to 5.45. The results showed that the overall daily activities of patients were improved. It can be observed that the planning and optimized management of a healthy city can greatly reduce chronic patients' pains, improve their daily activities, and encourage their normal lives. Overall, the finding reflects the importance of HCC.

3.2. *Community Health Services Coverage among Healthy Cities' Residents*

As previously completed, the chronic patients are evaluated on their self-control ability and the scores of daily activities. It is found that the proposed HCC-oriented community spatial planning can indeed improve chronicle patients' self-care ability and is thus effective. Therefore, this section will analyze the second part of the questionnaire, that is, the healthy city's resident's community health service coverage. The number of people who knew the community health service planning of healthy cities in the questionnaire survey is counted, and the proportion in the total number of respondents is calculated. Similarly, the results of the questionnaire will be drawn into a statistical graph, which is indicated in Figure 7.

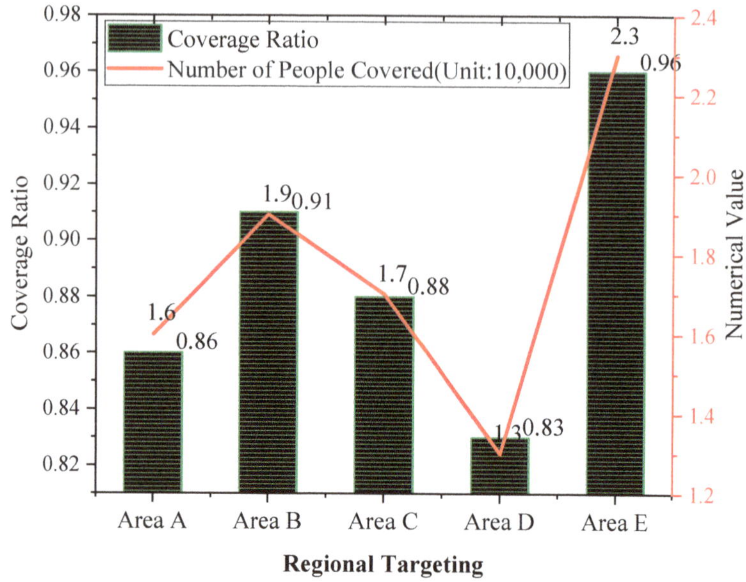

Figure 7. Healthy city residents' community healthy service coverage.

According to Figure 7, among regions A to E, the coverage of health services in community E is the highest, reaching 96%, and about 23,000 people can enjoy health security services, followed by region B, with a coverage rate of 91%, covering 19,000; followed by region C, with a coverage rate of 88%, including 17,000; followed by area A, with a coverage rate of 86%, covering about 16,000 people, and finally community D, with a coverage rate of only 83% and involving about 13,000 people. Theoretically, health security services in a healthy city should be a fundamental resident right. Nevertheless, so far, no community has reached 100% coverage. Therefore, more work is needed in HCC to reach 100% healthy service coverage, which is the basic requirement of a healthy city.

4. Discussion

The questionnaire survey on residents' health status and community health service comprehensively reveals the research results. First, the implementation of the proposed HCC-oriented community spatial planning can help patients alleviate pain, advance their self-control ability, and improve their daily activities. The proposed community structure has brought great benefits to residents' lives during the epidemic and sets an example for all communities to follow and practice. Second, the coverage of community health security is relatively high, reaching 88%. It indicates the vast majority of community health services are guaranteed. Due to varying factors, the investigation and cooperation of relevant departments and every citizen are required in HCC. HCC calls for more, and the following suggestions might help to accelerate HCC:

1. The provision of community health and medical personnel should be accelerated. Community health medical personnel are the foundation of the development of community health services. Without excellent personnel, there will be no quality service. General practice and community nursing programs can be established in medical colleges and universities to specifically train professionals suitable for community health [27].
2. A sound community health electronic information platform needs to be established. The informatization of community health service has played an important role in promoting the development and mode transformation of community health.

3. The income of employees should be increased, and a scientific performance appraisal system should be implemented. While implementing two lines of revenue and expenditure, a positive and effective performance appraisal system should be adopted. Otherwise, the efficiency of community health services will be inevitably affected [28].

In this way, healthy cities should be vigorously developed, and optimized management and health services in the community environment should be strengthened to prevent and control the epidemic better and safeguard residents' life and security.

5. Conclusions

The raging COVID-19 affects both China and countries worldwide, and it is said to coexist with humans for a long time. In this context, this work discusses how to develop a healthy city and optimize its management. Firstly, the theoretical sources of the healthy city are summarized together with its cognitive stages in human history and its main functions and completion standards. Secondly, a PSO is chosen to calculate the healthy city residents' health index. PSO can directly solve the best solution by combining the original high-level data characteristics. Finally, an HCC-oriented community spatial planning structure is proposed and tested through a questionnaire survey. The questionnaire survey tries to understand the patients' living ability and the community health security coverage under the influence of the pandemic. The following two results are obtained. (1) After the implementation of the proposed community structure, patients' daily activities have been greatly improved. (2) At present, the community residents' health security coverage has reached 88%. Then, corresponding recommendations are proposed based on the survey results. Last but not least, there are still some shortcomings in this experiment. The questionnaire questions are not comprehensive enough and need improvement. Therefore, in the follow-up research, the interview scope of the questionnaire will be expanded, and the generality of the experiment will be improved.

Author Contributions: Conceptualization, Y.Y. and Z.W.; Methodology, Y.Y. and Z.W.; Software, Z.W.; Validation, Y.H. and Z.W.; Formal analysis, Y.H. and H.W.; Investigation, Z.J. and H.W.; Resources, Y.H. and Z.W.; Data curation, Y.H. and H.W.; Writing—original draft, Z.J., Y.H. and H.W.; Writing—review & editing, Y.H.; Visualization, Z.J.; Supervision, H.W. and Z.W.; Project administration, H.W. and Z.W. All authors have read and agreed to the published version of the manuscript.

Funding: This research received no external funding.

Institutional Review Board Statement: Not applicable.

Informed Consent Statement: Not applicable.

Data Availability Statement: Not applicable.

Conflicts of Interest: The authors declare no conflict of interest.

References

1. Ardabili, S.F.; Mosavi, A.; Ghamisi, P.; Ferdinand, F.; Varkonyi-Koczy, A.R.; Reuter, U.; Rabczuk, T.; Atkinson, P.M. COVID-19 outbreak prediction with machine learning. *Algorithms* **2020**, *13*, 249. [CrossRef]
2. Gkiotsalitis, K.; Cats, O. Public transport planning adaption under the COVID-19 pandemic crisis: Literature review of research needs and directions. *Transp. Rev.* **2021**, *41*, 374–392. [CrossRef]
3. Costa, D.G.; Peixoto, J.P.J. COVID-19 pandemic: A review of smart cities initiatives to face new outbreaks. *IET Smart Cities* **2020**, *2*, 64–73. [CrossRef]
4. Louro, A.; Marques da Costa, N.; Marques da Costa, E. Sustainable urban mobility policies as a path to healthy cities—The case study of LMA, Portugal. *Sustainability* **2019**, *11*, 2929. [CrossRef]
5. Yang, J.; Luo, X.; Xiao, Y.; Shen, S.; Su, M.; Bai, Y.; Gong, P. Comparing the Use of Spatially Explicit Indicators and Conventional Indicators in the Evaluation of Healthy Cities: A Case Study in Shenzhen, China. *Int. J. Environ. Res. Public Health* **2020**, *17*, 7409. [CrossRef]
6. Singh, A.K.; Misra, A. Herd mentality, herds of migrants/people, and COVID-19 in India. *Diabetes Metab. Syndr.* **2020**, *14*, 497. [CrossRef]
7. Leporelli, E.; Santi, G. From psychology of sustainability to sustainability of urban spaces: Promoting a primary prevention approach for well-being in the healthy city designing. A waterfront case study in livorno. *Sustainability* **2019**, *11*, 760. [CrossRef]

8. De Jong, M.A.; Wagemakers, A.; Koelen, M.A. "We Don't Assume That Everyone Has the Same Idea About Health, Do We?" Explorative Study of Citizens' Perceptions of Health and Participation to Improve Their Health in a Low Socioeconomic City District. *Int. J. Environ. Res. Public Health* **2020**, *17*, 4958. [CrossRef]
9. Vujcic, M.; Tomicevic-Dubljevic, J.; Zivojinovic, I.; Toskovic, O. Connection between urban green areas and visitors' physical and mental well-being. *Urban For. Urban Green.* **2019**, *40*, 299–307. [CrossRef]
10. Sun, M.; Zhang, J. Research on the application of block chain big data platform in the construction of new smart city for low carbon emission and green environment. *Comput. Commun.* **2020**, *149*, 332–342. [CrossRef]
11. Ghosh, S. Smart homes: Architectural and engineering design imperatives for smart city building codes. In Proceedings of the 2018 Technologies for Smart-City Energy Security and Power (ICSESP), Bhubaneswar, India, 28–30 March 2018; pp. 1–4. [CrossRef]
12. Camero, A.; Alba, E. Smart City and information technology: A review. *Cities* **2019**, *93*, 84–94. [CrossRef]
13. Pozoukidou, G.; Chatziyiannaki, Z. 15-Minute City: Decomposing the new urban planning eutopia. *Sustainability* **2021**, *13*, 928. [CrossRef]
14. Zeng, C.; Huang, Y.; Yu, L.; Zeng, Q.; Wang, B.; Xu, Y. Long-Term Assessment of Rehabilitation Treatment of Sports through Artificial Intelligence Research. *Comput. Math. Methods Med.* **2021**, *2021*, 4980718. [CrossRef] [PubMed]
15. Zeyu, W.; Jia, L.; Mingyu, L.; Siting, Y.; Yutong, W.; Xin, C. Edge Computing and Blockchain in Enterprise Performance and Venture Capital Management. *Comput. Intell. Neurosci.* **2022**, *2022*, 2914936. [CrossRef]
16. Saini, J.; Dutta, M.; Marques, G. Fuzzy inference system tree with particle swarm optimization and genetic algorithm: A novel approach for PM10 forecasting. *Expert Syst. Appl.* **2021**, *183*, 115376. [CrossRef]
17. Dobrzański, L.A.; Dobrzańska-Danikiewicz, A.D.; Dobrzański, L.B. Effect of Biomedical Materials in the implementation of a long and healthy life policy. *Processes* **2021**, *9*, 865. [CrossRef]
18. Zeyu, W.; Yue, D.; Shouan, Z.; Zhongbang, W. Achieving sustainable development goal 9: A study of enterprise resource optimization based on artificial intelligence algorithms. *Resour. Policy* **2022**, *80*, 103212. [CrossRef]
19. Lee, M.; Park, S.; Yoon, K. Does Health Promotion Program Affect Local Resident' Emotions? *Int. J. Environ. Res. Public Health* **2019**, *16*, 549. [CrossRef]
20. Zeyu, W.; Shuting, Z.; Yuanyuan, Z.; Chuan, C.; Xiufang, D. Risk prediction and credibility detection of network public opinion using blockchain technology. *Technol. Forecast. Soc. Chang.* **2022**, *187*, 122177. [CrossRef]
21. Guo, A.; Yang, J.; Xiao, X.; Xia, J.; Jin, C.; Li, X. Influences of urban spatial form on urban heat island effects at the community level in China. *Sustain. Cities Soc.* **2020**, *53*, 101972. [CrossRef]
22. Bibri, S.E.; Krogstie, J.; Kärrholm, M. Compact city planning and development: Emerging practices and strategies for achieving the goals of sustainability. *Dev. Built Environ.* **2020**, *4*, 100021. [CrossRef]
23. Vieira, J.; Matos, P.; Mexia, T.; Silva, P.; Lopes, N.; Freitas, C.; Correia, O.; Santos-Reis, M.; Branquinho, C.; Pinho, P. Green spaces are not all the same for the provision of air purification and climate regulation services: The case of urban parks. *Environ. Res.* **2018**, *160*, 306–313. [CrossRef]
24. Blay-Palmer, A.; Santini, G.; Dubbeling, M.; Renting, H.; Taguchi, M.; Giordano, T. Validating the city region food system approach: Enacting inclusive, transformational city region food systems. *Sustainability* **2018**, *10*, 1680. [CrossRef]
25. Chetwynd, E. Critical Analysis of Reliability and Validity in Literature Reviews. *J. Hum. Lact.* **2022**, *38*, 392–396. [CrossRef]
26. Wang, S.; Gao, R.C.; Li, R.; Wu, G.C. Psychological status and correlated factors of primary medical staff during the COVID-19 outbreak in Hefei City, China. *J. Multidiscip. Healthc.* **2021**, *14*, 751. [CrossRef]
27. Perry, H.B.; Hodgins, S. Health for the people: Past, current, and future contributions of national community health worker programs to achieving global health goals. *Glob. Health Sci. Pract.* **2021**, *9*, 1–9. [CrossRef]
28. Moher, D.; Naudet, F.; Cristea, I.A.; Miedema, F.; Ioannidis, J.P.; Goodman, S.N. Assessing scientists for hiring, promotion, and tenure. *PLoS Biol.* **2018**, *16*, e2004089. [CrossRef]

International Journal of
Environmental Research and Public Health

MDPI

Article

The Impact of Government Governance on Regional Public Health and Development Measures from the Perspective of Ecological Environment

Tao He [1,†], Lulu Liu [1,2,†] and Manyi Gu [3,*]

1 School of Political Science & International Relations, Tongji University, Shanghai 200092, China
2 Party Committee Propaganda Department & United Front Work Department, Southwest Medical University, Luzhou 646000, China
3 School of Humanities and Management, Southwest Medical University, Luzhou 646000, China
* Correspondence: gumanyi@swmu.edu.cn
† These authors contribute equally to this work.

Abstract: In order to further improve the satisfaction of public health safety, this paper discusses the impact of government governance on the satisfaction of regional public health safety and discusses the effectiveness of government public health governance and development countermeasures. From the perspective of ecological environment protection, combined with the survey data of national urban public health safety satisfaction in the last two years, this paper performs an in-depth empirical analysis on the relationship between government governance, public health governance efficiency, public credibility and regional public health safety satisfaction, as well as the impact mechanism. Through the analysis, it is found that the efficiency of government governance directly affects the satisfaction of regional residents with public health safety. With the help of the intermediary effect test, the significant level standard error of the indirect effect is greater than 1.96, and the confidence interval does not include 0, which proves that the intermediary effect exists. On this basis, the strategy of improving the satisfaction of regional public health security is further analyzed.

Keywords: government governance; public health; ecological environment

Citation: He, T.; Liu, L.; Gu, M. The Impact of Government Governance on Regional Public Health and Development Measures from the Perspective of Ecological Environment. *Int. J. Environ. Res. Public Health* **2023**, *20*, 3286. https://doi.org/10.3390/ijerph20043286

Academic Editors: Yuan Li, Hongxiao Liu, Tong Wu and Paul B. Tchounwou

Received: 5 December 2022
Revised: 12 January 2023
Accepted: 5 February 2023
Published: 13 February 2023

1. Introduction

In recent years, with the gradual deterioration of the global ecological environment, eco-environmental protection has become a hot topic of general concern all over the world. Affected by this, the improvement of the quality of regional public health has also received general attention from all walks of life. The government plays an important role in regional public health governance [1].

With the implementation and continuous expansion of China's basic public health services in 2009, scholars have successively investigated the awareness rate and satisfaction of residents' public health services around the country. Many research results show that Chinese residents' satisfaction with public health services shows the characteristics of uneven geographical distribution [2]. Although many achievements have been made in the construction of Chinese government efficiency, relatively speaking, empirical research on the factors affected by government efficiency is not sufficient, and most of these factors focus on the impact of regional and even national macro policy implementation, including talent gathering, innovation and entrepreneurship, business environment and other fields.

Some scholars have carried out research on the impact and role of administrative efficiency on the willingness of high-level talent agglomeration, and through empirical testing, it is found that administrative efficiency, as one of the dimensions of administrative efficiency, has a significant positive effect on high-level talent agglomeration. Government regulation mainly includes economic regulation and social regulation. Economic regulation is mainly aimed at the problem of information asymmetry in reality, preventing the irrational allocation of resources, which can ensure the fair use of demanders. Social regulation is based on the perspective of national security and public interests to implement mandatory regulation on the social environment, natural resources, security and other aspects. The influence of the government on the demand of urban community health service mainly lies in the integration of market regulation and community management through policy guidance, financial support, strengthening publicity and other strategies and methods; following the basic laws of the medical service market; doing a good job of "distribution according to demand"; and transforming the potential health service demand of urban community residents into the actual distribution demand. This is in order to maximize the utilization of urban community health service resources to better meet the needs of both supply and demand, so as to realize their macro-control role and perform their regulatory functions.

Based on the global data from 1995 to 2022, some scholars found that government efficiency has a significant positive effect on national innovation capacity. In terms of exploring the impact of government effectiveness on the individual level, the relevant empirical research is less. Among them, some scholars have systematically studied the impact and mechanism of government effectiveness on residents' well-being. Based on the practical needs of the development of the Zhuhai government and the theory of social exchange, they used a structural equation model to discuss in detail the mechanism of a city's government effectiveness on residents' well-being. Thus, it reveals the positive impact of government effectiveness on residents' well-being and further clarifies the regulatory role of urban belonging in the impact of government effectiveness on residents' well-being, as well as the intermediary role of residents' living conditions in this impact process [3]. In order to achieve the fragile ecological environment, it is necessary to accurately identify the key environmental factors affecting the economic development of each province, giving full play to the role of government governance measures, and realize the benign economic development of each province.

2. Empirical Research

The relationship between the main variables in this study is shown in Figure 1:

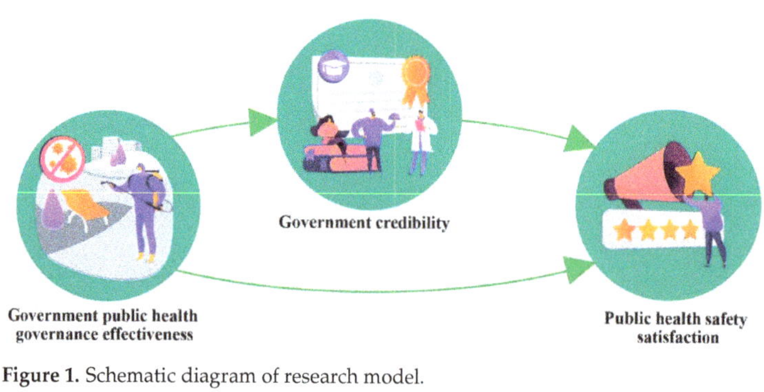

Figure 1. Schematic diagram of research model.

2.1. Research Data Sources

The data used in this study are from the 2017 national urban public security survey data in the urban public security database of the China University of Mining and Technology. In terms of the demographic characteristics of the respondents, statistics show that among all the residents who completed the questionnaire (excluding missing values), men accounted for 50.0% and women accounted for 50.0% [4]. After removing the missing values, the basic information of all residents who filled in the questionnaire in this data survey is shown in Table 1.

Table 1. Statistical table of residents' basic information filled in the questionnaire.

Variable	Category	Frequency	Frequency%	Variable	Category	Frequency	Frequency%
Political outlook	Member of the Communist Party of China	1668	18.0	Age	18–29 years old	4453	48.0
	Democratic party's	228	2.5		30–44 years old	2680	28.9
	Communist Youth League member	2770	29.9		45–59 years old	1588	17.1
	Masses	4580	49.4		Over 60 years old	544	5.9
Gender	Male	4632	50.0	Identity	Civil servant	342	3.7
	Female	4639	50.0		Personnel of public institutions	1061	11.4
Degree of education	Primary school and below	342	3.7		Clerk	2031	21.9
	Junior high school	1139	12.3		Migrant workers	485	5.2
	High school (vocational and technical secondary school)	2371	25.6		Student	2414	26.0
	University (junior college)	4868	52.5		Professional	1396	15.1
	Graduate and above	546	5.9		Retired personnel	532	5.7
					Other	996	10.7
Nation	Han nationality	8119	87.6	Religious belief	Nothing	7781	83.9
	Zhuang nationality	166	1.8		Buddhism	775	8.4
	Manchu	184	2.0		Taoism	95	1.0
	Hui nationality	240	2.6		Christianity	198	2.1
	Miao nationality	63	0.7		Islamism	214	2.3
	Uygur ethnic group	56	0.6		Catholicism	38	0.4
	Tujia nationality	37	0.4		Other	157	1.7
	Yi nationality	26	0.3	Household registration type	City of this city	4816	51.9
	Mongolian	60	0.6		Rural areas of the city	1357	14.6
	The Zang or Tibetan people	207	2.2		Foreign cities	1719	18.5
	Other	113	1.2		Rural areas outside the city	1369	14.8
Personal monthly income	Below CNY 2000	2734	9.5	Personal monthly income	CNY 5001–8000	1325	14.3
	CNY 2001–3500	2109	22.7		CNY 8001–12,500	430	4.6
	CNY 3501–5000	2288	24.7		More than CNY 12,500	202	2.2

2.2. Variable Measurement

2.2.1. Exploratory Factor Analysis

Before conducting the evaluation of the index, it was necessary to determine whether the index could be used for the analysis of the index itself. This can be checked by two key indicators: the KMO value of sampling quality and the result of Bartlett's test of sphericity [5]. In this study, KMO and Bartlett's tests were performed on different measures of public health satisfaction, and the results of the analysis are shown in Table 2.

Table 2. KMO and Bartlett's test of the scale.

KMO Sampling Suitability Quantity		**0.811**
Bartlett's Sphericity Test	Approximate chi square	30,531.954
	Freedom	21
	Significance	0.000

The final result of the factor analysis is shown in Table 3. The cumulative contribution rate of the two components is 73.558%, that is, the cumulative interpretation of the total variance is 73.558%, which is greater than the 0.6 suggested by hair. The observation variables xn1, xn2, xn3 and xn4 have a high load on factor 1. As these observation variables mainly reflect the subjective perception of the surveyed urban residents on the effectiveness of government public health governance, factor 1 is named as the effectiveness factor of government public health governance [6].

Table 3. Total variation and rotated component matrix of factor interpretation extracted from the scale.

	Question Item	Ingredients 1	2
Government public health governance effectiveness (GX)	Prevention effectiveness (xn1)	0.835	0.106
	Regulatory effectiveness (xn2)	0.889	0.076
	Specification efficiency (xn3)	0.881	0.069
	Emergency management efficiency (xn4)	0.861	0.128
Government credibility (GX)	Government expectation (gx1)	0.102	0.812
	Government confidence (gx2)	0.090	0.859
	Government trust (gx3)	0.080	0.828
	Variance explanatory quantity	43.273%	30.285%
	Cumulative variance interpretation		73.558%

In addition, the internal quality of each aircraft model can also be checked. The internal quality of the facet can be measured and verified by the reliability and mean difference of the measured values. According to Table 4, the average difference between the two different abilities is 0.696 and 0.558: both of them are higher than the standard 0.50, and the reliability is 0.901 and 0.790. A value of 0.60 indicates that the quality of the model is reasonable [7].

Table 4. Test results of internal structure suitability of the model by confirmatory factor analysis.

Potential Variables	Observation Variables	Standardized Load	Index Reliability	Measurement Error	Composite Reliability	Mean Variance Extraction
XG	Preventive effectiveness	0.757	0.573	0.427	0.901	0.696
	Regulatory effectiveness	0.892	0.796	0.204		
	Normative efficiency	0.827	0.684	0.316		
GX	Emergency management efficiency	0.856	0.733	0.267	0.790	0.558
	Government expectations	0.696	0.487	0.513		
	Government confidence	0.814	0.663	0.337		
	Government trust	0.723	0.523	0.477		

2.2.2. Normality Test of Sample Data

In this paper, spss26.0 was used to test the normality of each variable in the data model. The results of the clinical trials are shown in Table 5.

Table 5. Normality test results of the scale.

	Number of Cases	Average Value	Standard Deviation	Skewness	Kurtosis
Xn1	9273	5.10	2.674	0.147	−0.948
Xn2	9273	5.02	2.807	0.173	−1.050
Xn3	9273	4.92	2.691	0.192	−0.959
Xn4	9273	5.19	2.652	0.067	−0.959
my	9273	3.35	0.888	−0.223	−0.009
Gx1	9273	3.19	0.780	−0.283	0.611
Gx2	9273	3.30	0.817	−0.321	0.381
Gx3	9273	3.35	0.815	−0.366	0.364

It can be seen that the skewness coefficients and kurtosis coefficients of all observed variables in the sample data meet the test criteria, indicating that the data conform to the characteristics of normal distribution. Therefore, we can use the structural equation model to further analyze and verify the research hypothesis [8].

3. Current Situation of Regional Public Health Supply Structure under Government Governance

3.1. Main Structure of Public Health Supply

It can be seen from the data in Figure 2 below that the health spending of the Chinese government has not changed significantly in recent years. The distribution of government health spending has decreased from 2018 to 2021, from 30.7% in 2018; the lowest share of health spending in 2021 was 29.96%, and in 2022 increased to 30.45%, but remained lower than the government's 2018 medical expenses.

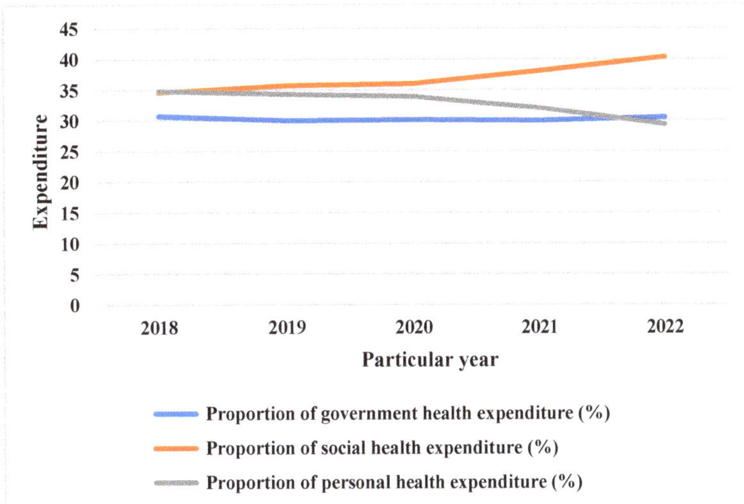

Figure 2. Main structure of public health expenditure in China.

3.2. Public Health Expenditure Structure

From the perspective of China's public health care, Table 6 shows that China's health care spending has increased year by year, from CNY 132.02 billion in 2013 to CNY 119.531 billion in 2018. In 2022, the price pattern also shows the pattern of financing. This shows that the increase in public health spending is mostly influenced by the budget. In 2013, 1.9% of medical and health care costs came from the central government, and 98.1% was public expenditure; meanwhile, in 2022, the central government accounted for only 0.7% of medical and health spending, while public spending accounted for 99.3% [9].

Table 6. Structure of government expenditure on health care in China from 2013 to 2022 unit: CNY 100 million, percentage%.

Particular Year	Total Medical and Health Expenditure	Central Expenditure	Local Expenditure	Central Proportion	Local Proportion
2018	6459	71.3	6358.2	1.1	98.9
2019	7245.1	74.3	7170.8	1	99
2020	8279.9	76.7	8203.2	0.9	99.1
2021	10,176.8	90.1	10,086.7	0.9	99.1
2022	11,953.1	84.4	11,868.7	0.7	99.3

3.3. Supply Structure of Public Health in the East, Middle and West

From the point of view of the region, the per capita health in the eastern region is higher than other regions, which is the reason for the regional disparity differences in population health in the eastern region. Spending on health per person in the eastern region is lower than other regions, which affects per capita health spending; in other regions in the east, there is little difference in the scale of health spending per capita. The difference in the index of public health expenditure per capita between the regions is small and insignificant; the level of health expenditure per person in the eastern region is higher than that of the middle region, and the level of health of the population in the eastern region is higher than on the middle ground [10].

To identify regional differences in China's health spending, the Theil index in the eastern region gradually decreased from 2018 to 2020, as can be seen from Theil index line in Figure 3 and the price changes in Table 7. It shows that the difference in health care costs of the population in the east is gradually decreasing. From 2018 to 2022, the TAIR index of the eastern region is the largest, and the power line of the TAIR index is at the top.

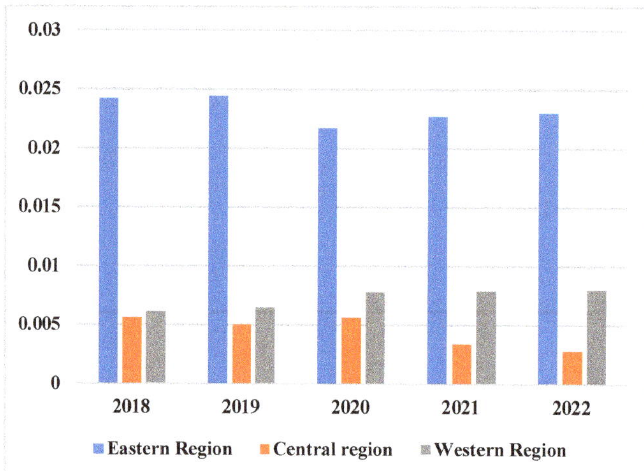

Figure 3. Change value of public health Theil index of fiscal expenditure in eastern, central and western regions.

Table 7. Changes of Theil index of public health in eastern, central and western regions.

Intra Group Theil Index	2018	2019	2020	2021	2022
Eastern region	0.0242	0.0244	0.0217	0.0227	0.0230
Central region	0.0056	0.0050	0.0056	0.0034	0.0028
Western region	0.0061	0.0065	0.0078	0.0079	0.0080

4. Empirical Analysis of the Impact of Government Public Health Governance Efficiency and Government Credibility on Public Health Safety Satisfaction

4.1. Descriptive Statistical Analysis

4.1.1. Public Health Safety Satisfaction Profile

The overall evaluation of urban residents' satisfaction with public health and safety is shown in Table 8. In general, the average score of the public health safety evaluation of the interviewed residents is 3.350, which is above the medium level [11]. However, under the macro social background of the increasing living standards of Chinese residents and the increasing needs and expectations of public safety, it is still questionable whether the above medium public health safety evaluation indicates that the public health safety situation has been widely recognized by urban residents.

Table 8. Overall satisfaction with public health and safety (unit: %).

Project	Public Health Safety Evaluation Score					Mean Value
	Very Poor	Relatively Poor	Commonly	Better	Very Nice	1–5
Public health safety satisfaction	2.5	12.1	42.1	34.8	8.5	3.350

4.1.2. Overview of Government Public Health Governance Effectiveness

The overall situation of urban residents' perception of government public health governance efficiency is shown in Table 9. Overall, the average score of the overall effectiveness of public health governance is 5.059, which is only at the medium level. The respondents scored the highest on the sub item of emergency management efficiency 5.191; the average score of normative effectiveness was the lowest at 4.919 [12]. This shows that the effectiveness of government public health governance at the subjective perception level of urban residents has great room for improvement, both on the whole and in specific sub items.

Table 9. Perceived effectiveness of government public health governance (unit: %).

Project	Rating Score of Worry Degree (Extremely Worried–Not Worried at all)										Mean Value
	1	2	3	4	5	6	7	9	10	1–10	5.059
Preventive effectiveness	12.1	7.3	11.7	11.2	15.8	10.0	9.6	9.9	5.0	7.2	5.104
Regulatory effectiveness	14.9	8.1	10.6	10.8	13.8	9.6	9.2	9.2	5.3	8.2	5.024
Normative efficiency	14.5	7.7	11.6	11.7	14.6	10.6	8.7	9.4	5.1	6.2	4.919
Emergency management efficiency	11.7	7.0	10.6	11.1	15.2	11.4	10.4	10.2	6.1	6.4	5.191

4.1.3. Overview of Government Credibility

The overall perception of urban residents on the evaluation of government credibility is shown in Table 10. In general, the average government credibility is 3.280, less than 3.5. Although the level of government credibility is above the medium level, it is not blindly optimistic, and there is still a risk of a weak foundation. The average scores of the surveyed residents on the degree of realization of the government's expectations, confidence and trust were 3.191, 3.297 and 3.351, respectively. The specific sub items of government credibility and the evaluation of public health safety satisfaction show the same distribution characteristics of large in the middle and small at both ends [13].

Table 10. Perception of government credibility (unit:%).

Project	Evaluation Score of Government Credibility					Mean Value (1–5)
	1	2	3	4	5	
Government credibility						3.280
Expected degree	2.7	11.2	53.6	29.2	3.3	3.191
Confidence level	2.5	10.4	47.0	35.1	5.0	3.297
Degree of trust	2.2	9.7	44.4	38.2	5.5	3.351

4.1.4. Correlation Analysis between Variables

By averaging the measurement items of government public health governance efficiency and government credibility, respectively, the correlation between the three main research variables can be obtained. The results are shown in Table 11. There is a positive correlation between the effectiveness of government public health governance, government credibility and urban residents' satisfaction with public health safety at a significance level of 0.01 [14].

Table 11. Correlation analysis between variables.

Variable	Relevance	XN	GX	MY
Government public health governance effectiveness (XN)	Pearson correlation	1		
Government credibility (GX)	Pearson correlation Sig (double tailed)	0.215 ** 0.000	1	
Public health safety satisfaction (MY)	Pearson correlation Sig (double tailed)	0.318 ** 0.000	0.345 ** 0.000	1

Note: ** $p < 0.01$.

4.2. Main Effect Test

4.2.1. Overview of Structural Equation Models

A structural equation model is composed of a measurement model and structural model. The measurement equation is:

$$X = \Lambda_x \xi + \delta \tag{1}$$

$$Y = \Lambda_y \eta + \varepsilon \tag{2}$$

The structural equation is:

$$\eta = B\eta + \Gamma\xi + \xi \tag{3}$$

Equation (1) is the measurement equation of the external latent variable, and ξ is the external latent variable; Equation (2) is the measurement equation of the internal dependent latent variable, and η is the internal dependent latent variable; Equation (3) is a structural equation, which describes the relationship between internal latent variables η and each other, and Γ describes the influence of external latent variables ς on internal latent variables η.

4.2.2. Empirical Test and Results

In this study, because public health safety satisfaction is an explicit variable, while the government's public health governance efficiency and government credibility are regarded as potential variables, and the structural model of path analysis includes both explicit and potential variables, the path analysis of the mixed model is constructed. The overall model fitness test results of the overall relationship model are shown in Table 12.

Table 12. Overall model fitness test results of overall relationship model.

Test Statistics	Adaptation Standard	Inspection Result Data	Model Adaptation Judgment
	Absolute fitness index		
GFI value	>0.90	0.997	Yes
AGFI value	>0.90	0.994	Yes
RMR value	<0.05	0.026	Yes
RMSEA value	<0.05 (good) <0.08 (reasonable)	0.023	Good
	Value added fitness index		
NFI value	>0.90	0.997	Yes
RFI value	>0.90	0.995	Yes
CFI value	>0.90	0.998	Yes
	Simple fitness index		
X^2/df value	1 < x/df value < 3, good; 3 < x2/df value < 5, acceptable; 5 < x2/df value, poor	5.987	Poor
PNFI value	>0.50	0.570	Yes
PGFI value	>0.50	0.443	No

It can be seen that, except for the x^2/df value and PGFI value, the other statistical test quantities meet the overall model fitness standard. As mentioned in Section 3, the problem of too large an x^2/df value caused by too large a sample size is also ignored here. Therefore, the fitness index of the model generally meets the fitness standard, indicating that the model has a good fitness. Table 13 shows the test results of the overall relationship model [15].

Table 13. Test results of the overall relationship model.

Relationship between Variables	Non-Standardized Estimation Results				Standardized Regression Coefficient
	Non-Standardized Regression Coefficient	Standard Error	Critical Ratio Value	Significance	
xn1←XN	1				0.802
xn2←XN	1.103	0.013	83.511	......	0.844
xn3←XN	1.098	0.013	83.805	......	0.875
xn4←XN	0.999	0.012	80.252	......	0.808
gx1←GX	0.704			......	0.704
gx2←GX	1.2	0.021	58.317	......	0.807
gx3←GX	1.075	0.019	56.995	......	0.724

Note: The symbol "←" indicates the action direction between variables.

Table 14 shows the hypothesis test results. It can be seen that the standardized regression coefficients of "government public health governance efficiency" on "government credibility" and "public health safety satisfaction" are 0.232 and 0.251, respectively, and the standardized regression coefficients of "government credibility" on "public health safety satisfaction" are 0.326, both reaching a significant level of 0.001. The value of standardized regression coefficient is positive, indicating that the influence path relationship between the three core variables is positive, which is consistent with the original theoretical hypothesis [16].

Table 14. Research assumptions and test results.

Relationship Between Variables	Non-Standardized Path Coefficient	Standardized Path Coefficient	Corresponding Assumption	Hypothesis Test Results
MY←XN	0.104	0.251 ***	H1	Support
GX Y←XN	0.059	0.232 ***	H2	Support
MY Y←GX	0.527	0.326 ***	H3	Support

Note: *** $p < 0.001$; The symbol "←" indicates the action direction between variables.

This means that the first three hypotheses of this study have been verified. Specifically:

First, the operation of the government's public health care has a positive effect on the satisfaction of the public's health security. The sign of the path coefficient of the influence of public health management on the results of urban residents' satisfaction with public health safety is positive and reaches a significant level of 0.001, indicating that the effectiveness of public health management has a positive effect on the assessment of health. In terms of the satisfaction of residents in cities with public health safety, hypothesis H1 is confirmed [17].

Second, the effectiveness of government public health care has a positive effect on trust in government. The sign of the path coefficient of the effect of health management on public trust is positive and reaches a significant level of 0.001, indicating that public health management has a positive effect on public confidence according to citizens, meaning that hypothesis H2 is confirmed.

4.3. Intermediary Effect Test

4.3.1. Overview of Mediation Effect

Mediation means that there are m intervening variables in the causal relationship mechanism between the independent variable X and the dependent variable Y: that is, the independent variable X can affect the difference between Y and M mediating variables [18]. In this case, m is the average difference between m affecting both the independent variable X and the variable Y. A schematic diagram of the mediation process is shown in Figure 4.

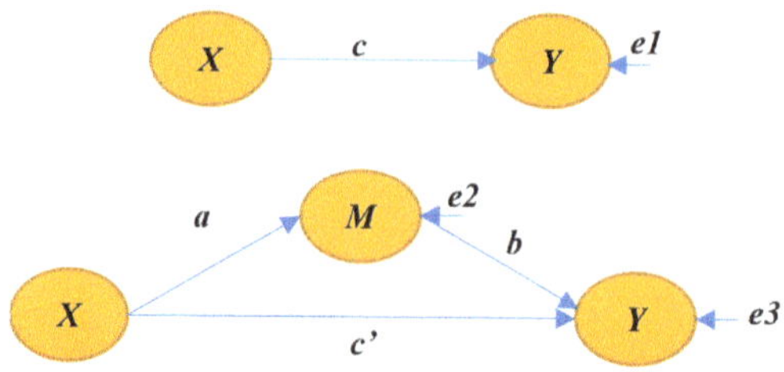

Figure 4. Schematic diagram of intermediary effect.

When all variables are centralized or standardized, the specific relationship between the three variables can be described by the following equation:

$$Y = cX + e_1 \tag{4}$$

$$M = aX + e_2 \tag{5}$$

$$Y = c'X + bM + e_3 \tag{6}$$

The intermediary effect is equal to the coefficient product *ab*, that is, the indirect effect. The total effect is the sum of the direct effect and the indirect effect. The specific relationship is as follows:

$$c = c' + ab \tag{7}$$

The mediating effect can be divided into the complete mediating effect and the partial mediating effect. In the complete intermediary effect, X can only affect Y through the intermediary variable m, but it cannot directly affect Y: that is, at this time, the coefficient c' is 0, $c = ab$. In some mediating effects, X can not only directly affect Y, but also affect Y through the mediating variable M.

4.3.2. Intermediary Effect Test

The bootstrap method is one of the common methods to test the mediation effect. It does not need samples to obey the hypothesis of positive distribution and does not rely on theoretical standard error. It has high statistical effect, and is also considered the most ideal mediation effect test method at present. Bootstrap is a nonparametric statistical method that estimates the variance of statistics and then estimates the interval. Its basic core idea is to use resampled sample data to calculate statistics and estimate sample distribution [19]. The test results are shown in Table 15.

Table 15. Intermediary effect test results.

Effect Type	Non-Standardized Test Results		Bias-Corrected		Percentile	
	Coefficient	Standard Error	95% Confidence Interval	*p*	95% Confidence Interval	*p*
Total effect	0.143	0.005	0.133, 0.153	0.001	0.133, 0.153	0.001
Indirect effect	0.034	0.002	0.030, 0.039	0.001	0.030, 0.039	0.001
Direct effect	0.109	0.005	0100, 0.118	0.001	0.100, 0.118	0.001

Table 15 shows that government trust plays a mediating role in the effect of government health care on public health on urban residents' satisfaction with public services. The health Z value of the significance level of the indirect effect is equal to the value of the unstandardized coefficient of the indirect effect/standard error of the indirect effect = 0.034/0.002 > 1.96. In addition, the confidence interval for the direct effect does not include 0, thus confirming the existence of a moderated effect. The planting time for the direct effect does not include 0, so there is still an indirect effect [20]. Through the analysis, it is found that the efficiency of government governance directly affects the satisfaction of regional residents with public health security. With the help of the mediation effect test, the significant level standard error of the indirect effect is greater than 1.96, and the confidence interval does not include 0, demonstrating the existence of the mediation effect. On this basis, strategies to improve regional public health safety satisfaction were further analyzed.

5. Development Countermeasures of Government Governance to Improve the Satisfaction and Quality of Public Health Security from the Perspective of Ecological Environment

5.1. Improve the Existing Public Health Administrative Norms

First, governance plays an important role in improving the legal system of public health, establishing and forming the legal department of public health, improving the status of public health law in China's legal system, and making public health law an independent department of the legal system. Second, it solves the basic theoretical problems related to public health law (characteristics, essence, origin, legal basic rights and obligations of health) and determines the basic categories of public health law to adjust various social relations based on the right to life and health [21]. This is to explore the relationship between public health law and other legal norms, and to study the relationship between public health legal rights and obligations, and the related legal responsibilities, so as to provide a theoretical basis for the establishment of the basic theoretical system of health law. Third, it is of great significance to strengthen the construction of a socialist public health legal system. (1) In terms of public health legislation, through the study of the legal system of public health, we can clearly find the defects and deficiencies of the current legal system of public health, so as to scientifically formulate public health legislation strategies; gradually implement public health legislation planning; realize legislative intent, legislative value and legislative effectiveness; and provide theoretical support for the construction of public health rule of law. (2) It is of great significance to guide public health administrative law enforcement and judicial practice, improve the health level and awareness of the safeguarding rights of the whole population, and protect citizens' health rights and interests. Only by enhancing the understanding and recognition of the national public health legal system can the public increase their knowledge, understanding of, and ability to abide by the law [22].

5.2. Formulate the Basic Law of Public Health and Standardize Government Functions

The formal rationality of law can serve as a powerful footnote to the systematization requirements of public health law. The so-called formal rationality of law refers to the systematization and scientization of legal rules controlled by reason and the formalization of the process of law making and application. According to Max Weber, a famous German legal sociologist, formal rationality is the highest level of rationality pursued by law. He believes that theoretically, the last stage of legal development is the systematic development of professional jurists on the basis of literature and formal logic training [23]. However, due to the lack of command and coordination, separate public health laws and regulations are isolated from each other. This situation is not conducive to the unified regulation and management of the field of public health, nor to the implementation of a single public health law. People involved in legislation at different places often start from the perspective of their own departments and regions, causing the relevant laws and regulations to be one-sided and limited, rather than overall and long term. Therefore, it is necessary to formulate a unified basic law of public health.

5.3. Give Full Play to the Government's Supervision of Public Health Services

The government's supervision function in the field of public health is particularly important under the current market economy. Since the reform and opening up, under the conditions of the market economy, due to the profit seeking nature of the market, many bad businesses, in order to pursue interests, have taken advantage of problems such as untimely government supervision and loopholes in supervision, waiting for the opportunity to produce and sell fake and shoddy food and drugs, causing serious social harm, as well as causing public opinion to falter regarding the government's supervision ability. China has had a market economy for more than 30 years [24]. With the rapid economic development, weak supervision of market behavior has occurred from time to time, especially in the field of public health. The field of public health is closely related to all aspects of society; whether it is food safety or drug safety, environmental protection or ecological balance, it is related to the lives and safety of the public. It is also the strictest set of regulatory measures that public health functional departments must implement. From production license, production to transportation, sales and other links, professional supervisors must be sent to supervise the whole process, in order to be open and transparent and ensure that businesses can conduct market behavior in accordance with relevant professional standards [25].

6. Conclusions

In the new era of social transformation, various public health events emerge in an endless stream; this leads to a great threat to public health and causes people to question the government's governance capacity in the field of public health. In this case, the government should aim at building a service-oriented government, give full play to its initiative in performing public health functions, and strive to reform and improve China's public health undertakings under the guidance of public health law.

Author Contributions: Software, T.H.; Resources, T.H.; Writing—original draft, M.G.; Writing—review & editing, L.L. All authors have read and agreed to the published version of the manuscript.

Funding: This work is supported by the: 1. Center for American Studies. Project name: Research on the Comparison and Enlightenment of Chinese and American governments in the disposal of COVID-19, Project No.: ARC2021007; 2. Luzhou Federation of Social Sciences–Social Governance Innovation Research Center. Project name: Research on the path to improve the level of grass-roots governance based on the integrated development of urban and rural areas, Project No.: SHZLYB2106.

Institutional Review Board Statement: Not applicable.

Informed Consent Statement: Not applicable.

Data Availability Statement: The labeled data set used to support that findings of this study are available from the corresponding author upon request.

Conflicts of Interest: The authors declare no conflict of interest.

References

1. Balisi, S.; Madisa, M. COVID-19 preventative measures: Lessons for public health policy. *Open J. Soc. Sci.* **2021**, *9*, 15. [CrossRef]
2. Sheard, D.J.; Clydesdale, G.; Maclean, G. Governance structure and public health provision. *J. Health Organ. Manag.* **2019**, *33*, 426–442. [CrossRef]
3. Lin, G. Book reviews: Urbanization, regional development and governance in China, by Jianfa Shen. *China Rev. Interdiscip. J. Greater China* **2019**, *19*, 169–171.
4. Mwoka, M.; Ajayi, A.I.; Kibunja, G.; Cheruiyot, C.; Ushie, B.A. Cocreated regional research agenda for evidence-informed policy and advocacy to improve adolescent sexual and reproductive health and rights in Sub-Saharan Africa. *Br. Med. J. Glob. Health* **2021**, *6*, e005571. [CrossRef] [PubMed]
5. Wolman, H. Looking at regional governance institutions in other countries as a possible model for u.s. metropolitan areas: An examination of multipurpose regional service delivery districts in British Columbia. *Urban Aff. Rev.* **2019**, *55*, 321–354. [CrossRef]
6. Pan, X.; Li, M.; Guo, S.; Pu, C. Research on the competitive effect of local government's environmental expenditure in China. *Sci. Total Environ.* **2020**, *718*, 137238. [CrossRef]
7. Huang, Z. Value analysis of sports industry policy from the perspective of national governance. *Mod. Manag. Forum* **2021**, *5*, 189. [CrossRef]
8. Wang, Y.; Zhao, Y. Is collaborative governance effective for air pollution prevention? A case study on the Yangtze river delta region of China. *J. Environ. Manag.* **2021**, *292*, 112709. [CrossRef]

9. Li, C.; Wu, X.; Sheridan, S.; Lee, J.; Wang, X.; Yin, J.; Han, J. Interaction of climate and socio-ecological environment drives the dengue outbreak in epidemic region of China. *PLoS Negl. Trop. Dis.* **2021**, *15*, e0009761. [CrossRef]
10. Panjaitan, R.B.; Sumartono, S.; Saleh, C. The role of central government and local government and the moderating effect of good governance on forest fire policy in Indonesia. *Benchmarking* **2019**, *26*, 147–159. [CrossRef]
11. Lu, T. Study on Effectiveness of Government Governance in the Cities with Serious Air Pollution: A case study of Zheng-Zhou City. *Meteorol. Environ. Res.* **2019**, *10*, 20–23+30.
12. Titah, R.; Maclean, D. A systematic literature review of empirical research on the impacts of e-government: A public value perspective. *Public Adm. Rev.* **2022**, *82*, 23–38.
13. Yun, N.S.; Lee, S.K. The effect of public office values on government trust: Focusing on the effects of regulating the level of citizenship. *Restaur. Bus.* **2019**, *118*, 366–374. [CrossRef]
14. Cheong, Y.M.; Choi, S.O. The effect of e-government on digital divide. *J. Korea Technol. Innov. Soc.* **2019**, *22*, 942–961. [CrossRef]
15. Liargovas, P.; Apostolopoulos, N. Economic governance of a populist government: From the fear of grexit to continued instability. *Southeast. Eur. L'Europe Du Sud-Est* **2021**, *45*, 48–63. [CrossRef]
16. Zhang, F.; Feng, Y.; Liao, Z.; Xiaogang, W.; Zhang, L. The effect of ecological factors on Fagaceae species richness in Sichuan-Chongqing region, China. *Chin. J. Appl. Environ. Biol.* **2020**, *26*, 689–696.
17. Ryabova, L.N.; Zalygina, I.A. Criterion for the assessment of the ecological condition of bottom sediments (on the example of the Brest region). *Proc. Natl. Acad. Sci. Belarus Chem. Ser.* **2019**, *54*, 455–466. [CrossRef]
18. Zheltukhina, V.I.; Kulikova, M.A.; Kolesnichenko, E.Y.; Manokhina, L.A. Assessment of the ecological state of the Maysky village reservoirs, belgorod region, based on the analysis of morphological features of silver carp Carassius Gibelio. *IOP Conf. Ser. Earth Environ. Sci.* **2021**, *848*, 012160. [CrossRef]
19. Ivanova, D. On the issue of ecological problems of the Baikal region forests: An economic aspect. *Bull. Baikal State Univ.* **2019**, *29*, 24–31. [CrossRef]
20. Ma, L.; Liu, Z.; Huang, X.; Li, T. The impact of local government policy on innovation ecosystem in knowledge resource scarce region: Case study of Changzhou, China. *Sci. Technol. Soc.* **2019**, *24*, 29–52. [CrossRef]
21. Zhang, X.; Rane, K.; Kakaravada, I.; Shabaz, M. Research on vibration monitoring and fault diagnosis of rotating machinery based on internet of things technology. *Nonlinear Eng.* **2021**, *10*, 245–254. [CrossRef]
22. Huang, R.; Zhang, S.; Zhang, W.; Yang, X. Progress of zinc oxide-based nanocomposites in the textile industry. *IET Collab. Intell. Manuf.* **2021**, *3*, 281–289. [CrossRef]
23. Chen, J.; Liu, J.; Liu, X.; Xu, X.; Zhong, F. Decomposition of Toluene with a Combined Plasma Photolysis (CPP) Reactor: Influence of UV Irradiation and Byproduct Analysis. *Plasma Chem. Plasma Process.* **2021**, *41*, 409–420. [CrossRef]
24. Ajay, P.; Jaya, J. Bi-level energy optimization model in smart integrated engineering systems using WSN. *Energy Rep.* **2022**, *8*, 2490–2495.
25. Sharma, K.; Chaurasia, B.K. Trust Based Location Finding Mechanism in VANET Using DST. In Proceedings of the Fifth International Conference on Communication Systems & Network Technologies, Gwalior, India, 4–6 April 2015; pp. 763–766.

International Journal of
*Environmental Research
and Public Health*

MDPI

Article

Exploring the Relationship between Urban Street Spatial Patterns and Street Vitality: A Case Study of Guiyang, China

Junyue Yang [1,2], Xiaomei Li [1,2,*], Jia Du [1] and Canhui Cheng [2]

1 College of Architecture and Urban Planning, Guizhou University, Guiyang 550025, China
2 School of Architecture and Urban Planning, Chongqing University, Chongqing 400044, China
* Correspondence: xmli2@gzu.edu.cn

Abstract: Understanding how street spatial patterns are related to street vitality is conducive to enhancing effective urban and street design. Such analysis is facilitated by big data technology as it enables more accurate methods. This study cites data from street view imagery (SVI) and points of interest (POI) to assess street vitality strength after the classification of street spatial and vitality types to explore the relationship between street spatial patterns and street vitality with a further discussion on the layout features of street vitality and its strength in various street spatial patterns. First, street spatial patterns are quantified based on SVI, which are further classified using principal component analysis and cluster analysis; POI data are then introduced to identify street vitality patterns and layout, and the strength of street vitality is evaluated using spatial overlay analysis. Finally, relevance analysis is explored to cast light on the relationship between street vitality layout and street spatial patterns by overlaying street spatial pattern, street vitality types, and street vitality strength in the grid cells. This paper takes the urban area of Guiyang, China, as an example and the analysis shows that a pattern is discovered in Guiyang regarding the layout of street vitality types and vitality strengths across different street spatial patterns; compact street spaces should be prioritized in designing street space renovation; and cultural leisure vitality is most adaptive to street spatial patterns. Based on big data and using grids to understand the intrinsic relationship between street spatial patterns and the type and strength of street vitality, this paper brings more options to urban street studies in terms of perspective and methodology.

Keywords: street spatial patterns; street vitality; street design; big data; China

Citation: Yang, J.; Li, X.; Du, J.; Cheng, C. Exploring the Relationship between Urban Street Spatial Patterns and Street Vitality: A Case Study of Guiyang, China. *Int. J. Environ. Res. Public Health* **2023**, *20*, 1646. https://doi.org/10.3390/ijerph20021646

Academic Editors: Yuan Li, Hongxiao Liu and Tong Wu

Received: 25 November 2022
Revised: 11 January 2023
Accepted: 11 January 2023
Published: 16 January 2023

1. Introduction

Streets are the main spaces of public life in cities, carrying diversity and vitality [1]. At the same time, streets help to shape a city by determining the way people move and stay [2]. Peoples' impressions of the urban space are directly impacted by the spatial patterns of streets. In addition, a large number of scholars believe that mixed street vitality and high street vitality strength are more conducive to sustainable development [3,4]. However, the spatial forms of streets in urban areas are different, and the distribution and development trends of streets with different space morphology are different [5]; notably, the very different space morphology of urban streets has different influences on the urban, social, economic, cultural, and natural environment [6]. Streets with different spatial forms provide people with different ways of living and activities [7–9]. Remarkably, mixed land use is one of the characteristics of urban street spaces [10], which means that the urban street space contains rich urban functions and multiple urban vitality, and the vitality of a street is closely related to the physical space form of the street [11].

In this context, a large number of studies on the vitality and spatial patterns of urban streets have emerged, but the assessment of street vitality and the classification of street spatial patterns are largely based on manual survey methods, such as questionnaires, field statistics, and measurements [12–14]. In recent years, new perspectives emerging from big-data-based computer analytics and geographic information science and technology are facilitating space analysis of urban streets [15–17]. Compared with the traditional manual collection of information and data, urban big data that promise fast and free-of-charge access show a huge advantage in that big data do not only carry objective physical information but also preferences and characteristics regarding peoples' activities, making it highly applicable and effective in the analysis of the vitality and spatial patterns of urban streets. Many urban street studies have already been using multiple sources of big data such as street view imagery (SVI) and points of interest (POI).

Existing studies of street vitality are currently dominated by vitality evaluation. Liu et al., used POI data as the basis for urban vitality identification, which was combined with the features of land use in surrounding areas to calculate the space–time layout of vitality and analyze the influencing factors of layout evolution [18]. Zeng et al. used big data to create spatially explicit indices to identify spatial patterns of vitality in cities and developed a graded assessment of vitality [19]. Chen et al. cited POI and Weibo check-in geographic markers to analyze vitality distribution and indices [20]. These attempts to study urban vitality have demonstrated the feasibility of analytical methods that use big data to reflect the characteristics of population activity as a means to obtain an accurate distribution of urban vitality [21]. They equally prove the validity of geo-information data in the analysis of the spatial pattern and vitality of streets [22]. However, current studies on urban vitality tend to neglect the interaction of vitality and built-up environment [23] and few have provided a categorization framework for vitality understanding, which limits the creative value of vitality analysis for urban design [24].

Street space morphology studies, on the one hand, evaluate the quality of street space by analyzing peoples' perceptions of street space from the public viewpoint [25,26]. Wu et al. formed an SVI-based random forest model trained on perceptual samples to assess the visual quality of street space [27]. Hu et al. accessed POI data to identify street functions, which were further integrated into an SVI semantic segmentation analysis of street space to illustrate the impacts of function variance on street spatial patterns [28]. On the other hand, many studies focus on the relationship between such patterns and the speed of the vehicles passing through [29] them, as the street patterns change drivers' perceptions of street length and the density of buildings and consequently, their speed [30]. The influencing variables on street patterns, in turn, include urban traffic volumes and traffic demand [31]. Differences in spatial patterns of the streets and their design speed also influence their vitality [32]. Evidence also shows that street spatial patterns have implications for urban microclimate and pollution, most of which are SVI-based analyses [33–35]. In general, SVI is an important source of data for street space studies [36], and the mutual relations of street spatial patterns and vitality are supported by a wide range of studies.

There also exist many studies which act as support to the analysis of street vitality and spatial patterns, while we can find equally abundant evidence for the relationship between the two. Studies on how the environmental features of buildings improve street vitality can be dated back to the 1960s [37]. Xu et al. proposed a framework to assess street vitality, which includes street pattern, street business type, and street accessibility [38]. Long et al. discovered a positive correlation between street function mix and its 'walk score' [39]. Jiang et al. based the analysis of a human-scale investigation of vitality vis à vis environmental features of buildings on the various types of LBS (location-based service) data, such as bus cards, taxis, mobile phones, and taxi trajectory [40]. Xia et al. studied city vitality in daytime and at night with the help of nighttime lighting data and extended the analysis to the spatial relationship between land use intensity and urban vitality [41].

This study emphasizes the interactive relationship between the spatial patterns of streets, vitality types, and vitality strength (the more vitality types in the same area, the higher the vitality strength). It explores the differences in street spatial patterns, which are believed to be a perspective to the spatial layout features of street vitality types and strength, so as to contribute to enriching the methodology and angles towards urban street space renovation design and the shaping of street spatial vitality.

Street spatial patterns embody the feature of the area by integrating the function of the buildings and the morphological elements. As such, streets and buildings, the functions the buildings have, and the vitality of such functions are closely interconnected [42]. Streets are the core space for public life in cities and should highlight the integration of street spatial patterns and the vitality that takes the function of the street as the foundation [43]. For a long time, planners and designers have been working to create streets with a unique identity and high vitality [44], which can only be accomplished after understanding the interactions between the spatial pattern, vitality type, and strength of streets, allowing the full release of the potential in street spatial patterns and vitality during the urban and street design process [45]. Hence, the main concerns of this study are:

- Exploring the classification of urban street spatial patterns;
- Exploring methods for determining street vitality types and vitality strength;
- Analyzing the inter-relationship between the spatial pattern, vitality type, and vitality strength of streets.

This paper proposes a classification method of street spaces and an evaluation method of street vitality based on multi-sourced data based on ArcGIS. Such methodology and quests into a perspective on urban street analysis that integrates street spatial pattern and vitality serve as compliments to the existing system of urban street analysis and are more suitable for the urban design of intensive city centers.

2. Research Scope and Data Processing

2.1. Research Scope

The scope of this paper is the city center of Guiyang, Guizhou, China. To be more specific, it is north of Beijing Road, south of the southern section of the central ring road; west of Huaxi Avenue, and east of Guiyang Ring Highway. The urban streets in the abovementioned scope are considered research subjects, which are cut by intersections. The total number of streets concerned in this study is 1153.

Guiyang is located in the mountain area of southwest China, and cities of its kind typically lack flat lands [46]. As a result of changing landscapes, streets in central Guiyang show high density and complex spatial patterns [47]. Moreover, as a city with a long history, Guiyang has many buildings and areas of historical importance. Our team has calculated the number of historical and cultural sites protected at the provincial- and municipal-level and the result is 39 in the study area alone. For example, the Qianling Mountain monuments and stone inscriptions (4,260,000 square meters), the ancient buildings of Guizhou Normal University (694,667 square meters) and the former residence of Mao Guangxiang and the site of warehouses for Forbidden City relics (404,290 square meters) are the three main sites. There are also seven more sites with areas larger than 10,000 square meters and seventeen larger than 1000 square meters. Among the 39 sites in the research area, 25 of them are adjacent to streets, meaning that these law-protected sites are key driving forces behind the changes in street spatial patterns in central Guiyang. Although facing huge demand due to a large population, the central area of the city has maintained diversity in spatial patterns. At the same time, as urbanization speeds up, the street functions in the central area continue to be more integrated and denser, resulting in a highly diverse combination of street function and spatial pattern. Therefore, the central area of Guiyang is an ideal research scope for the understanding of the relationship between the spatial pattern, vitality type, and vitality strength of urban streets.

2.2. Data Processing

A total of 1153 central points, latitude, and longitude coordinates of all streets in the research scope were collected. The starting point, ending point, and central points of street sections were used as anchors to crawl SVI data in Tencent Map, which resulted in three street photos for each section. A total of 3459 Tencent Streetview pictures of central Guiyang were collected, as shown in Figure 1. The OpenStreetMap (OSM) and satellite map of Guiyang were imported into ArcGIS to sketch all 1153 street sections in the research scope. The street centerlines were simplified to generate the street network in ArcGIS. In addition, POI data of the research scope were gathered using online tools, resulting in 95,522 entries of 11 categories.

Figure 1. Tencent Streetview pictures of downtown Guiyang (selected).

3. Research Methodology

3.1. Street Space Classification

3.1.1. Establishing Street Space Indicators

From the perspective of humans' perceptions of the enclosing state of street space, this paper draws on the street skeleton variable system proposed by Harvey Chester Wollaeger in 2014 [48] to include indicators which are street width, street length, total building numbers, building height, building cross-section ratio, street wall continuity, ratio of total building numbers to street length, and height variance. The definition and importance of each indicator is introduced in Table 1. The research classifies street space types by evaluating the enclosing state of street space. In view of the difference between the two sides of the street, 8 indicators are decomposed into 11 sub-indicators in Table 2. The detailed explanation of each spatial index is shown in Figure 2.

Table 1. Spatial definition and importance of indicators.

No.	Category	Spatial Definition and Importance
1	Street Width	Distance between the edges of the street view. The width is the field of view for a user.
2	Street Length	Length of the centerline of a block. Street length is the depth of the street view for a user.
3	Total Building Numbers	Total number of buildings along the street. It reflects the granularity of the street view.
4	Building Height	Average building height along the street. It is an important indicator of the street enclosing degree and determines the user's range of vision.
5	Building Cross-Section Ratio	Average building height to width ratio. It describes the interaction of these dimensions.
6	Street Wall Continuity	Ratio of street frontage building width to street length. It describes the interaction between street buildings.
7	Ratio of Total Building Numbers to Street Length	The count of buildings along both sides of a segment standardized by centerline length. It reflects the density of the block.
8	Height Variance	The difference between the tallest building and the lowest building along the street. It describes the continuity of street enclosing degree.

Source: Harvey Chester Wollaeger, 2014 [48].

Table 2. Quantitative Street Space Indicators.

No.	Category	Sub-Category	
1	Street Width	Street Width	
2	Street Length	Street Centerline Length	
3	Total Building Numbers	Total Building Numbers	
4	Building Height	Average Building Height	Higher Side Lower Side
5	Building Cross-Section Ratio	Average Building Height to Width Ratio	Higher Side Ratio Lower Side Ratio
6	Street Wall Continuity	Ratio of Street Frontage Building Width to Street Length	More Continuous Side Less Continuous Side
7	Ratio of Total Building Numbers to Street Length	Ratio of Total Building Numbers to Street Length	
8	Height Variance	Difference between High End and Low End of the Building	

Source: Harvey Chester Wollaeger, 2014 [48].

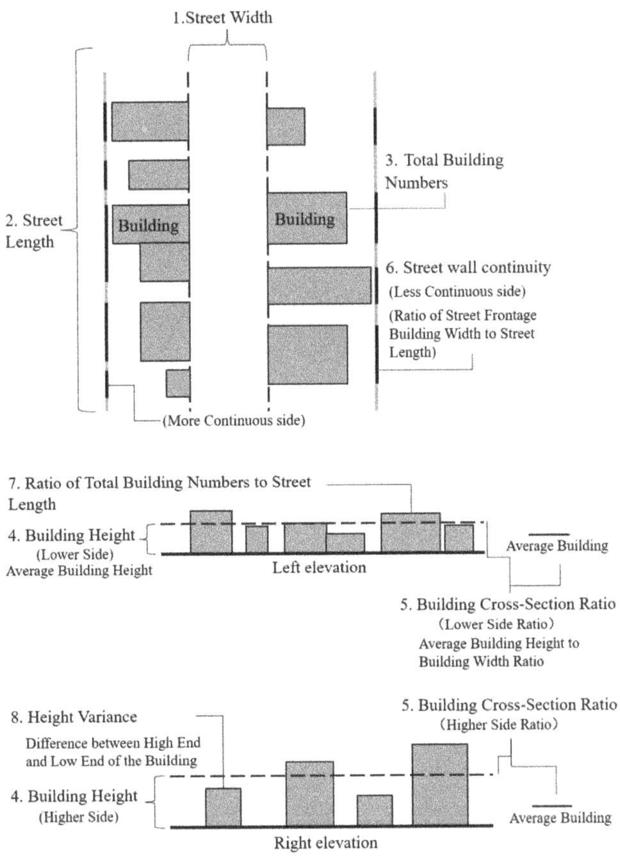

Figure 2. Quantitative street space indicator interpretation map [48].

3.1.2. Collection of Street Space Data Set

As shown in the indicator table, this study requires six data sets, namely, street length, street width, higher building height, lower building height, building width, and the number of buildings. Street length is calculated using the Calculate Geometry tool in ArcGIS and street width is calculated based on high-resolution satellite images.

As for building heights, widths, and numbers, four undergraduates with urban and rural planning and design backgrounds were recruited to make estimations based on the street view pictures and input into the software. They started with the collective measurement using a shared method to collect street building data from street view pictures and then worked individually to import data into ArcGIS street network attributes table. The imported data were converted into indicator data using the aggregated statistics tool in ArcGIS. In total, this took six days. Specifically, every undergraduate was required to complete 48–49 streets each day, and the import time of each street was controlled within 10 min; hence, each student was supposed to work less than 8 h a day.

The calculation of the mean and standard deviation for each indicator data set was completed as shown in Table 3.

A total of 7398 buildings were counted in the research scope. Most of the streets were two lanes in both directions in terms of average width. The mean heights of buildings fell in the range of 10.32–19.7, and most of the buildings had one floor or a couple of floors. Street wall continuity, which in general was high at 59% in dense places and 42% in sparse places, reflects the proportion of the length of the street versus building width.

This was based on the deviation calculation formula:

$$CV = \frac{SD}{Mean} \times 100\%$$

In the formula, CV is the deviation value, SD is the standard deviation, and Mean is the mean. The CV values for each indicator were calculated (within a reasonable range of 5–35%) and it was found that the CV values for the street space indicators within the data collection area were all higher than the maximum skewness value of 35%, indicating that the street spaces in downtown Guiyang are significantly different with a wide range of street types.

3.1.3. Street Spatial Pattern Identification

1. Principal Component Analysis (PCA)

PCA is conducted in SPSS on street space data, yielding a matrix as shown in Table 4. The criteria for the spatial classification of streets can be derived from the proportion of the components corresponding to the indicators. The data obtained were divided into four principal components with a total contribution of 64.86%, cumulatively.

In the component matrix, we selected data with component indicators greater than 0.6. In the first principal component, the indicators that met the criteria were building height (lower and higher) and street wall continuity (1.2); in the second principal component, it was the number of buildings; in the third and fourth principal components, there were no values that met the selection criteria. Hence, the three categories of building height, street wall continuity, and the number of buildings were used as the basis for classifying the street space for analysis in the next step.

2. Cluster Analysis

Building height, street wall continuity, and the number of buildings were used as the basis for cluster analysis of street spaces, which was conducted using Ward's method (sum of squares of deviations). The indicators were grouped into four categories according to the results. As breakdown data show in Table 5, they corresponded to the four types of street space: closed, compact, semi-open, and open.

Table 3. Street space indicators of core area in downtown Guiyang.

Building Numbers		Street Centerline Length (m)	Street Width (m)	Building Height (m) (Lower Side)	Building Height (m) (Higher Side)	Cross-Section Ratio 1	Cross-Section Ratio 2	Street Wall Continuity 1	Street Wall Continuity 2	Building Numbers/ Street Length	Height Variance (m)
7398	SD	148.75	12.84	10.32	19.7	0.66	1.26	0.42	0.59	0.12	7.77
	Mean	121.10	9.81	9.93	18.74	5.59	9.16	1.83	0.32	0.97	8.55
	CV (%)	122.83	130.88	139.27	105.12	11.80	13.80	22.90	184.43	12.37	90.87

Source: Authors' calculations.

Table 4. Street Spatial Component Matrix.

	1	2	3	4
Building Numbers	0.372	0.650	0.428	0.110
Street Width	0.362	0.191	−0.459	−0.026
Building Height (Lower Side)	0.728	0.158	−0.050	−0.410
Building Height (Higher Side)	0.810	−0.265	−0.116	0.073
Street Centerline Length	0.152	0.549	0.594	0.224
Cross-Section Ratio 1	0.555	−0.627	0.310	0.058
Cross-Section Ratio 2	0.531	−0.697	0.279	0.185
Street Wall Continuity 1	0.662	0.336	−0.056	−0.348
Street Wall Continuity 2	0.670	0.105	−0.009	0.533
Building Numbers Street Length	0.110	0.125	−0.410	0.566
Height Variance	0.594	0.266	−0.297	0.189

Source: Authors' calculations.

Table 5. Classification of street space in downtown Guiyang.

No.	Type	Number of Streets (Sections)	Building Numbers	Height Range (m)	Street Wall Continuity
1	Closed	593	8	10.80–24.08	0.59–0.71
2	Compact	34	10	8.98–15.32	0.57–0.71
3	Semi-open	459	6	8.98–18.9	0.33–0.54
4	Open	64	6	7.35–13.24	0.31–0.43

Source: Authors' calculations.

- Closed streets are characterized by high continuity of street walls and multi-floor or high buildings. The street space looks narrow from the human eye's perspective;
- Compact streets are similar to closed streets in street wall continuity and building numbers, but most buildings in this type of street are ground-floor buildings or podiums of high-rise buildings. The buildings are lower than those in closed streets;
- Semi-open streets are occupied by fewer buildings and characterized by large gaps in the continuity of the street walls on both sides of the street, with large building spacing and independent distribution;
- Open streets are spacious, with squares, public buildings, or vacant lots appearing on both sides of the street. The wall continuity is low, and the obstruction of views is fewer.

3.2. Street Vitality Identification

3.2.1. Street Vitality Categories

Urban street vitality is represented by the categories and frequency density of POI data. Based on the street type classification method of Better Street Plan in San Francisco [49] and the urban space characteristics of Guiyang city, this paper divides street vitality into four categories based on Baidu POI data categories, namely, residential, commercial, cultural leisure, and management and medicine (see Table 6). The residence category includes the buildings for people to live in and services, such as housekeeping, moving companies, wedding, elderly care, auto repair, renovation, car wash, and parking, among others; the commercial category includes facilities for personal, enterprise, and government consumption, such as enterprises, financial services, office buildings, accommodation, shopping, and catering services; cultural leisure refers to those facilities for entertainment, culture, and learning, among other purposes, such as tourism attractions, historical sites, education centers, research buildings, sports centers, and art performance buildings; management and medicine refer to the facilities required for the operation of the city and for the health care of residents, such as government buildings, municipal services, welfare centers, hospitals, clinics, and pharmacies.

Table 6. Street functions based on Baidu POI.

Street Vitality Category	Baidu POI Category
Residencial	Residential buildings, utility services
Commercial	Enterprises, shopping, catering, and accommodation services
Cultural leisure	Science, education, culture, sports, leisure, and scenic beauty
Management and medicine	Government and health care

Source: Collated by the author.

3.2.2. Street Vitality Categories and Levels

This paper draws upon the San Francisco street type classification method and quantitative identification method of urban functional area [49]; the research divides the research area into grids of 500 m × 500 m. POI data are then cited to determine the vitality type of each grid, which is quoted as the vitality type of the streets in the grid [50]. This paper uses the 500 m × 500 m scale because the target area includes many historical sites, which occupy large areas. Furthermore, the local residence facilities are largely big, closed communities. Consequently, the street spaces that are the target of design and renovation scatter in such large objects. The 500 m × 500 m scale allows the grids to include street spaces that are possible for renovation and design, and is fit for maintaining a complete street spatial experience. Normally, 500 m is the distance for a 10–30 min walk, allowing pedestrians to have a more comfortable and complete street spatial experience and that of the street vitality.

Firstly, the frequency density of every POI category was calculated according to the following formula:

$$f_i = \frac{n_i}{\sum_{i=1}^{4} n_i} (i = 1, 2, 3, 4)$$

In the formula, f_i denotes the frequency density of type i POI in the grid and n_i is the number of type i POI in the grid.

Then, the frequency density of each POI category within each grid was compared to the total frequency density to determine the dominant POI category and thus the type of street vitality within that grid. The formula to calculate total frequency density of POI is as follows:

$$F_i = \frac{N_i}{N} (i = 1, 2, 3, 4)$$

In the formula, F_i denotes the total frequency density of type i POI; N_i is the number of type i POI; and N represents the number of all POI.

By overlaying the four vitality types, more vitality types appear in a grid, the higher the vitality strength of the grid that is considered, and the grid vitality strength is graded according to this method. Hence, streets can be classified into four levels: grids with four vitality types are high vitality grids, which correspond to high vitality streets; grids with three vitality types are relatively high vitality grids, which correspond to relatively high vitality streets; grids with two or one vitality type(s) are relatively low and low vitality grids, which correspond to relatively low and low vitality streets.

This paper overlays the maps of vitality grids with street spatial pattern layouts to calculate the proportion of different patterns in each grid. The pattern with the highest proportion becomes the pattern of the grid. As a result, the layout of street vitality types is generated by counting the number of grids with a certain type of street vitality under a certain spatial pattern. Then, the vitality strength level grids are overlaid with the grids of street spatial patterns to show the numbers of grids with a certain type of vitality under a certain spatial pattern, so as to generate the vitality strength layout of various spatial patterns.

4. Research Conclusion

4.1. Street Spatial Pattern Classification

The four street types are connected in the ArcGIS street network map according to the classification of indicator data to form a visual analysis model to obtain a spatial distribution of street types in downtown Guiyang (see Figure 3).

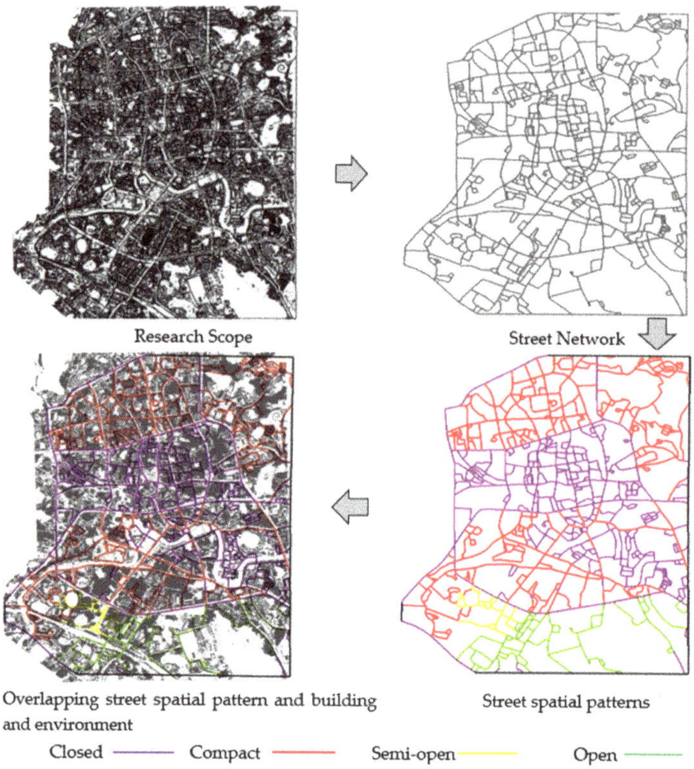

Figure 3. Process of street space classification.

The street spatial patterns in downtown Guiyang are concluded as follows: compact (540) > closed (483) > open (95) > semi-open (29); in terms of distribution, the closed street spaces are located in the city center, while the closed streets are surrounded by compact streets. The two types of streets are in typical commercial, office, and living areas and most of them have two narrow lanes for two-way traffic. The semi-open and open street spaces are located close to hills, squares, large public buildings, and express roads. Such a distribution structure is closely related to the type of urban street vitality and the degree of street vitality.

4.2. Street Vitality Classification and Leveling Results

As shown in Figure 4, the layout of street vitality types is cultural leisure vitality (47 grids), accounting for 54.02% of the total; management and medicine (46), 52.87%; and residential and commercial are both 40, accounting for 45.98%, respectively. In terms of the layout features of the street vitality types, streets that boast residential and management and medicine features are more evenly distributed, while commercial vitality is more concentrated in the west and south-west of the study area, and cultural leisure vitality is more seen in the center and north-east regions.

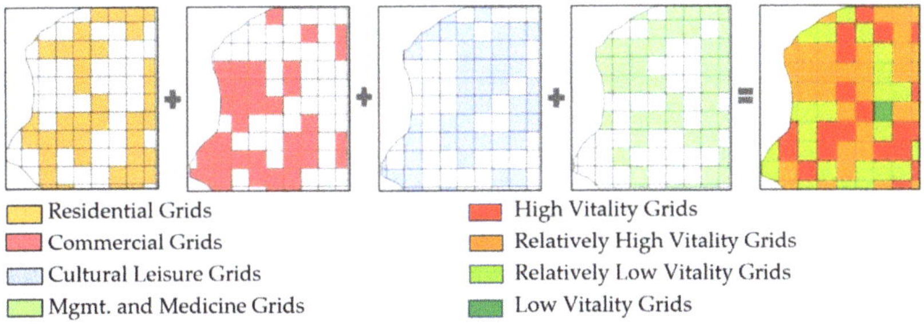

Figure 4. Identification of the type and level of grid vitality based on POI data.

The street vitalities can also be classified according to their levels. The percentage of the grids considered to have high vitality levels is 25.29%; it is 44.83% with relatively high vitality, 28.74% with relatively low vitality, and 1.15% with low vitality levels. An investigation of space distribution of vitality levels shows that most of the study area is covered by grids with high and relatively high vitality levels that stretch out in patches.

4.3. Vitality Types and Levels in Different Street Spatial Patterns

As shown in Figure 5, commercial vitality occupies the largest share in open street spaces at 35%, followed by lifestyle, which stands at 25.00%. In semi-open street spaces, the four vitality types are equally mixed with residential, commercial, and management and medicine accounting for 28.57%, respectively, and cultural leisure accounting for 14.29%, which is relatively lower. It is noteworthy of our attention that in compact and closed street spaces, the most widely distributed vitality type is cultural leisure, with 32.69% and 31.96%, respectively, whereas in compact spaces, others (28.85%) outperform those of closed spaces (24.74%). The same pattern is observed for the residential category, which accounts for 21.15% of vitality in compact spaces and 15.46% in closed spaces. The proportion of commercial vitality is higher in closed street patterns (27.83%) than in compact street patterns (17.31%).

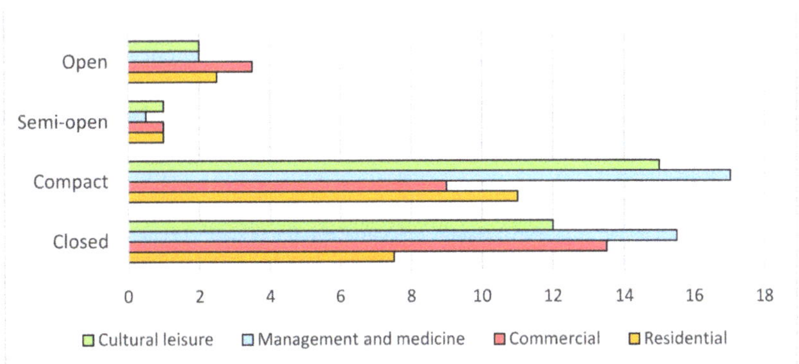

Figure 5. Layout of vitality types against street space patterns.

In general, compact street spaces enjoy the highest level of vitality. Among compact streets, 33.33% are high vitality streets and 43.33% are relatively high vitality streets. Closed street patterns are the second highest, with high vitality accounting for 25.00% and relatively high vitality accounting for 45.45%; in open street spatial patterns, the vitality level is the lowest, with 75.00% of the streets considered to be relatively low vitality streets (see Figure 6).

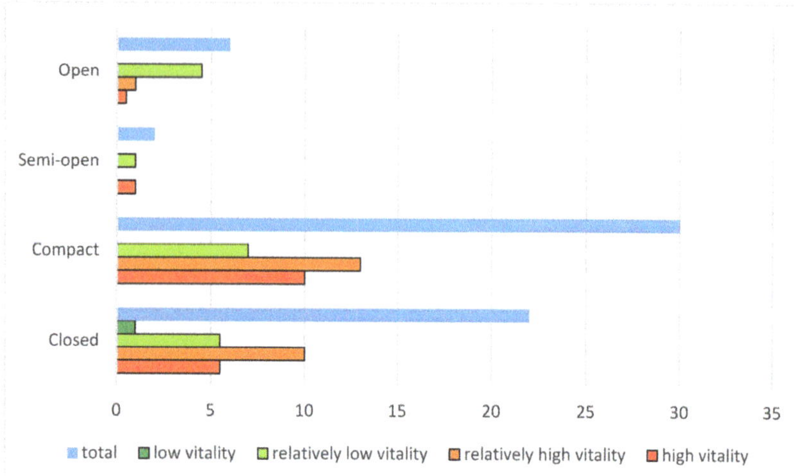

Figure 6. Layout of vitality levels against different street space patterns.

5. Discussions

Effectiveness and Universality of Research Method and Data Processing

This paper proves that the space data in SVI are feasible for the quantified classification of urban street spaces, avoiding uncertainty in spatial pattern classification. Meanwhile, the POI data help to accurately identify street vitality types and layout features. The two quantitative methods for street analysis provide a sound foundation for the quest into the relationship between urban street space and vitality. Moreover, a new method and perspective are introduced for urban street research when grids are used to breakdown the spatial layout of street spaces and vitality types.

6. Conclusions

Driven by quick urbanization, the street vitality types in the central area streets of Guiyang continue to overlap and the street space form is undergoing rapid transformation. Hence, Guiyang is in urgent need of clarifying the relationship between street space form, street vitality type, and street vitality strength, so as to provide guidance for shaping street space scientifically and efficiently.

6.1. Regularity Found in Layout of Vitality Type and Strength in Various Spatial Patterns

In terms of vitality types, an obvious difference is found in Guiyang regarding the layout of different vitality types in different spatial patterns. The commercial vitality has the widest layout in closed street spaces; residence, cultural leisure, and management and medicine are mostly found in compact spaces; and in semi-open street spaces, all vitality types show the lowest distribution.

In terms of vitality strength, closed and compact street spaces are stronger than semi-open and open. Compact street spaces registered as the highest strength level, whereas the semi-open spaces registered as the lowest.

6.2. Compact Street Spaces Should Be Prioritized in Designing Street Space Renovation

Among the four street spatial patterns in central Guiyang, compact street spaces show the highest level of vitality and most diverse vitality types, also making them the best street patterns for dense and rich vitality. As a result, it is recommended that more emphasis should be given to compact street spaces in urban and street design in Guiyang. To be specific, on one hand, this would strengthen the refined design of existing compact street spaces to give full play to the street vitality carried in such street spaces and, on the other, it would intentionally cultivate new compact street spaces in the city to promote urban vitality.

6.3. Cultural Leisure Vitality Is Most Adaptive to Street Spatial Patterns

In central Guiyang, it can be found that cultural leisure vitality is the most widely distributed vitality type in its street spaces, especially in compact and closed patterns where cultural leisure outnumbers others extensively. This shows that cultural leisure vitality in central Guiyang streets can adapt to different spatial patterns and co-exist with them, which is highly conducive to both the mental and physical health of the Guiyang people [51].

Author Contributions: Conceptualization, J.Y. and X.L.; methodology, J.Y. and J.D.; software, J.Y. and C.C.; validation, X.L., J.D. and J.Y.; formal analysis, J.Y.; investigation, J.Y. and J.D.; data curation, C.C.; writing—original draft preparation, J.Y.; writing—review and editing, X.L.; visualization, J.Y. and C.C. All authors have read and agreed to the published version of the manuscript.

Funding: This research was funded by the National Natural Science Foundation of China Regional Science Foundation Project, Number 52068006 and Guizhou Natural Science Foundation, Number [2019] 2850.

Institutional Review Board Statement: Not applicable.

Informed Consent Statement: Not applicable.

Data Availability Statement: All data used in this study are presented in the manuscript.

Conflicts of Interest: The authors declare no conflict of interest.

References

1. Wang, F.; Tan, C.; Li, M.; Gu, D.; Wang, H. Research on Traffic Design of Urban Vital Streets. *Sustainability* **2022**, *14*, 6468. [CrossRef]
2. El-Shimy, H.G.; Ragheb, R.A. Sustainable Urban Street Design: Evaluation of El-Moaz Street in Cairo, Egypt. *Procedia Environ. Sci.* **2017**, *37*, 689–698. [CrossRef]
3. Kim, L.Y.; Moon, T.H. The Effect of Compact City's Spatial Factors on Social Sustainability. *Hous. Stud.* **2010**, *18*, 51–72.
4. Artmann, M.; Kohler, M.; Meinel, G.; Gan, J.; Ioja, I.-C. How smart growth and green infrastructure can mutually support each other—A conceptual framework for compact and green cities. *Ecol. Indic.* **2019**, *96*, 10–22. [CrossRef]
5. Chung, J.-Y. A Study on the Urban Development Direction of Overseas and Domestic Sustainable Compact City. *J. Archit. Inst. Korea Plan. Des.* **2011**, *27*, 297–306.
6. Halas, M.; Kladivo, P.; Roubinek, P. Concept of a Compact-City: Contribution to Research and Administration. In Proceedings of the 16th International Colloquium on Regional Sciences, Brno, Czech Republic, 19–21 June 2013; pp. 140–146.
7. Bardhan, R.; Kurisu, K.; Hanaki, K. Does compact urban forms relate to good quality of life in high density cities of India? Case of Kolkata. *Cities* **2015**, *48*, 55–65. [CrossRef]
8. Yao, Y.; Pan, H.; Cui, X.; Wang, Z. Do compact cities have higher efficiencies of agglomeration economies? A dynamic panel model with compactness indicators. *Land Use Policy* **2022**, *115*, 106005. [CrossRef]
9. Soon, L.J. A Study on the Present Situation and Improvement Plan of Compact City in Japan—Focused on Practical Case of Toyama City in Japan. *J. Northeast Asian Econ. Stud.* **2018**, *30*, 145–171.
10. Abdullahi, S.; Pradhan, B.; Mansor, S.; Shariff, A.R.M. GIS-based modeling for the spatial measurement and evaluation of mixed land use development for a compact city. *GISci. Remote Sens.* **2015**, *52*, 18–39. [CrossRef]
11. Jalaladdini, S.; Oktay, D. Urban Public Spaces and Vitality: A Socio-Spatial Analysis in the Streets of Cypriot Towns. *Procedia Soc. Behav. Sci.* **2012**, *35*, 664–674. [CrossRef]
12. Zarin, S.Z.; Niroomand, M.; Heidari, A.A. Physical and Social Aspects of Vitality Case Study: Traditional Street and Modern Street in Tehran. *Procedia Soc. Behav. Sci.* **2015**, *170*, 659–668. [CrossRef]
13. Wang, M.; He, Y.; Meng, H.; Zhang, Y.; Zhu, B.; Mango, J.; Li, X. Assessing Street Space Quality Using Street View Imagery and Function-Driven Method: The Case of Xiamen, China. *ISPRS Int. J. Geo-Inf.* **2022**, *11*, 282. [CrossRef]
14. Alhasoun, F.; Gonzalez, M. Streetify: Using Street View Imagery and Deep Learning for Urban Streets Development. In Proceedings of the 2019 IEEE International Conference on Big Data (Big Data), Los Angeles, CA, USA, 9–12 December 2019.
15. Niu, C.; Tan, K.; Jia, X.; Wang, X. Deep learning based regression for optically inactive inland water quality parameter estimation using airborne hyperspectral imagery. *Environ. Pollut.* **2021**, *286*, 117534. [CrossRef]
16. Nadeem, M.; Aziz, A.; Al-Rashid, M.A.; Tesoriere, G.; Asim, M.; Campisi, T. Scaling the Potential of Compact City Development: The Case of Lahore, Pakistan. *Sustainability* **2021**, *13*, 5257. [CrossRef]

17. Kotharkar, R.; Bahadure, P.; Sarda, N. Measuring Compact Urban Form: A Case of Nagpur City, India. *Sustainability* **2014**, *6*, 4246–4272. [CrossRef]
18. Liu, S.; Zhang, L.; Long, Y. Urban vitality area identification and pattern analysis from the perspective of time and space fusion. *Sustainability* **2019**, *11*, 4032. [CrossRef]
19. Zeng, C.; Song, Y.; He, Q.; Shen, F. Spatially explicit assessment on urban vitality: Case studies in Chicago and Wuhan. *Sustain. Cities Soc.* **2018**, *40*, 296–306. [CrossRef]
20. Pan, C.; Zhou, J.; Huang, X. Impact of Check-In Data on Urban Vitality in the Macao Peninsula. *Sci. Program.* **2021**, *2021*, 7179965. [CrossRef]
21. Gong, Z.; Ma, Q.; Kan, C.; Qi, Q. Classifying Street Spaces with Street View Images for a Spatial Indicator of Urban Functions. *Sustainability* **2019**, *11*, 6424. [CrossRef]
22. Li, Q.; Cui, C.H.; Liu, F.; Wu, Q.R.; Run, Y.D.; Han, Z.G. Multidimensional Urban Vitality on Streets: Spatial Patterns and Influence Factor Identification Using Multisource Urban Data. *ISPRS Int. J. Geo-Inf.* **2022**, *11*, 2. [CrossRef]
23. Wey, W.; We, W.L. Urban Street Environment Design for Quality of Urban Life. *Soc. Indic. Res.* **2016**, *126*, 161–186. [CrossRef]
24. Von Schönfeld, K.C.; Bertolini, L. Urban streets: Epitomes of planning challenges and opportunities at the interface of public space and mobility. *Cities* **2017**, *68*, 48–55. [CrossRef]
25. Xia, Y.; Yabuki, N.; Fukuda, T. Sky view factor estimation from street view images based on semantic segmentation. *Urban Clim.* **2021**, *40*, 100999. [CrossRef]
26. Zhang, L.; Ye, Y.; Zeng, W.; Chiaradia, A. A Systematic Measurement of Street Quality through Multi-Sourced Urban Data: A Human-Oriented Analysis. *Int. J. Environ. Res. Public Health* **2019**, *16*, 1782. [CrossRef] [PubMed]
27. Wu, B.; Yu, B.; Shu, S.; Liang, H.; Zhao, Y.; Wu, J. Mapping fine-scale visual quality distribution inside urban streets using mobile LiDAR data. *Build. Environ.* **2021**, *206*, 108323. [CrossRef]
28. Hu, F.; Liu, W.; Lu, J.; Song, C.; Meng, Y.; Wang, J.; Xing, H. Urban Function as a New Perspective for Adaptive Street Quality Assessment. *Sustainability* **2020**, *12*, 1296. [CrossRef]
29. Pospelov, P.; Kostsov, A.; Martiakhin, D. Results of studies on design speed limitations on urban streets of the Russian Federation. *Transp. Res. Procedia* **2018**, *36*, 617–621. [CrossRef]
30. García-Ramírez, Y.; Paladines, L.; Verdesoto, C.; Torres, P. Car Drivers Do Not Choose Their Speed in Urban Environments: Speed Models in Tangent Streets. In Proceedings of the 7th International Conference on Vehicle Technology and Intelligent Transport Systems—VEHITS, Online, 28–30 April 2021; pp. 421–428. [CrossRef]
31. Prelovskaya, E.; Levashev, A. Modern Approach of Street Space Design. *Transp. Res. Procedia* **2017**, *20*, 523–528. [CrossRef]
32. Deshpande, R.; Gartner, N.H.; Zarrillo, M.L. Urban Street Performance: Level of Service and Quality of Progression Analysis. *Transp. Res. Rec.* **2010**, *2173*, 57–63. [CrossRef]
33. Sharifi, A. Resilient urban forms: A review of literature on streets and street networks. *Build. Environ.* **2019**, *147*, 171–187. [CrossRef]
34. Hu, C.-B.; Zhang, F.; Gong, F.-Y.; Ratti, C.; Li, X. Classification and mapping of urban canyon geometry using Google Street View images and deep multitask learning. *Build. Environ.* **2020**, *167*, 106424. [CrossRef]
35. Huang, Y.-D.; Ren, S.-Q.; Xu, N.; Luo, Y.; Sin, C.H.; Cui, P.-Y. Impacts of specific street geometry on airflow and traffic pollutant dispersion inside a street canyon. *Air Qual. Atmos. Health* **2022**, *15*, 1133–1152. [CrossRef]
36. Su, T.; Sun, M.; Fan, Z.; Noyman, A.; Pentland, A.; Moro, E. Rhythm of the streets: A street classification framework based on street activity patterns. *EPJ Data Sci.* **2022**, *11*, 43. [CrossRef]
37. Jacobs, J. *The Death and Life of Great American*; Random House: New York, NY, USA, 1961.
38. Xu, X.; Xu, X.; Guan, P.; Ren, Y.; Wang, W.; Xu, N. The Cause and Evolution of Urban Street Vitality under the Time Dimension: Nine Cases of Streets in Nanjing City, China. *Sustainability* **2018**, *10*, 2797. [CrossRef]
39. Long, Y.; Zhao, J.; Li, S.; Zhao, Y.; Xu, L. The Large-Scale Calculation of "Walk Score" of Main Cities in China. *New Archit.* **2018**, *3*, 4–9. [CrossRef]
40. Jiang, Y.; Han, Y.; Liu, M.; Ye, Y. Street vitality and built environment features: A data-informed approach from fourteen Chinese cities. *Sustain. Cities Soc.* **2022**, *79*, 103724. [CrossRef]
41. Xia, C.; Yeh, A.G.O.; Zhang, A. Analyzing spatial relationships between urban land use intensity 2 and urban vitality at street block level: A case study of five Chinese megacities. *Landsc. Urban Plan.* **2020**, *193*, 103669. [CrossRef]
42. Lee, J.H. A Study on Typology and Property of the Method of Spatial Composition in the Block of Street. *J. Archit. Inst. Korea Plan. Des.* **2005**, *21*, 177–186.
43. Kim, J. The adoption of Western urban design theories in urban developments in Korea from the perspective of building-street relationship—Focusing on urban development cases in the Seoul metropolitan area. *GRI Rev.* **2017**, *19*, 1–24.
44. Dumbaugh, E.; King, M. Engineering Livable Streets: A Thematic Review of Advancements in Urban Street Design. *J. Plan. Lit.* **2018**, *33*, 451–465. [CrossRef]
45. Ding, W.; Tong, Z. An approach for simulating the street spatial patterns. *Build. Simul.* **2011**, *4*, 321–333. [CrossRef]
46. Yang, J.; Xing, Z.; Cheng, C. How Urban Fringe Expansion Affects Green Habitat Diversity? Analysis from Urban and Local Scale in Hilly City. *J. Environ. Public Health* **2022**, *2022*, 8566686. [CrossRef] [PubMed]
47. Kang, M. Three Inherent Conflicts over Compact City: Social Welfare vs Individual Interest. *J. Korean Reg. Dev. Assoc.* **2012**, *24*, 27–40.

48. Harvey, C. Measuring Streetscape Design for Livability Using Spatial Data and Methods. Master's Thesis, University of Vermont and State Agricultural College, Burlington, VT, USA, 2014.
49. San Francisco Planning Department. San Francisco Better Streets Plan Policies and Guidelines for the Pedestrian Realm. 2010. Available online: https://sfplanning.org/sites/default/files/archives/BetterStreets/docs/Better-Streets-Plan_Final-Adopted-10-7-2010.pdf (accessed on 5 January 2023).
50. Jun, C.; Zhan, G.; Wei, S. Livable Street Assessment Index System and Data Adaptation Analysis. In *Intelligent Planning-Ecological Habitat-Quality Space, Proceedings of the 2019 China Annual Conference on Urban Planning Informatization, Shenzhen, China, 15 November 2019*; Guangxi Science and Technology Press: Nanning, China, 2019; pp. 167–175. [CrossRef]
51. Kang, I.-H. The Strategy of Compact City in Japan Local Governments. *Korea Assoc. Policy Stud.* **2018**, *27*, 221–244.

International Journal of
Environmental Research and Public Health

Article

Exploring Urban Green Space Optimization of the Urban Walking Life Circle in Fuzhou, China

Huili Xie, Xinke Wang, Xiaoting Hu, Zhiyong Shi, Hong Lin, Xiangqun Xie, Lingxiu Chen, Hongxia Dai, Jiao Zhang, Mengjie Xu and Xingzhao Liu *

College of Landscape Architecture and Art, Fujian Agriculture and Forestry University, Fuzhou 350002, China
* Correspondence: xzliu@fafu.edu.cn

Abstract: The spatial distribution of urban green spaces (UGS) is closely related to the health of residents and the ecological pattern of cities. Exploring the equity of UGSs plays an important role in urban planning and also provides guidance for urban development. Taking the main urban area of Fuzhou City as an example, this study uses network big data and census data to pinpoint the population demand, evaluates the accessibility and equity of UGS within the basic living circle, neighborhood living circle and daily living circle of residents at the scale of residential and sub-districts. Based on the G2SFCA model, we also quantify the actual effective UGS's service capacity. Then, using the scale and travel range as the entry point, we further discuss the similarities and differences under different scales and different travel ranges. Finally, optimization strategies are proposed for the construction status. The results show that: (1) The spatial allocation of urban green space resources varies significantly, and there is a serious inequity in the spatial distribution of urban green space under pedestrian conditions; (2) The results of UGS accessibility, equity, and service capacity in Fuzhou at both residential and sub-district scales are consistent; (3) Urban construction should be multi-level overall planning, combined with local economic and social development factors in accordance with local conditions to take measures. The results of the study can provide a scientific reference for the optimization of the spatial distribution of UGS.

Keywords: urban green space; life circle; different research scales; equity; G2SFCA

Citation: Xie, H.; Wang, X.; Hu, X.; Shi, Z.; Lin, H.; Xie, X.; Chen, L.; Dai, H.; Zhang, J.; Xu, M.; et al. Exploring Urban Green Space Optimization of the Urban Walking Life Circle in Fuzhou, China. *Int. J. Environ. Res. Public Health* **2023**, *20*, 1180. https://doi.org/10.3390/ijerph20021180

Academic Editors: Hongxiao Liu, Tong Wu and Yuan Li

Received: 21 December 2022
Revised: 4 January 2023
Accepted: 6 January 2023
Published: 9 January 2023

1. Introduction

Urban green space (UGS) is an important part of urban natural ecosystems, which has the functions of purifying the environment [1], regulating microclimate [2], alleviating the urban heat island effect [3], and maintaining biodiversity and other ecosystem services [4]. It has many positive effects on human physiological and psychological health [5]. The justice of UGS resource allocation is not only conducive to enhancing the human well-being, but also promotes social justice and harmony [6]. Therefore, exploring the equity of UGS has an important role to play in urban planning and also provides guidance for urban development.

Most previous studies on UGS equity have started from accessibility, discussing whether residents have access to urban green space services within a certain spatial range. Accessibility was first proposed by Hansen in 1959 [7], and the existing measurement methods mainly include the buffer zone method, the minimum proximity method, the network analysis method, and the two-step floating catchment area method. Compared with other accessibility analysis methods, the two-step floating catchment area method takes into account the interaction between supply and demand, which was first proposed by R. John et al. [8] and it has been adopted by a large number of scholars since then. However, this method not only ignores the actual resistance of road network formation, but also does not consider the radius difference between the supply and demand side of the service [9,10]. To solve this problem, a series of improved algorithms have been derived from the two-step floating catchment area method. Shalini K et al. established the enhanced two-step floating catchment area model (E2SFCA) [11]; Polzin et al. proposed the expanded kernel density two-step floating catchment area (KD2SFCA) method and validated it in the Portuguese region [12]; and Weiji Xu [13], Xiaoyan Zhou et al. [14] used the gravity floating catchment area method (Gravity-2SFCA) to measure UGS accessibility. Xiaotong Xue et al. [15] used the Gaussian floating catchment area method (G2SFCA) to measure the urban accessibility of green space supply in different communities in Xuzhou. In previous studies, the demand of residents was mainly represented by population distribution data or census data, which is difficult to accurately match to the demand of the population. However, high-density cities lack space for development and construction. Accurate matching of supply and demand and urban micro renewal are new trends in urban planning.

Previous research scales mainly focused on macro-scales [16], but the evaluation of the mezzo-scale or micro-scale was insufficient, which led to many problems such as too-large units, inaccurate data, and so on. The existing studies lack comparative studies between different scales. In order to evaluate the equity of UGS more scientifically and accurately, this study takes the main area of Fuzhou City as an example, combines the data of the 7th Chinese census and network big data to locate the supply and demand, and adds a Gaussian decay coefficient on the traditional two-step floating catchment area (G2SFCA) method to measure the accessibility of UGS at different scales under various travel distances of residents in the main urban area of Fuzhou. At the same time, the Gini coefficient and Lorenz curve are introduced to evaluate the overall social equity of UGS allocation, and the effective coverage of UGS and the locational entropy method are used to discuss the real capacity of service supply of UGS. Previous studies lacked commonalities and diversities among different research scales and different travel scopes, so this study will also continue to improve and explore the spatial differentiation of UGS accessibility, equity and actual supply capacity of services at different research scales and different travel ranges. Then we will make planning proposals for urban construction in Fuzhou based on the results, with a view to providing a reference for the optimization of the spatial layout of UGS.

2. Materials and Methods

2.1. Study Area

Fuzhou, the capital of Fujian Province, is located at 25°15′~26°39′ N latitude and 118°08′~120°31′ E longitude, with a total area of 11,968 km^2. It has a subtropical monsoon climate with suitable temperatures and warm and humid conditions. By the end of 2021, the total resident population in Fuzhou City was about 8.42 million, the regional GDP was 113.248 billion RMB, and the urbanization rate was 73.0% (access on http://tjj.fuzhou.gov.cn/zz/fztjnj/2022tjnj/zk/indexch.htm (accessed on 8 March 2022)). In this study, the main urban area of Fuzhou City was selected as the study area, with an area of about 311.8 km^2 as of the end of 2021. It includes the whole area of Cangshan District, Gulou District and Taijiang District, and six sub-districts in Jin'an District, for a total of 39 sub-districts (Figure 1).

2.2. Data Source and Preprocessing

2.2.1. Urban Green Space

UGS is the green infrastructure of the city, as well as a basic spatial guarantee for sustainable urban development [17]. Previous studies mostly focus on park green space and mostly take the geometric center of green space as the service supply point, which ignores the service value of other types of UGS and is not in line with the actual situation. The Fuzhou Urban Green Space System Plan (2016–2020) divided UGS into five categories: park green space, square green space, protective green space, regional green space and attached green space. Since the protective green spaces focus more on ecological protection function and the attached green spaces were not developed externally leading to the specificity of the service population, only park green space, square green space and regional green space were included in this study. This study extracted 85 UGSs with a total area of about 76,537.6 ha with reference to the "Urban Green Line of Fuzhou City (2016–2020)", published by the Fuzhou Landscape Bureau and Google remote sensing image data. We assumed that residents have access to the entrance of the UGS within a certain walking range that is served by the UGS. According to previous studies [13], the geometric centers of green areas with a scale of less than 5 ha were extracted and used as the center of the network analysis; for green areas with a scale larger than 5 ha, the actual entrances and exits of the green areas were determined by Baidu map poi data and field research (Table 1). The park green space dataset was established (Figure 2a).

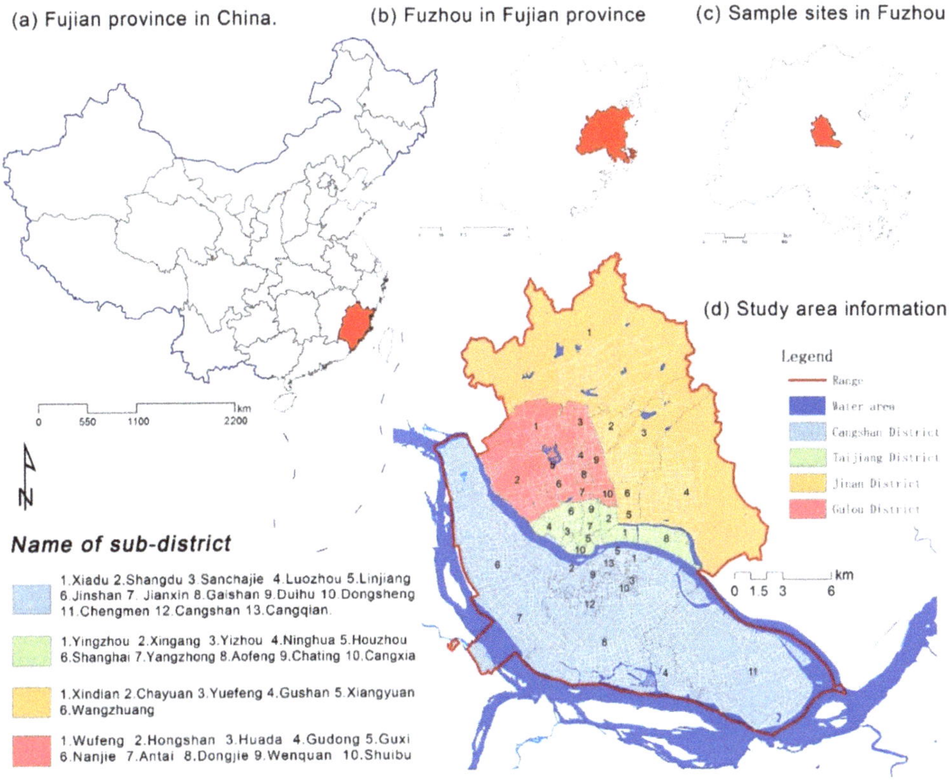

Figure 1. Location and information of study area.

Table 1. Classified statistics of UGS.

Type	Count	Ratio/%	Area/ha	Ratio/%
>5 ha	43	50.59	2000.30	96.19
<5 ha	42	49.41	79.24	3.81
Total	85	100.00	2079.54	100.00

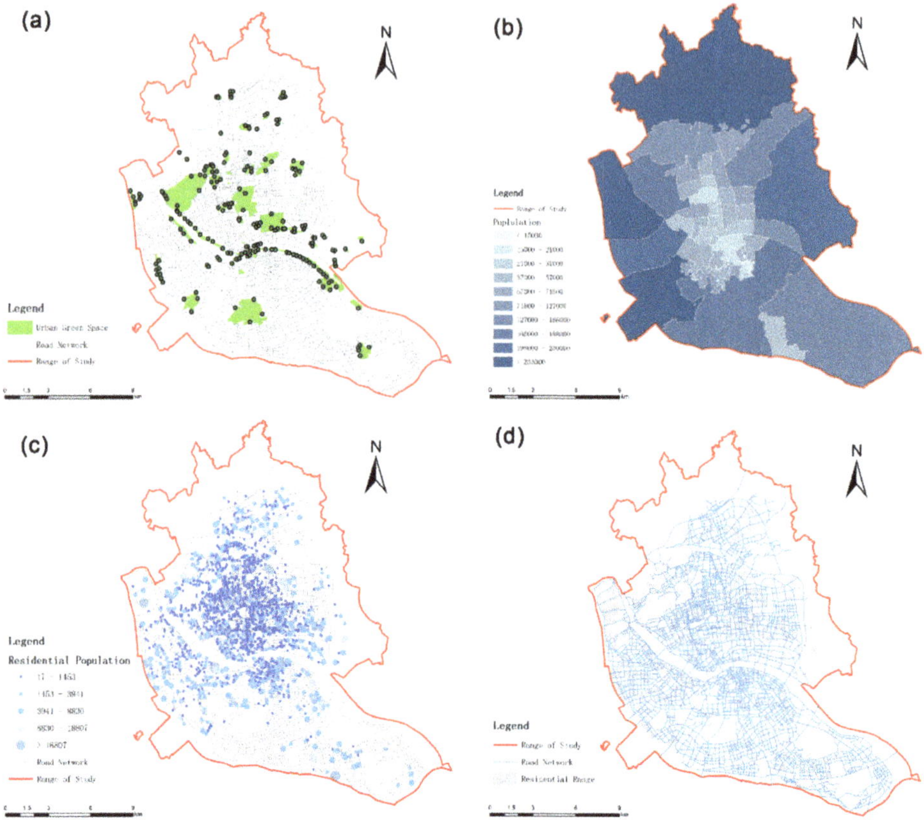

Figure 2. (**a**) Distribution of UGS in the study area. (**b**) Population distribution in sub-district scale. Information on the road network in the study area. (**c**) Population distribution in residential scale. (**d**) Information on the road network in the study area.

2.2.2. Residential Population Data

The population can represent the demand for urban green space to a certain extent. To obtain more accurate research results, this study used residential areas as the measurement unit for analysis. The outline of residential areas within the study area was obtained from Baidu Map (https://lbsyun.baidu.com/, accessed on 29 December 2021). The geometric center of each residential area was used as the service demand point, and a total of 2020 residential points were finally obtained. This study assumed that the population was evenly distributed within the settlement area [18]. The population of all settlements in each sub-district was calculated based on the population data of the sub-district (Figure 2b), and the proportion of each settlement in the total area of the settlement in its street (Figure 2c, Table 2). The population data of sub-districts were obtained from the data of the seventh national census published by the Fuzhou City Bureau of Statistics (http://tjj.fuzhou.gov.cn, accessed on 20 December 2021). The formula is as follows.

$$D_k = \frac{RA_K}{RA} \times SP, \tag{1}$$

where Dk is the population in the kth residential area; RAk is the area of kth residential area; RA is the total residential area of the sub-district where the residential area is located; SP is the total population of that sub-district. Thirty residential areas were randomly selected as samples and, by surveying the number of households in these residential areas, the calculation results of the above formula were tested according to the average household size of 3.5 persons (the results of the seven popular population statistics in Fuzhou City). The average accuracy rate was at 76%, so the present estimation was considered to be able to estimate the residential population well.

Table 2. Data display for residential.

Number	District	Sub-District	Population	Number	District	Sub-District	Population
A1	Cangshan	Jinshan	3166	A6	Jinan	Xiangyuan	1254
A2	Jinan	Xindian	663	A7	Taijiang	Shanghai	409
A3	Gulou	Huada	1024	A8	Jinan	Wangzhuang	282
A4	Gulou	Antai	618	A9	Taijiang	Chating	1393
A5	Cangshan	Duihu	1305	A10	Gulou	Gxi	1296

2.2.3. Road Network

Open Map Street platform was used to obtain open-source data (www.openstreetmap.org, accessed on 4 February 2022) and extract the road network data in and around the study area. The data were combined with Google Maps' high-accuracy images to correct the data, screen out railways, highways and other roads that are not suitable for pedestrians. Finally, we obtained a road network that better matched the real situation (Figure 2d).

2.3. Methodology

Firstly, we used multi-source data to build a database of UGSs and population. After that, calculated the accessibility of residents under pedestrian condition based on G2SFCA model, and further explored the fairness of accessibility. We also quantified the service capacity of UGS using service coverage and locational entropy. Finally, we identify the inequitable areas of UGS service provision in Fuzhou and make reference suggestions for urban construction (Figure 3).

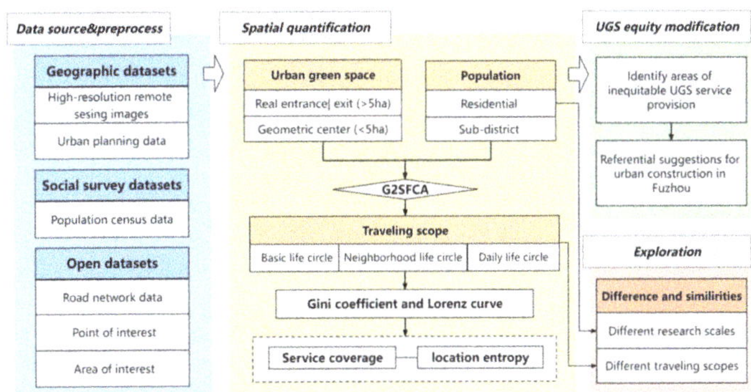

Figure 3. Technology Route.

2.3.1. G2SFCA-Based Accessibility Evaluation

The Gaussian two-step floating catchment area (G2SFCA) method is an accessibility research method optimized from the traditional two-step floating catchment area method. The distance decay coefficient is added to the traditional calculation, which is more suitable for the actual situation while the calculation results are more reliable. In this study, residents were set to reach UGS by walking, and the walking speed was determined to be 5 km/h according to previous studies [13]. The 5-min walking distance is the basic activity range of residents [19], and 15 min is the best travel time for residents to use the UGS service [20,21], while the ultimate travel time is 30 min [2,22]. So, this study delineated the 5-min living circle, 15-min living circle and 30-min living circle as the travel range (Table 3) to further quantify the residents' received services by UGS under different activity ranges (Table 3).

Table 3. Traveling scopes.

Type	5-min Life Circle	15-min Life Circle	30-Min Life Circle
Distance	500 m	1000 m	2000 m

The spatial accessibility evaluation based on G2SFCA was divided into two main steps. In the first step, the UGS point data was used as the starting point and the residential points as the destination points for analysis, and the residential points k within the search space range d_0 were filtered out from the UGS_j. The number of all populations within d_0 was aggregated and assigned weights according to the distance decay law using the Gaussian function [22]. These weighted populations were summed and aggregated to calculate the supply–demand ratio Rj.

$$R_j = \frac{S_j}{\Sigma_{k\in\{d_{jk}\leqslant d_0\}}{}^{G(d_{jk})D_k}} \tag{2}$$

where D_k is the population number of each residential k, d_{kj} is the actual road network distance between locations k and j. For the UGSs with multiple entrances, the road network distance from the settlement to the nearest entrance was selected, while the settlement k is in the search field (i.e., $d_{kj} \leq d_0$); S_j is the area of urban green space j; $G(d_{kj})$ is a Gaussian decay function considering the spatial friction problem. Its expression as follows.

$$G\left(d_{kj}\right) = \frac{e^{-\frac{1}{2}\times(\frac{d_{kj}}{d0})^2} - e^{-\frac{1}{2}}}{1-e^{-\frac{1}{2}}}\left(d_{jk}<d_0\right). \tag{3}$$

In the second step, the analysis was performed with residential points as the starting point and UGS point data as the destination point. All UGSs j in the search domain d_0 were found separately, and the supply and demand ratios R_j of these UGSs were summed up on the basis of Gaussian decay function to obtain k the spatial accessibility Ak based on the distance cost, whose larger value indicates a higher degree of accessibility. The formula is as follows:

$$A_k = \Sigma_{j\in\{d_{kj}\leqslant d_0\}}{}^{G(d_{kj})R_j} \tag{4}$$

The accessibility results calculated by the G2SFCA have the meaning of UGS capita in a wide sense. Therefore, it can be compared with UGS per capita as a basis for evaluating accessibility, especially for horizontal comparison of different units in the region [23]. The UGS per capita is the ratio of urban green space to total population, and the measured green space per capita in the study area is 6.95 m^2/person.

2.3.2. Accessibility-Based Spatial Equity Evaluation

The Lorenz curve method is often used to explore the justice of national income in distribution. Recently, the Lorenz curve has been widely used in the fields of urban public resource allocation and spatial layout of green space to reflect the justice level. The value of Gini coefficient is taken between 0~1 [24]. The smaller the Gini coefficient, the more reasonable the level of spatial distribution of resources among urban residents and the fairer its spatial distribution. In this study, the Lorenz curve was introduced to analyze the justice level of the spatial distribution of UGS in the main urban area of Fuzhou. The calculation formula is as follows.

$$G = 1 - \frac{1}{n}\left(2\Sigma_{i=1}^{n-1}W_i + 1\right) \tag{5}$$

where G is the Gini coefficient, n is the total number of sub-districts (or residential areas) in the study area, ' i' is the i-th sub-district (residential area) after the result of the ratio of the cumulative value of accessibility of each sub-district (residential area) to the number of people in the sub-district (residential area) is ranked from smallest to largest, $i = 1, 2, \ldots, n$; Wi is the proportion of urban green space accessibility per capita from sub-district (residential area) 1 to sub-district (residential area) i to the total accessibility per capita in the whole area.

2.3.3. Accessibility-Based Performance Evaluation of UGS Services

(1) UGS service coverage

Referring to the studies of Shuang Niu et al. [25] and Zilai Tang et al. [26], and we selected the ratio of the effective service area of UGS in a spatial unit (the sum of population) to the area of the spatial unit (population) as a quantitative indicator of the service level of UGS, i.e., the effective service area ratio and the effective service population ratio. The formulas are as follows:

$$T_j = \frac{\sum S_n}{S_j} = \frac{S_1 + S_2 + S_3 + \ldots + S_n}{S_j}, \tag{6}$$

$$U_j = \frac{\sum P_n}{P_j} = \frac{P_1 + P_2 + P_3 + \ldots + P_n}{P_j}, \tag{7}$$

where T_j is the UGS effective service area ratio of unit j; U_j is the UGS effective service population ratio of unit j; n represents the presence of n residential areas in sub-district j.

(2) Location entropy method

The locational entropy represents the ratio of UGS services received by the population living in a study unit to the per capita UGS resources of the residents in the whole study area. It can be used to express the spatial distribution pattern of the justice of the per capita distribution of resources in different study units. The higher the locational entropy value, the higher the level of resource allocation equipped in the study unit. If the locational entropy of the study unit is greater than 1, it indicates that the per capita level of UGS resources receiving services in the study unit is higher than the overall level of the whole study area; if the locational entropy is less than 1, it indicates that the per capita level of urban green space resources enjoyed in the study unit is lower than the overall level. The calculation formula is as follows:

$$AALQ_j = \frac{Aj}{\left(\frac{S}{P}\right)}, \tag{8}$$

where A_j denotes the per capita UGS enjoyed by the population of unit j within the travel threshold, i.e., the actual accessibility of UGS obtained using the two-step floating catchment area method, S denotes the total UGS within the study area, P denotes the total population within the study area, and the ratio of S to P is the per capita UGS resources in the study area.

3. Results

3.1. Evaluation of Spatial Accessibility of UGS

3.1.1. Residential Scale

Based on the G2SFCA method, the spatial accessibility of UGSs within the 5-min life circle, 15-min life circle and 30-min life circle of residential areas in the study area was analyzed. The results were graded by the natural breakpoint method into five grades: very low, lower, middle, higher and very high (Table 4).

Table 4. Accessibility grading statistics of residential areas under different traveling scopes.

Level	5-min Life Circle	Ratio	15-min Life Circle	ratio	30-min Life Circle	Ratio
Inaccessible	1689	83.6%	1417	70.1%	907	44.9%
Very low	286	14.2%	496	24.6%	918	45.4%
Lower	27	1.3%	66	3.3%	129	6.4%
Medium	10	0.5%	21	1.0%	56	2.8%
Higher	5	0.2%	17	0.8%	8	0.4%
Very high	3	0.1%	3	0.1%	2	0.1%
Accessible	331	16.4%	603	29.8%	1113	55.1%
Higher than the UGS per capita	190	57.4%	322	53.4%	540	48.5%

A total of 1689 residential areas in the main urban area of Fuzhou are not accessible within the 5-min life circle, i.e., 83.6% of the residential areas in the study area cannot be served by UGSs; only 5.7% of the accessible residential areas are rated as higher and very high in accessibility, among which those with very high accessibility are located in the Xiangyuan sub-district and the Yingzhou sub-district. The accessibility of only 190 settlements is greater than the per capita UGS in the study area, accounting for 57.4% of the total accessible settlements. The mean value of the accessibility evaluation result is 40.2, while the standard deviation is 953.7, with obvious spatial variation (Table 5).

Table 5. Statistical results of accessibility for different traveling scopes.

	5-min Life Circle	15-min Life Circle	30-min Life Circle
Average	40.2	35.8	32.2
Standard deviation	953.7	337.3	186.0

When the search radius was set to 15 min living circle, 70.1% of the settlements were inaccessible. The settlements with grade of inaccessible or very low in theHongshan sub-district, the Xindian sub-district, the Jianxin sub-district and the Gushan sub-district were the most numerous, with the number of settlements with both classifications exceeding 5% of the total number of settlements; settlements with very high accessibility were located in the Hongshan, Wenquan and Xindian sub-districts. There are 322 settlements with accessibility greater than the per capita green area in the study area, accounting for 53.4% of the total number of accessible settlements. The mean value of the settlement accessibility evaluation results is 35.8 in the 15-min life circle, and the standard deviation is 337.3, which also shows obvious spatial differentiation (Table 5).

When the search scope was expanded to the 30 min living circle, 55.1% of the residential spots in the study area were accessible; among them, 540 residential areas were accessible more than the UGS area per capita, accounting for 48.5%, which is a significant improvement in accessibility compared to the first two cases. However, there are still cases of inaccessible settlements. These are concentrated in the incompletely developed areas at the edge of the study area and in the relatively sparsely constructed areas of the Sanchajie and Duihu sub-districts.

According to the results of UGS accessibility at the scale of residential area, the spatial separation of UGS occurs in the study area under the search radii of 500 m, 1000 m and 2000 m (Figure 4). The smaller the activity circle, the more green space service blind areas; even within the 30 min living circle, the percentage of residential areas within the service blind area reaches 44.9%. These indicate that the current UGS arrangement can hardly meet the population demand.

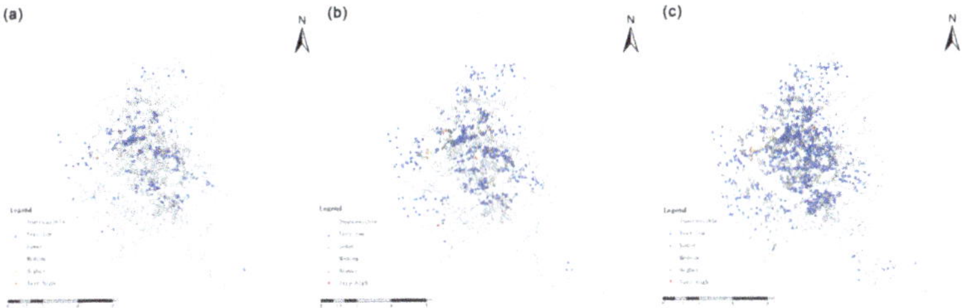

Figure 4. G2SFCA-based residential scale accessibility under different traveling scopes (**a**) 5-min life circle. (**b**) 15-min life circle. (**c**) 30-min life circle.

3.1.2. Sub-District Scale

The accessibility analysis results of sub-district scale based on G2SFCA are shown in Figure 5: within the 5-min life circle, the Wangzhuang, Xiangyuan and Yingzhou sub-districts, which are located in the center of the study area, have the highest accessibility, while Sanchajie sub-district is beyond the accessibility space and has inaccessibility. In the case of the 15-min life circle, the number of sub-districts with high accessibility increases, accounting for one-third of the total. The accessibility of the Hongshan sub-district ranks first. Further expanding the search threshold to the 30 min living circle, the sub-districts with the highest accessibility are the Xindian sub-district and the Cangxia sub-district.

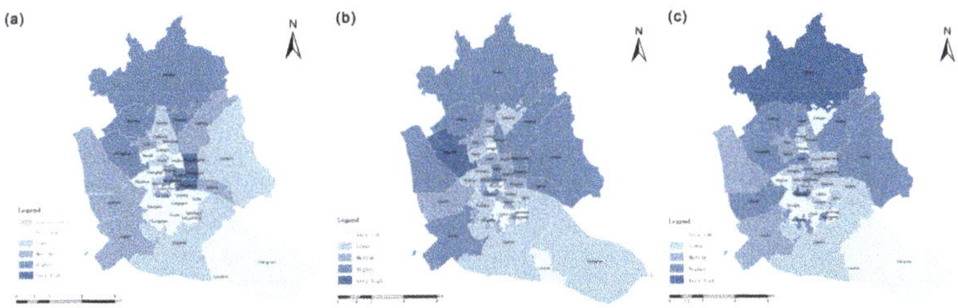

Figure 5. G2SFCA-based sub-district scale accessibility under different traveling scopes (**a**) 5-min life circle. (**b**) 15-min life circle. (**c**) 30-min life circle.

The overall accessibility of the study area showed as high in the northwest and low in the southeast under all three activity range circles. There are four sub-districts with unchanged grades, from higher, middle, lower to very low accessibility corresponding to Wufeng, Guxi, Gudong, and Luzhou sub-districts, respectively. The accessibility level of Shangdu sub-district and Chengmen sub-district has been at a very low or lower level. Eighteen percent of the sub-districts have changed significantly, with the increase of travel range, the accessibility levels of the Antai sub-district, Cangxia sub-district and Changshan sub-district have continued to become higher, while the accessibility levels of four sub-districts, namely Chayuan sub-district, Yingzhou sub-district, Wangzhuang sub-district and Xiangyuan sub-district, have been reduced.

3.2. Evaluation of Spatial Equity of UGS

The Gini coefficient and Lorenz curve express the overall social equity performance of UGS allocation (Figure 6). The calculated results of the Gini coefficient under the residential scale are all greater than 0.9. The UGS resource allocation varies greatly among residential areas and the social inequity is significant. The spatial allocation of UGS in the study area at the sub-district scale has a large gap in each sub-district, and all three spatial thresholds show a large difference in the allocation of UGS resources; especially at the search radius of the 15-min life circle, the Gini coefficient achieves 0.723, indicating the serious social inequity. Both research scales reveal significant differences in the allocation of UGS and serious social inequity.

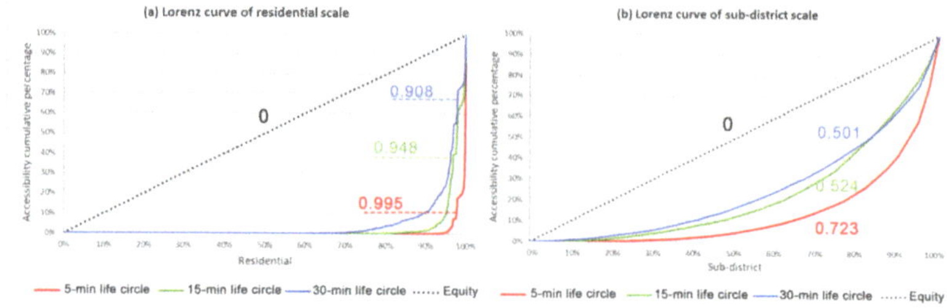

Figure 6. Gini coefficient and Lorenz curve under different scales. (**a**) Residential scale. (**b**) Sub-district scale.

3.3. Service Performance Evaluation of UGS

3.3.1. Residential Scale

The location entropy reflects the specific pattern of "spatial matching" between the allocation of UGS and the resident. Additionally, the sub-district location entropy value grade generally represents the evaluation level of the actual per capita UGS allocation. Therefore, after measuring the fairness in general, the spatial distribution pattern of UGS social equity can be analyzed using the locational entropy method (Figure 7). With reference to previous literature, the locational entropy value is divided into five grades: very low, lower, medium, higher, and very high. The locational entropy grading under the two research scales is counted separately [27].

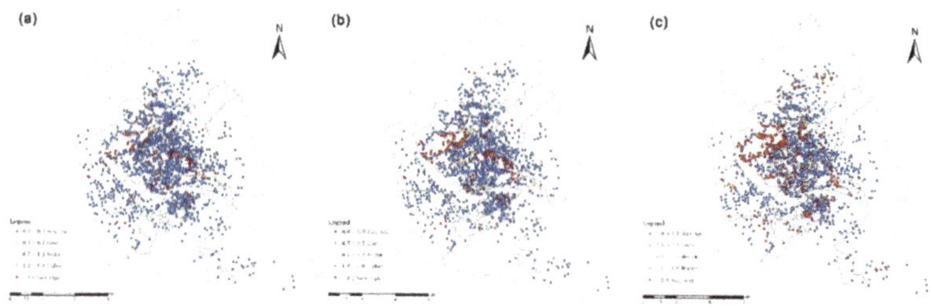

Figure 7. Locational entropy at residential scale under different traveling scopes (**a**) 5-min life circle. (**b**) 15-min life circle. (**c**) 30-min life circle.

Based on the locational entropy method, it was calculated that 9.5%, 16.0% and 26.8% of the settlements had a higher per capita acceptance level of UGS than the overall level of the study area at 500 m, 1000 m and 2000 m search radii, respectively. A hierarchical classification was made for each residential area. As can be seen from Table 6, the total number of settlements with very low and lower locational entropy tends to decrease as the spatial threshold increases, while the number of settlements with higher and very high locational entropy increases significantly. This indicates that, within a certain range, an increase in the actual choice of the travel distance of residents will result in more residents accepting the UGS. However, even at the travel limit distance (2000 m) about 70.6% of residential areas receive a lower or very low level of UGS services.

Table 6. Classification statistics of residential location entropy under different traveling scopes.

	Very Low	**Lower**	**Medium**	**Higher**	**Very High**	**Above-Average**
Type	0–0.5	0.5–0.75	0.75–1.2	1.2–2.0	>2.0	>1.0
5-min life circle	1802	16	20	30	152	191
15-min life circle	1624	39	55	33	268	323
30-min life circle	1351	75	74	68	451	541

3.3.2. Sub-District Scale

The effective service area ratio is the ratio of UGS service coverage area to total sub-districts within the study area, which reflects the effective service of UGS in the spatial dimension. Under the three living circle activity ranges, Antai sub-district, Chating sub-district, Dongsheng sub-district and Ninghua sub-district all show a high level, while Aofeng sub-district, Cangxia sub-district, Duihu sub-district and Xiadu sub-district all show a low level. With the increase of activity range, the UGS service coverage of each sub-district was improved, but only Sanchajie sub-district stopped increasing after the travel range reached 1000 m. The results show spatial consistency with accessibility (Figure 8a).

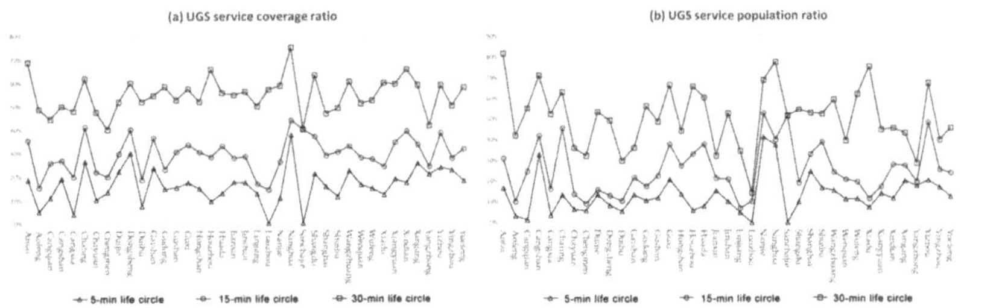

Figure 8. UGS service capability under different traveling scopes. (**a**) UGS service coverage ratio. (**b**) UGS service population ratio.

The effective service population ratio is the ratio of the actual number of populations covered by the UGS service range in the spatial unit to the total population, which reflects the actual service of UGS from an objective perspective. None of the three living circle ranges can make the UGS service cover the whole population; with the increase of travel range, the population covered by each sub-district as a whole also tends to increase (Figure 8b).

The results of the sub-district scale locational entropy calculation were matched with the UGS service coverage of each sub-district under different search thresholds to derive the actual equity spatial pattern of UGS distribution. Within the 5-min life circle and the 15-min life circle, all sub-districts in the study area have a locational entropy of less than 1 and are at a low grading level; the per capita acceptance level of UGS is also lower than the overall level (Figure 9).

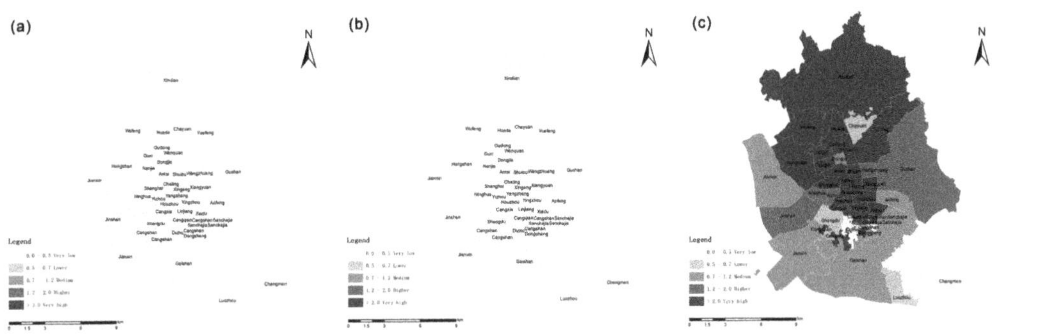

Figure 9. Locational entropy at sub-district scale under different traveling scopes (**a**) 5-min life circle. (**b**) 15-min life circle. (**c**) 30-min life circle.

4. Discussion

4.1. Similarities and Differences at Different Study Scales

Accessibility at both the residential scale and the sub-district scale showed a spatial distribution pattern of being high in the northwest and low in the southeast. When residential areas are used as the evaluation unit, the residential areas with higher and very high accessibility ratings are mainly distributed in Gulou District and Taijiang District, while the residential areas without accessibility are scattered in all parts of the study area. However, they are mainly concentrated in the underdeveloped areas at the edge of the study area and in some areas with relatively sparse road construction, such as Sanchajie sub-district and Taihu sub-district.

The Gini coefficient shows social inequity at both scales, with inequity more pronounced at the residential scale. This is because the distribution of UGSs is more concentrated in the central part of the study area, where residences have more options and the per capita of UGS is much higher than the average, while some residential areas are not accessible. This creates a serious polarization and an extremely inequitable distribution of UGSs. The evaluation unit of sub-districts is evaluated at the macro level. Each sub-district includes several residential areas, which weakens the characteristics of local area differences.

The results of locational entropy at both evaluation scales indicate that UGS services vary widely among regions in the study area, and there is an obvious spatial mismatch between resources and population. In addition, both scales show an overall spatial distribution pattern of high in the northwest and low in the southeast, but differ in the spatial distribution. When residential areas are used as the evaluation unit, more than 70% of the residential areas receive fewer UGS services than the overall level. At the sub-district scale, more than eight sub-districts are less than the overall level of the study area, accounting for 20.5% of the total number of sub-districts. The difference is that the UGS services of the study area are all at a poor level and are severely under-served at the sub-district scale when the living circle range is small. Additionally, all sub-districts cannot enjoy equitable UGS services within 1000 m. In contrast, some zones can be served by abundant UGS resources at the residential scale. In the sub-district with a low value of locational entropy grading at the macro-scale, there are high value residential units distributed at the micro-scale; it is similar in the sub-district units with good UGS service, containing residential areas with insufficient green space service.

4.2. Similarities and Differences at Different Traveling Scopes

Spatial divergence between UGS and population occurs under all three travel scopes. The smaller the activity circle, the more blind areas of UGS services there are. Compared with the 5-min life circle and the 15-min life circle, the values of accessibility result in the 30-min activity circle changing more gently, the number of spatial units with higher accessibility increasing and the service blind areas decreasing. The results show that the spatial accessibility of urban green space will change when the travel range changes. The longer the distance people travel within a certain range, the higher the accessibility; the smaller the variability of green space accessibility between spatial units, the more people will enjoy urban green space services, and the more balanced the distribution of green space resources will be.

In the evaluation of the equity of UGS, all three activity ranges show inequity. As the activity circle range increases, the more the Gini coefficient calculation results converge to 0, the closer the Lorenz curve is to the absolute equity line; that is, the spatial distribution of UGS service supply in Fuzhou is relatively more equal.

With the increase of activity range, the level of UGS service supply in Fuzhou City is significantly improved. When the travel range of residents increases, the effective service area and the population coverage of UGS increases, and the more UGS services residents enjoyed, the better the supply of UGS services.

4.3. Application of UGS Planning in Fuzhou

In order to scientifically build an urban green space system and achieve the purpose of equitable supply of services and precise matching of supply and demand, this study mainly proposes the following optimization suggestions in view of the existing problems of UGS allocation in the main urban area of Fuzhou.

(1) The per capita level of UGS resources in Chengmen sub-district, Duihu sub-district and Sanchajie sub-district is much lower than the average level. They are at a disadvantageous level in both scales even under the extreme travel scope. This is mainly due to the incomplete construction, the lack of concentrated distribution of UGS resources, and the inadequate road system. They should be further evaluated and prioritized for measures in urban construction.

(2) For the high population density, the dense distribution of residential areas and land constraints of Chayuan sub-district and Dongjie sub-district, the existing urban layout should make full use of abandoned space to build pocket parks, set up street side green space and corner green space to improve the efficiency of space utilization.

(3) Walking is the primary choice of daily travel for residents. Not only should the construction of UGS areas be strengthened, but also the planning of slow-moving transportation systems should be paid attention to. Due to the sparse road network and inconvenient traffic in Shangdu sub-district, some residents are inconvenienced in using the UGS. The accessibility of UGS for residents should be improved by increasing UGS. Building green corridors connecting with surrounding green space, strengthening road network construction, opening up cut-off roads, and improving road network continuity seem to be enforceable.

(4) In the context of building a livable city in the new era, urban renewal and construction should be tailored to local conditions and strategies should be adopted for different situations. Different regions are not consistent in terms of population structure, economic structure and development status. We should avoid using a single indicator as the basis for spatial optimization when conducting optimization. Urban planners should coordinate and unify with government with considering the city, region and street as a whole at multiple levels. A more scientific and effective construction management system should be formed by satisfying both macroscopic rigid regulations and a microscopic grasp of construction details.

(5) The 5 min, 15 min and 30 min walkable living circles defined in this study determine the spatial evaluation range based on time. They correspond to the basic, neighborhood and daily social and travel distances of residents, respectively. There are also cases of urban spatial planning that emphasize the time-scale as the boundary [28], such as a maximum 5-min walk to all amenities and public transport in Copenhagen [29], the 15-min city in Paris [30], the 20-min neighborhood in the United States [31] and Australia [32], the 20-min town in Singapore [33], and the 15-min convenient living circle in China [28], which have transformed the concepts of urban planning and construction in the past. However, due to the differences in the time delineation of the living circle and the choice of transportation in different countries, a unified spatial scope division has not yet been formed. With the development of intelligent technology, the planning mode of living circle series will undergo new changes. UGS, which has great ecological and social benefits, should bear the brunt; the new planning methods deserve further attention and exploration.

4.4. Limitation

Due to the limitations to obtaining data, this study assumed that the population is evenly distributed within the residential area. The number of building floors was not considered. The result calculated from the area weight and the census data is an approximate expression of the number of people in the settlement. In reality, the population distribution is not uniform and should be related to the number of building floors. The accuracy of the research results will be greatly improved if the population data can be accurately obtained by using big data. As a service provider, the higher the quality of the physical environment of the urban green space itself, the more residents can be attracted to visit. Although there have been some studies on the assessment of UGS quality in the past, there is no uniform assessment standard and further discussion should be had. The real meaning of equity should include more consideration of the demand-side state, taking into account the distribution of population, social attributes, and the ability to obtain green space services, etc., which means people-oriented planning. To ensure UGS construction meets the diverse needs and preferences of different social groups is the meaning of UGS equity evaluation.

5. Conclusions

This study used web big data, based on G2SFCA, to measure the accessibility of UGS for residents of 39 sub-districts and 2020 residential areas in Fuzhou at different travel scopes, to identify the inequities in the spatial distribution of UGS at the different scales, respectively, and to quantitatively evaluate the actual service of UGS. Finally, scientific proposals were provided for the urban construction and improvement of the urban green space system in Fuzhou. The main conclusions of this study are as follows:

1. Fuzhou city has significant variations in the allocation of urban green space resources with serious social inequities. The accessibility of UGS shows a spatial distribution pattern of being high in the northwest and low in the southeast.
2. Within the walking limit travel distance, with the expansion of residents' travel range, the accessibility degree of UGS in Fuzhou increases, the distribution of UGS resources is more balanced, therefore the higher the extent of equity, and the level of urban green space service supply is significantly improved.
3. The evaluation results of UGS accessibility, equity, and green space services in the main city of Fuzhou are consistent overall at the residential scale and sub-district scale. However, at smaller travel ranges, the residential scale has an advantage in identifying access to green space resources.
4. The city construction should be planned holistically from multiple levels and take into account the population structure, economic structure and development status, and so on. We should take measures according to local conditions. The construction of UGS should also be further explored and investigated with reference to the planning model of new living circles.

Author Contributions: Conceptualization, H.X. and X.L.; methodology, H.X. and X.L.; software, H.X.; validation, X.W., X.H. and Z.S.; formal analysis, H.L.; investigation. X.X. and L.C.; resources, H.X.; data curation, H.D., J.Z. and M.X.; writing—original draft preparation, H.X. and X.L.; writing—review and editing, H.X., X.L. and X.W.; visualization, H.X.; supervision, X.L.; project administration, X.L.; funding acquisition, X.L. All authors have read and agreed to the published version of the manuscript.

Funding: This study was supported by the National Natural Science Foundation of China (32071578) and the Science and Technology Innovation Project of Fujian Province (KY-090000-04-2021-012) and Subject Cross-Integration Project of College of Landscape Architecture, Fujian Agriculture and Forestry University (No. YSYL-xkjc-2).

Institutional Review Board Statement: Not applicable.

Informed Consent Statement: Not applicable.

Data Availability Statement: The data presented in this study are available on request from the author. The data are not publicly available due to privacy. Images employed for the study will be available online for readers.

Conflicts of Interest: The authors declare no conflict of interest.

References

1. Hartig, T.; Mitchell, R.; de Vries, S.; Frumkin, H. Nature and Health. *Annu. Rev. Public Health* **2014**, *35*, 207–228. [CrossRef] [PubMed]
2. Wang, L.; Jia, J.; Lu, Y.; Jing, Z.W.; Yao, Y.L. Dynamic Response of Green Space Configuration and Cooling Efficiency in Changchun City. *Chin. Landsc. Archit.* **2022**, *38*, 44–49. [CrossRef]
3. Bowler, D.E.; Buyung-Ali, L.; Knight, T.M.; Pullin, A.S. Urban Greening to Cool Towns and Cities: A Systematic Review of the Empirical Evidence. *Landsc. Urban Plan.* **2010**, *97*, 147–155. [CrossRef]
4. Nagendra, H.; Gopal, D. Tree Diversity, Distribution, History and Change in Urban Parks: Studies in Bangalore, India. *Urban Ecosyst.* **2011**, *14*, 211–223. [CrossRef]
5. Gan, J.; Yang, W.G.; Wang, L. Spatial Characteristics of Urban Green Space under Different Health Impact Pathways. *Landsc. Archit.* **2020**, *27*, 95–100. [CrossRef]
6. Iraegui, E.; Augusto, G.; Cabral, P. Assessing Equity in the Accessibility to Urban Green Spaces according to Different Functional Levels. *ISPRS Int. J. Geo-Inf.* **2020**, *9*, 308. [CrossRef]
7. Hansen, W.G. How Accessibility Shapes Land Use. *J. Am. Inst. Plan.* **1959**, *25*, 73–76. [CrossRef]
8. Radke, J.; Mu, L. Spatial Decompositions, Modeling and Mapping Service Regions to Predict Access to Social Programs. *Ann. GIS* **2000**, *6*, 105–112. [CrossRef]
9. Xing, L.; Liu, Y.; Wang, B.; Wang, Y.; Liu, H. An Environmental Justice Study on Spatial Access to Parks for Youth by Using an Improved 2SFCA Method in Wuhan, China. *Cities* **2020**, *96*, 102405. [CrossRef]
10. Yue, B.J.; Lin, A.W.; Sun, C. The Analysis of Low-Income People's Park Green Space Accessibility Based on 2SFCA in Wuhan. *Mod. Urban Res.* **2017**, *8*, 99–107.
11. Kanuganti, S.; Sarkar, A.K.; Singh, A.P. Evaluation of Access to Health Care in Rural Areas Using Enhanced Two-Step Floating Catchment Area (E2SFCA) Method. *J. Transp. Geogr.* **2016**, *56*, 45–52. [CrossRef]
12. An Extended Kernel Density Two-Step Floating Catchment Area Method to Analyze Access to Health Care—Pierre Polzin, José Borges, António Coelho. 2014. Available online: https://journals.sagepub.com/doi/abs/10.1068/b120050p (accessed on 6 December 2022).
13. Xu, W.J.; Fang, S.M.; Liu, C.Y. Spatial equity of public green in Wuhan central districts based on gravity two-step floating catchment area method. *Resour. Sci.* **2017**, *39*, 430–440.
14. Zhou, X.Y.; Li, X.W.; Hou, M.L. Accessibility and Equity of Public Green Space in Wuhan from the Perspective of Residential Quarter. *Geomat. World* **2020**, *27*, 124–129.
15. Xue, X.T.; Yong, X.Q.; Li, X.; Hu, Z.N. Evaluation of spatial equity of urban green space based on internet big data for central area of Xuzhou city. *Ecol. Sci.* **2019**, *38*, 140–148. [CrossRef]
16. Rao, Y.; Zhong, Y.; He, Q.; Dai, J. Assessing the Equity of Accessibility to Urban Green Space: A Study of 254 Cities in China. *Int. J. Environ. Res. Public Health* **2022**, *19*, 4855. [CrossRef]
17. Yin, H.W.; Kong, F.H.; Zong, Y.G. Accessibility and equity assessment on urban green space. *Acta Ecol. Sin.* **2008**, *7*, 3375–3383.
18. Wu, J.S.; Si, M.L.; Li, W.F. Spatial equity analysis of urban green space from the perspective of balance between supply and demand: A case study of Futian District, Shenzhen, China. *Chin. J. Appl. Ecol.* **2016**, *27*, 2831–2838. [CrossRef]
19. Wang, L.F.; Guo, X.H. Green Space Justice from the Perspective of Living Circle. *J. Chin. Urban For.* **2021**, *19*, 83–88.
20. Li, Q.; Liu, X.Z.; Lian, X.X.; Li, S.N.; Li, J.X. Research on Park Green Space Accessibility of Nantai lsland in Fuzhou City, China, Based on Space Syntax. *J. Guangxi Norm. Univ.* **2021**, *39*, 181–195. [CrossRef]
21. Li, Q.; Peng, K.; Cheng, P. Community-Level Urban Green Space Equity Evaluation Based on Spatial Design Network Analysis (SDNA): A Case Study of Central Wuhan, China. *Int. J. Environ. Res. Public Health* **2021**, *18*, 10174. [CrossRef]
22. Jin, H.X.; He, G.; Huang, Q.S. Research on the Accessibility of Urban Green Space Through Environmental Justice View. A Case Study of Hangzhou, China. *J. Northwest For. Univ.* **2022**, *37*, 261–267.
23. Gong, D.; Sun, Y.Y.; Xie, M.M. Evaluation of green space accessibility based on improved Gaussian two-step floating catchment area method:A case study of Shenzhen City, China. *Prog. Geogr.* **2021**, *40*, 1113–1126.
24. Wang, H.; Wei, X.; Ao, W. Assessing Park Accessibility Based on a Dynamic Huff Two-Step Floating Catchment Area Method and Map Service API. *ISPRS Int. J. Geo-Inf.* **2022**, *11*, 394. [CrossRef]
25. Niu, S.; Tang, X.M. Research on the Equity Measurement of Park Green Space Distribution in High-density Urban Areas—A Case Study of Huangpu District, Shanghai. *Chin. Landsc. Archit.* **2021**, *37*, 100–105. [CrossRef]
26. Tang, Z.L.; Gu, S. Social Performance Evaluation of Public Green Space Distribution in Central Shanghai: From Geographical Equity to Social Equity Urban Plan. *Forum* **2015**, 48–56. [CrossRef]
27. Wan, Y.; Bi, C.; Zhang, J.G.; Zhao, B. Spatial Equity of Urban Parks in the Central Urban Area in Nanjing Based on Different Transportations. *J. Beijing For. Univ.* **2020**, *19*, 47–54. [CrossRef]
28. Logan, T.M.; Hobbs, M.H.; Conrow, L.C.; Reid, N.L.; Young, R.A.; Anderson, M.J. The X-Minute City: Measuring the 10, 15, 20-Minute City and an Evaluation of Its Use for Sustainable Urban Design. *Cities* **2022**, *131*, 103924. [CrossRef]
29. State of Green. 5 Minutes to Everything. In *State of Green. 2 August 2016.*. Available online: https://stateofgreen.com/en/partners/ramboll/news/5-minutes-to-everything/ (accessed on 5 December 2022).
30. Moreno, C.; Allam, Z.; Chabaud, D.; Gall, C.; Pratlong, F. Introducing the "15-Minute City": Sustainability, Resilience and Place Identity in Future Post-Pandemic Cities. *Smart Cities* **2021**, *4*, 6. [CrossRef]

31. Capasso Da Silva, D.; King, D.A.; Lemar, S. Accessibility in Practice: 20-Minute City as a Sustainability Planning Goal. *Sustainability* **2020**, *12*, 129. [CrossRef]
32. Bikeability and the 20-Min Neighborhood: How Infrastructure and Destinations Influence Bicycle Accessibility—Nathan McNeil. 2011. Available online: https://journals.sagepub.com/doi/abs/10.3141/2247-07 (accessed on 6 December 2022).
33. Land Transport Authority. *Land Transport Master Plan 2040*; Land Transport Authority: Singapore, 2019.

International Journal of
Environmental Research and Public Health

MDPI

Article

Study on Chinese Farmland Ecosystem Service Value Transfer Based on Meta Analysis

Liangzhen Nie [1], Bifan Cai [1], Yixin Luo [1], Yue Li [2], Neng Xie [1], Tong Zhang [1], Zhenlin Yang [1], Peixin Lin [1] and Junshan Ma [1,*]

1 College of Landscape Architecture, Zhejiang A & F University, Hangzhou 311300, China
2 College of Landscape Architecture, Nanjing Forestry University, Nanjing 210037, China
* Correspondence: majunshan@zafu.edu.cn

Abstract: An analytic database was built based on meta-regression analysis (MRA) method, mainly including ecosystem service type, farmland division, cultivated land type, value assessment method, and farmland characteristics. The feasible weighted least square (FWLS) method was adopted to comprehensively investigate the seventy observations from empirical studies. The results indicate that: (1) except the negative impact of farmland area on farmland value, such factors as paddy field, good soil conservation function, mainly providing agricultural products, and using market value method for assessment all produce positive effect on the promotion of farmland value. (2) In meta-regression analysis, the average transfer error is 36.74%, and the median transfer error is 14.59%. (3) Under the A1B, A2, B1, and B2 scenarios of IPCC SRES, it is discovered from calculation that the value changes under different scenarios have some differences, in which, the total value rises significantly under A2 scenario and will reach to 15,220 billion yuan until the year of 2100; while the total value loss is the greatest under B1 scenario and will fall to 6320 billion yuan until the year of 2100. Finally, this paper gives some suggestions for scholars to deeply study the service value of farmland ecosystem as well as for the government to formulate differentiation policies.

Keywords: ecosystem services; value transfer; farmland; meta-regression analysis

Citation: Nie, L.; Cai, B.; Luo, Y.; Li, Y.; Xie, N.; Zhang, T.; Yang, Z.; Lin, P.; Ma, J. Study on Chinese Farmland Ecosystem Service Value Transfer Based on Meta Analysis. *Int. J. Environ. Res. Public Health* **2023**, *20*, 440. https://doi.org/10.3390/ijerph20010440

Academic Editors: Hongxiao Liu, Tong Wu and Yuan Li

Received: 26 October 2022
Revised: 21 December 2022
Accepted: 22 December 2022
Published: 27 December 2022

1. Introduction

Nowadays, the increasingly severe natural resource depletion, ecology destroying, and environmental pollution have attracted people's attention to the environment [1]. China is the world's largest developing country, and its long-term extensive way of development has resulted in many ecological environment problems, which show sharp conflict to people's growing demands for comfortable environment. Around the world, the farmland ecosystem is always one of the most important ecosystems [2]. Up to the 25 August 2021, according to the public data of Ministry of Natural Resources of the People's Republic of China, Chinese cultivated land area is 1.43×10^6 square kilometer, covering 14.90% of national territorial area, mainly distributed in Northeast China, Huanghai-Huaihai-Haihe region and Yangtze Plain, Middle, and Lower. Farmland ecosystem not only provides crops [3–5] and recreational tourism [3,6] for people, but also plays an important role in gas regulation [3,7,8], soil conservation [3,4,6,9], water conservation [3,10,11], and soil nutrient circulation [3,4], with huge environmental and economic benefits [12]. Recently, the main methods to evaluate farmland ecosystem service value include shadow project approach, market valuation approach, surrogate market approach, and carbon tax approach [11,13–15]. Before 2003, the ecosystem service value assessment in China mainly focused on forest [16–18], wetland [19], landscape [20], and river ecosystems [21,22], especially for forest and wetland ecosystems, but there were few contents about assessment of farmland ecosystem service value. In 2003, Zhao Rongqing [23] et al. initially defined and classified farmland ecosystem service functions. After that, China started to study the farmland ecosystem service value of different or specific region at different time periods [3,24,25].

At present, research on the application of Meta-analytic value transfer methods in the field of ecosystem service valuation is mainly two-fold: first, to develop a meta-analytic value transfer model with general applicability for ecosystem service value prediction within a broad ecosystem service valuation framework, and second, to develop a meta-analytic value transfer model for a specific policy context [26].The applied research of meta-analysis value transfer method in farmland ecosystem service value assessment field mainly focuses on the former aspect. In the research of ecosystem service value transfer, meta-analysis method can take effect in three points: (1) Integrate and evaluate the ecosystem service value assessment study; (2) build meta-regression model for evaluating the influential factors of ecosystem service value changes; (3) use the meta-regression model built to predict ecosystem service value [27]. Meanwhile, there are very few applied research on meta-analysis value transfer method. According to the current literature retrieval condition, overseas and domestic scholars have already launched the applied research of meta-analysis value transfer method in China's forest ecosystem service value assessment [28], study of dry farmland ecosystem's active and passive restoration measures [29], land use changes and ecosystem service value transfer in Changbai Mountains region [30], study of agroforestry system's grain yield [31], and assessment of wetland ecological protection value in Changbai Mountain region [32]. However, there is no applied research of farmland ecosystem service value assessment in China yet.

IPCC SRES has four broad scenarios: A1, A2, B1, and B2. Scenario A describes a world that tends toward economic development, while Scenario B tends toward social and environmental development; Scenario 1 has a more globalized world with frequent economic, cultural, and technological interactions among countries, while Scenario 2 has a less globalized world. Specifically, the A1 scenario depicts a future that is called the "Golden Age of the Economy. Scenario A2 is called "Cultural Diversity". People are concerned with regional independence and environmental issues are not a priority. Populations continue to grow, trade barriers exist, and technological progress is slow. Scenario B1 is similar to "sustainable development": there is a high level of globalization, stable populations, international cooperation, a preference for more efficient and cleaner energy sources, and rapid development of energy-efficient technologies. "Regional Solutions". People are concerned about regional independence, but also about environmental protection. Some regions would use clean coal as fuel, others would use clean energy sources such as wind and solar. Among them, A1B fuel use is more balanced. The different scenarios are based on different drivers, such as economy, energy use and temperature. Several specific drivers are represented in these four scenarios:

1. Economic development: globalization is better than non-globalization (1 series > 2 series), but focusing on the economy does not necessarily yield better benefits than focusing on the society/environment (A2 < B2).
2. The Human Development Index (HDI) is an index proposed by the United Nations Development Programme (UNDP) that includes three dimensions: human health and longevity, knowledge and education, and economic development. A2 is basically the worst scenario, with B2 slightly worse.
3. Temperature rise: Focus on more economic warming (A > B), less globalization warming (1 < 2)

In light of the above problems, we collected the value assessment results of empirical research literature about farmland ecosystem service value assessment in China, established a value transfer database, built a meta-analysis value transfer model of farmland ecosystem service in China by using meta-analysis and multiple regression analysis method, and calculated farmland ecosystem value change condition in China between 2010 and 2100 based on these. First, an influential factor analysis and a model error evaluation were conducted though the regression results. Then, on this basis, according to the A1B, A2, B1, and B2 scenarios of IPCC SRES, the farmland value change condition in China between

2010 and 2100 was calculated, and the applicability and development prospect of Meta-analysis value transfer method in farmland ecosystem service value assessment field in China were discussed.

2. Research Method and Data Processing

2.1. Database Establishment

The monetizing valuation of the entire biosphere's ecosystem service value made by Costanza et al. [33] has become an important milestone for the entire ecosystem service value assessment system in future. Now, there are three sources for ecosystem service value in China: Costanza et al. [33], The Millennium Ecosystem Assessment (MA) [34], and Xie Gaodi et al. [35]. This paper referred to the value assessment procedure of existing literature and the MA reports, and established a meta-analysis farmland ecosystem service value assessment system (see Figure 1).

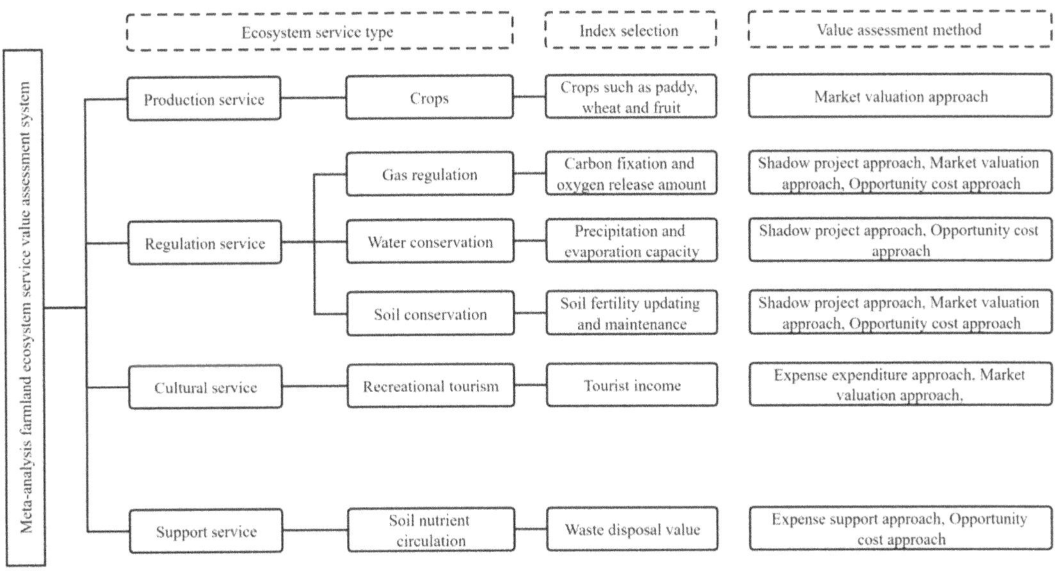

Figure 1. Meta-analysis farmland ecosystem service value assessment system.

Based on the farmland ecosystem service value assessment system in Figure 1, the research information such as the title, author, research time, research region, farmland type, evaluation method, ecosystem service type, and ecosystem service value were collected uniformly from the literature and then input into the Excel table. At last, a total of 26 literature and 70 value observations were included in the meta-analysis database. It can be seen from the geographical location distribution of study sample site (Figure 2) that the evaluated farmland ecosystems were mainly distributed in the major grain-producing area such as Huanghai-Huaihai-Haihe region and Yangtze Plain, Middle, and Lower, while very few study samples were distributed in the non-major grain-producing area such as the central and western regions of China.

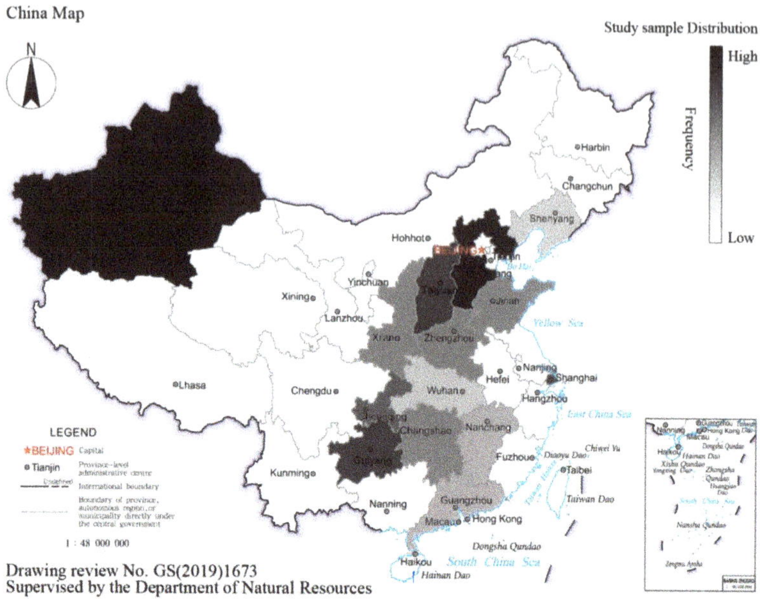

Note: The darker the color in this area, the higher the frequency of the study samples in this area; This map is made based on the drawing review No. GS(2019)1673 downloaded from the standard map service website of National Administration of Surveying.The base map is not modified.

Figure 2. Geographical location distribution of study samples.

2.2. Selection of Independent Variables

The study on ecosystem service value transfer at home and abroad generally considers the value assessment method of study samples, research region area, ecosystem service type, regional economic development level of research region, substitution effect of the same type of ecosystems, as well as the influence of other factors on ecosystem service value changes [36]. By reference to overseas and domestic influential factors considered, combined with the actual data collection condition in this paper, the independent variables used to build meta-regression model were selected:

Ecosystem service type: different types of services provided by an ecosystem are different, their importance are also differed, so the generated values are varied. The difference between different ecosystem service types may have impact on the value of farmland ecosystem.

Value assessment method: the ecosystem value assessment method can affect the assessment results to a great extent, especially the selection of service quality evaluation index and price parameters. This paper focuses on analyzing the influence of using different value assessment methods on the change of farmland ecosystem service value.

Farmland division: farmland is mainly divided into major crops producing region and non-major crops producing region. The crops producing region covers the northeast plain, Huanghai-Huaihai-Haihe plain, Yangtze River plain, with wide distribution scope and good production condition.

Farmland area: it is known from recently collected literature that ecosystems have boundary effect. Within the boundary effect, with the increase of farmland area, the unit area value of some farmland ecosystems will increase as well; but when it exceeds a certain critical threshold, the value of some farmland ecosystems may show decreasing scale benefits.

Number of beneficiaries: the number of beneficiaries of farmland ecosystem service reflects the demands or market size of farmland ecosystem service [37]. This paper defines the scope of farmland beneficiaries as within the administrative region. The population data within administrative regions originates from the Statistical Yearbook of each province and city.

Per capital GDP: the farmland ecosystem service value is closely related to the economic development level of the region in which the farmland is located, for example, there is a huge gap between the developed and under-developed regions in identification and market realization of farmland ecosystem service value [38]. This paper uses per capital GDP as the index to measure regional economic development state. According to data collection condition, the scale of the region is defined as within the administrative region. The per capital GDP data of the administrative region in which the farmland is located originate from the Statistical Yearbook of each province and city.

Cultivated land type: the level-1 cultivated land mainly includes paddy field and dry land. Dry land normally refers to the lands in which dry crops are planted without seasonal irrigation. Paddy field refers to the farmlands in which paddy and other aquatic plants are planted with seasonal ponding every year. Different cultivated land types may have impact on farmland ecosystem service value.

Count the number and mean of value observations corresponding to each variable, follow the statistical and metric data requirements to assign various information of independent variables, and calculate their mean values and standard deviations respectively. See the variable information of meta-regression model in Table 1 [3,14,25,39–60].

Table 1. Variable definitions and summary statistics [a] (in CNY¥ ha^{-1} yr^{-1}).

Variable Names	Variable Description	Mean	SD	N
Dependent variable				
Farmland value	Annual value per hectare in 2015 CNY¥ in logarithmic form	10.770	1.101	70
Independent variables				
Value evaluation approach				
Surrogate market approach	Baseline category [b]	0.143	0.352	10
Opportunity cost approach	If the opportunity cost approach is used for assessment, the value is set as 1, otherwise as 0	0.443	0.500	31
Carbon tax approach	If the carbon tax approach is used for assessment, the value is set as 1, otherwise as 0	0.143	0.352	10
Shadow project approach	If the shadow project approach is used for assessment, the value is set as 1, otherwise as 0	0.457	0.502	32
Market valuation approach	If the market valuation approach is used for assessment, the value is set as 1, otherwise as 0	0.529	0.503	37
Farmland ecosystem services				
Water conservation	Baseline category [b]	0.500	0.504	35
Crops	If the ecosystem service type is crops, the value is set as 1, otherwise as 0	0.786	0.413	55
Gas regulation	If the ecosystem service type is gas regulation, the value is set as 1, otherwise as 0	0.600	0.493	42
Soil conservation	If the ecosystem service type is soil conservation, the value is set as 1, otherwise as 0	0.457	0.502	32
Recreational tourism	If the ecosystem service type is recreational tourism, the value is set as 1, otherwise as 0	0.257	0.440	18
Soil nutrient circulation	If the ecosystem service type is soil nutrient circulation, the value is set as 1, otherwise as 0	0.414	0.496	29
Dry land	Baseline category [b]	0.643	0.483	45
Paddy field	If the cultivated land type is paddy field, the value is set as 1, otherwise as 0	0.371	0.487	26

Table 1. *Cont.*

Variable Names	Variable Description	Mean	SD	N
Non-major crops producing region	Baseline category [b]	0.700	0.462	49
Northeast region	If the crops' geographic region is northeast region, the value is set as 1, otherwise as 0	0.043	0.204	3
Huanghai-Huaihai-Haihe region	If the crops' geographic region is Huanghai-Huaihai-Haihe region, the value is set as 1, otherwise as 0	0.257	0.440	18
Yangtze Plain, Middle and Lower	If the crops' geographic region is Yangtze Plain, Middle and Lower, the value is set as 1, otherwise as 0	0.057	0.234	4
Farmland size	Area of Farmland site in logarithmic form	12.599	2.256	70
Number of beneficiaries	Numerical variables in logarithmic form	15.476	2.485	70
GDP per capita [c]	GDP per capita in logarithmic form	9.896	0.961	70

[a] Note: N = number of observations for each variable or variable level; SD = standard deviation. [b] Baseline category refers to that which is excluded for each categorical variable in order to avoid perfect collinearity. [c] Referring to year 2015.

2.3. Model Establishment

In meta-regression, the weighted regression model, panel data model or hierarchical linear model are usually used to explain individual research effect [61]. In weighted regression, every study has its independent weight, irrelevant to the number of value obtained from a single research. Panel data include fixed effect model and random effect model, in which the former assumes every study has fixed individual effect, and its interior estimator could only explain the influence inside the study. While the latter assumes every study has random individual effect, and it often uses the generalized least square estimation, able to explain the influence between studies and inside the study simultaneously. Hierarchical linear model mixes random and fixed models [62]. However, in meta-regression dataset of this paper, the estimated value variance change is not sufficiently explained inside single research, so panel data model or hierarchical model is not suitable for the dataset of this paper [62].

In the existing relevant meta-regression analysis, the majority adopts the ordinary least squares(OLS), with general expression as below:

$$\ln(y_i) = \alpha + \beta_s X_s + \beta_t X_t + \beta_p X_p + \beta_e X_e + \delta_i \quad (1)$$

In which, the dependent variable y is the value vector of farmland ecosystem service, in CNY ha^{-1} a^{-1}. α is a constant term, δ is a residue term, β is the regression coefficient matrix of independent variable, and X is the independent variable matrix, in which, X_s represents the variable of farmland ecosystem service value assessment method, X_t represents the characteristics of evaluated farmland, X_p represents farmland ecosystem service type, X_e represents the geographical environment characteristics around evaluated farmland.

The main reason for choosing a log-linear model in this paper is that the logarithmic transformation can reduce fluctuations in the original data, improve the accuracy of the fit and reduce heteroskedasticity [62,63]. Since the value assessment in literature was often based on different years, in order to make the data comparable, the paper used the value observation conversion method adopted by Kochi et al. [64] and Johnston et al. [65] for reference, and regulated the values of different assessment base years into the price level of 2015 through consumer price index (CPI); then, divided the value (CNY a^{-1}) of unified base year by farmland area (ha) of study region, obtained the unit area value (CNY ha^{-1} a^{-1}) of farmland ecosystems in different study regions, and took it as the dependent variable of meta-regression model.

3. Results Analysis

3.1. Meta-Regression Analysis

The meta-regression results are shown in Table 2. The Model(A) reports the general model covering all explanatory variables and makes estimation through OLS, but the B-P test results indicate the existence of heteroscedasticity ($p = 0.000$). The reason may be that the potential assumption of least square method is that different observations are irrelevant. In the established meta-analysis database, one literature can provide 11 value observations at the most, and 50% literature provide multiple observations. Since the observations from the same literature are not independent and different studies may be correlated, these could result in biased estimation simultaneously. In Model(B), we referred to the solutions to this problem in existing studies and used the weighted least square (WLS), in which some used the weighted least square taking the reciprocal of observation number as the weight [36,63,66], reducing the influence of sample correlation to a certain extent. But the demerit of WLS is that it assumes the covariance matrix of disturbing term is known, which is often an unrealistic hypothesis. In view of this, it only can be used after using sample data for uniform estimation, and this method is called the feasible weighted least square (FWLS) [67], see the results in Model (C). Moreover, we tested whether the regression model had serious multicollinearity problem by figuring out the variance inflation factor (*VIF*), which was defined as:

$$VIF = \frac{1}{1 - R_{\mathrm{i}}^2} \quad (2)$$

In which, R_{i} is the negative correlation coefficient of regression analysis performed by independent variable X_{i} on other independent variables. The greater the *VIF* is, the larger the possibility of multicollinearity exists between independent variables will be. Generally speaking, if *VIF* is more than ten, the regression model has serious multicollinearity. While the independent variable's VIF being less than ten is generally acceptable, this indicates there is no multicollinearity problem between independent variables [67]. It is known from Table 3 that the VIF of the independent variable-number of beneficiaries is greater than ten, so after we delete it from Model(D-G), the VIF values of all independent variables are tested to be less than ten. It can be seen from the results of Model (D) that the significance of farmland area has been improved, and the independent variable-number of beneficiaries is not significant in Model (A) and Model (B). This demonstrates that the model accuracy has been improved after the number of beneficiaries is deleted. In Model (E), the significance of opportunity cost method and per capital GDP declines slightly, but it is still statistically significant at a level of 5%, but the significance of such ecosystem service type variables as crops and gas regulation rise to some extent. This may be due to the influence of multicollinearity interference which can be improved after deleting the influential variables. Brander et al. [68] considers when evaluating a given ecosystem's service value, more than one assessment method is often used, and the value assessment method of a given ecosystem service is generally definite, for example, using shadow project approach to evaluate the value of water conservation [69], using opportunity cost method to evaluate the value of biodiversity, etc. [70] Therefore, the regression results can be affected. So, we deleted the value assessment method variables from Model (F) and Model (G), but its results showed that except the significance of farmland area did not change, the significance and regression coefficient of other variables had changed. This phenomenon also manifests that the value assessment method play a certain effect in regression model, so we should refer to existing studies to reserve the value assessment method variable. [36,64,66].

Table 2. Estimated meta-regression value transfer function.

Variable	Full Model				Reduced Model		
	Model (A)	Model (B)	Model (C)	Model (D)	Model (E)	Model (F)	Model (G)
Opportunity cost approach	−0.462 (0.296)	−0.288 (0.239)	−0.559 *** (0.194)	−0.326 (0.230)	−0.465 ** (0.190)	-	-
Carbon tax approach	−0.279 (0.318)	−0.170 (0.374)	0.068 (0.422)	−0.208 (0.395)	0.116 (0.358)	-	-
Shadow project approach	0.344 (0.294)	0.189 (0.257)	−0.084 (0.209)	0.262 (0.256)	−0.171 (0.184)	-	-
Market evaluation approach	0.870 *** (0.215)	1.326 *** (0.262)	0.623 *** (0.134)	1.338 *** (0.257)	0.627 *** (0.128)	-	-
Crops	0.224 (0.256)	0.443 * (0.257)	0.445 ** (0.168)	0.403 * (0.227)	0.464 *** (0.168)	0.164 (0.348)	0.403 * (0.225)
Gas regulation	0.237 (0.297)	0.212 (0.331)	0.490 (0.323)	0.188 (0.336)	0.495 * (0.284)	0.417 (0.367)	0.486 ** (0.189)
Soil conservation	0.444* (0.230)	0.152 (0.222)	0.872 *** (0.164)	0.169 (0.216)	0.923 *** (0.155)	0.411 (0.397)	1.577 *** (0.297)
Recreational tourism	−0.780 *** (0.249)	−0.654 ** (0.273)	−0.989 *** (0.343)	−0.629 ** (0.246)	−1.018 *** (0.301)	−0.540 (0.391)	−0.544 ** (0.240)
Soil nutrient circulation	0.255 (0.253)	0.403 (0.270)	−0.108 (0.230)	0.409 (0.269)	−0.152 (0.209)	−0.273 (0.294)	−0.419 *** (0.142)
Paddy field	0.734 *** (0.240)	0.996 *** (0.330)	0.649 *** (0.222)	1.013 *** (0.323)	0.590 *** (0.199)	0.747 * (0.444)	0.080 (0.315)
Northeast region	−0.222 (0.453)	−0.268 (0.685)	0.103 (0.403)	−0.220 (0.665)	−0.073 (0.403)	−0.368 (0.855)	−0.841 * (0.493)
Yangtze Plain, Middle and Lower	−1.056 *** (0.331)	−1.411 *** (0.360)	−1.147 *** (0.412)	−1.411 *** (0.357)	−1.141 *** (0.371)	−0.921* (0.544)	0.321 (0.528)
Huanghai-Huaihai-Haihe region	−0.388 (0.269)	−0.422 (0.418)	−0.189 (0.359)	−0.407 (0.422)	−0.218 (0.303)	−0.407 (0.396)	0.080 (0.248)
Farmland area	−0.215 ** (0.082)	−0.166 (0.104)	−0.149 *** (0.053)	−0.135 ** (0.053)	−0.181 *** (0.046)	−0.190 *** (0.067)	−0.298 *** (0.044)
Number of beneficiaries	−0.014 (0.087)	0.035 (0.099)	−0.034 *** (0.006)	-	-	-	-
Per capital GDP	−0.083 (0.090)	−0.149 * (0.075)	−0.106 *** (0.034)	−0.149 * (0.074)	−0.090 ** (0.035)	−0.040 (0.125)	0.047 (0.052)
Constant term	13.620 *** (1.112)	12.420 *** (0.853)	13.268 *** (0.357)	12.570 *** (0.771)	13.014 *** (0.323)	13.129 *** (1.669)	13.108 *** (0.640)
Number of observations	70	70	70	70	70	70	70
R^2	0.827	0.742	0.990	0.741	0.987	0.554	0.937
Adjusted R^2	0.775	0.664	0.986	0.670	0.983	0.469	0.925

Note: Figures in brackets are standard errors. Model (A) uses OLS. WLS is used for Model (B, D, F), using reciprocal of sample size as weights. FWLS is used for Model (C, E, G). Significance is indicated with ***, **, and * for 1%, 5%, and 10% statistical significance levels, respectively.

Table 3. MAPE of MRA models.

MAPE (%)	In-Sample MAPE		Out-of-Sample MAPE	
	(1) Model (D)	(2) Model (E)	(3) Model (D)	(4) Model (E)
Average MAPE	47.60	36.74	94.96	88.87
Median MAPE	42.18	14.59	47.97	23.04
Maximum	263.69	314.64	874.91	1802.72
Minimum	0.72	0.07	0.84	0.04

Note: Columns (1) and (3) use the WLS results from Model (D), Table 2. Columns (2) and (4) use the FWLS results from Model (E), Table 2. Columns (1) and (2) report the in-sample MAPE. Columns (3) and (4) report out-of-sample comparisons.

In Model (E), the ecosystem service type, value assessment approach, farmland division, cultivated land type, farmland area and per capital GDP in all can explain 98.3% of the value change of sample size farmland. In the regression results, the regression coefficient of virtual variables (ecosystem service type, value assessment method, farmland division, cultivated land type) reflects the deviation direction and degree of specific variables relative to control group; the regression coefficient of continuous variable represents elastic coefficient, i.e., the ratio of change rate of dependent variable to independent variable. A specific analysis of regression results is made as below:

1. Value assessment method: the regression coefficient of opportunity cost approach and market evaluation approach is statistically significant. This indicates when other influential factors remain unchanged, the value estimates obtained through opportunity cost approach and market valuation approach show significant difference with that obtained through surrogate market approach. The market valuation approach is higher than other value assessment method, while the value estimated obtained through opportunity cost approach is the lowest.
2. Ecosystem service type: in the six farmland ecosystem services, the regression coefficients of crops, gas regulation, soil conservation, and recreational tourism are statistically significant. This indicates when other influential factors remain unchanged, the values of crops, gas regulation, soil conservation, and recreational tourism services show significant difference with water conservation. The values of crops, gas regulation, and soil conservation are higher than other farmland ecosystem services, in which the value of soil conservation service is the highest, while the value of recreational tourism service is the lowest.
3. Cultivated land type: the regression coefficient of paddy field is positive and is statistically significant at the level of 1%. This indicates when other influential factors remain unchanged, the value estimates obtained when farmland ecosystem's cultivated land type is paddy field are evidently higher than that obtained from dry land.
4. Farmland division: in the four farmland division regions, the regression coefficient of Yangtze Plain Middle and Lower is statistically significant. This indicates when other influential factors remain unchanged, the farmland ecosystem service value of Yangtze Plain Middle and Lower is obviously lower than the service value of other regions.
5. Farmland area: the regression coefficient of farmland area is significantly negative. This indicates the per hectare value of farmland ecosystem has decreasing return to scale, but this effect will decrease geometrically with the increase of ecosystem area [71,72]. Taking the regression coefficient −0.181 of farmland area as an example, for a farmland ecosystem of 10 ha, if the area increases by 1%, its per hectare value will decrease by 1.81%; but for a farmland ecosystem of 1000 ha, if the area increases by 1%, its per hectare value only decreases by 0.018%. So, the total value of farmland ecosystem still increases with the increase of farmland area.
6. Per capital GDP: the regression coefficient of per capital GDP variable is statistically significant, showing negative correlation. This indicates when other conditions remain unchanged, if the per capital GDP of research region becomes higher, the unit area value of farmland ecosystem will be lower, and economic growth may lead to a recession of ecosystem function. When per capital GDP increases by 10%, the unit area value will reduce by 0.9%.

3.2. Value Transfer

Transfer error is used to test the consistency between model prediction value and value observations [12,28], equivalent to the mean absolute percentage error (*MAPE*), defined as:

$$MAPE = \sum [\frac{|V_{est} - V_{obs}|}{V_{obs}} \cdot 100]/n \quad (3)$$

where V_{est} is the transferred (predicted) farmland value from the MRA, V_{obs} is the farmland value as reported in a primary study, and n is the number of estimates. Generally speaking, the smaller the transfer error is, the higher the effectiveness of value transfer model will be [12,28].

The paper adopts the leave-one-out cross validation [63,68,73], i.e., successively select every observation as test set, take other observations as training set, and calculate the transfer error of corresponding observation in the test set respectively. Compared to the ordinary K-fold cross validation($k > 1$), though the calculation of leave-one-out cross validation is more tedious, its sample use ratio is the highest. It not only can accurately evaluate the model's global error, but also make it easier to analyze the change characteristics of error with observations.

We use MAPE to calculate the transfer error rate according to Equation (3). The transfer error rates are presented in Table 3, columns (1) and (2) for in-sample comparisons. Column (1) represents the transfer error of Model (D) in Table 2, and column (2) represents the transfer error of Model (E), with the average transfer errors of 47.60% and 36.74%, respectively. The results show that the regression results of FWLS are better than WLS results. We also report out-of-sample predictions in columns (3) and (4), with average errors of 94.96% and 88.87%, respectively, both of which are greater than in-sample errors. This condition conforms to the prediction [71]. The column (2) in Table 3 indicates that the median transfer error rate is 14.59%, in which, 74% of sample transfer error rate is less than 40%, and 9% of sample transfer error rate is greater than 100%. Therefore, it is necessary to be cautious due to that some transfer error rates are very large.

By sorting the value observations in ascending order, the change condition of model's predicted value and transfer error is obtained as shown in Figures 3 and 4. Figure 3 displays the change of predicted values of Model (D) and Model (E). It can be seen that when value observations are small, the model's predicted value is higher with large deviation degree; when value observations are large, the model's predicted value is lower with gradually descending deviation degree and data fluctuation range. The predicted value deviation and fluctuation range of Model (E) is obviously better than that of Model (D). Figure 4 indicates that with the increase of value observations, the transfer error shows a declining trend, and the transfer error of Model (E) is less than the results of Model (D) in most cases.

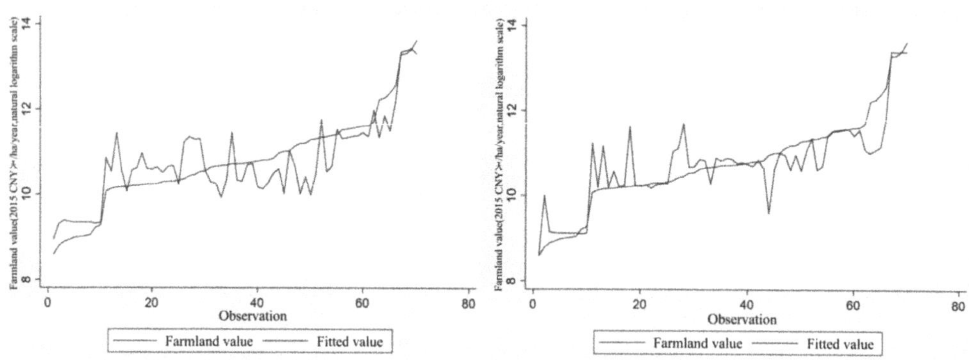

Figure 3. Value observations and predicted value (value observations are sorted in ascending order, with Model (D) on the left and Model (E) on the right).

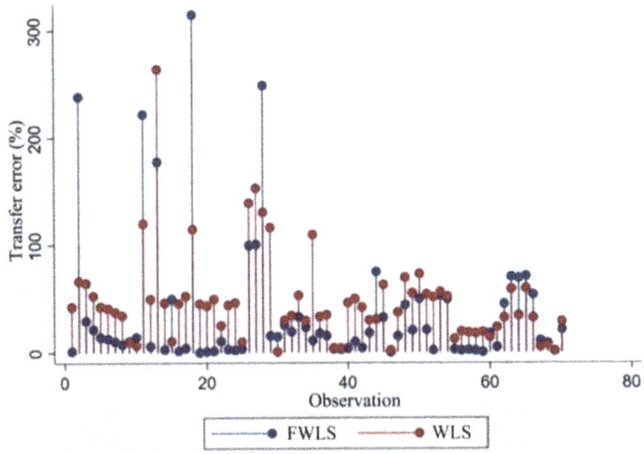

Figure 4. FWLS and WLS transfer error.

3.3. Value Change Assessment of China's Farmland Ecosystem between 2010 and 2100

This paper adopts the global 1 km land cover change dataset of 2010~2100 established by Li Xia et al. [2]. This dataset is built based on IMAGE module [74] and cellular automata techniques [75], and simulates the evolutionary process of different land use types between 2010 and 2100 according to the four scenarios of IPCC SRES: A1B, A2, B1, and B2, in which the baseline scenario is in the year of 2010. A1B belongs to the sub-scenario of various energy balance development in A1 scenarios. This dataset can be downloaded at https://geosimulation.cn/download/GlobalSimulation/ (accessed on 2 December 2021).

The paper extracted the dataset of China's land use (Farmland, Water, Urban, Forestry, Barren and Grassland) in the years of 2050 and 2100 under four scenarios, as shown in Figure 5. The meta-regression model results were used to calculate the change of China's farmland ecosystem service value under different scenarios, see the calculation results in Table 4. In 2010, the total farmland area in China was about 110 million hectare, and the total ecosystem service value was 8860 billion yuan. Under four scenarios, the change in total values of farmland area and farmland ecosystem service with time shows the same tendency. Under A2 and B2 scenarios, the total values of farmland area and farmland ecosystem service continuously increase, reaching to 15,220 billion yuan and 14,930 billion yuan till the year of 2100, respectively. Under A1B and B1 scenarios, both the total values of farmland area and farmland ecosystem decline after a rise, and the total farmland values will be only 7920 billion yuan and 6320 billion yuan in the year of 2100, respectively. In general, as China's farmland ecosystem is concerned, A2 scenario is the optimal development path, and the total service value loss of farmland ecosystem in B1 scenario is the maximum.

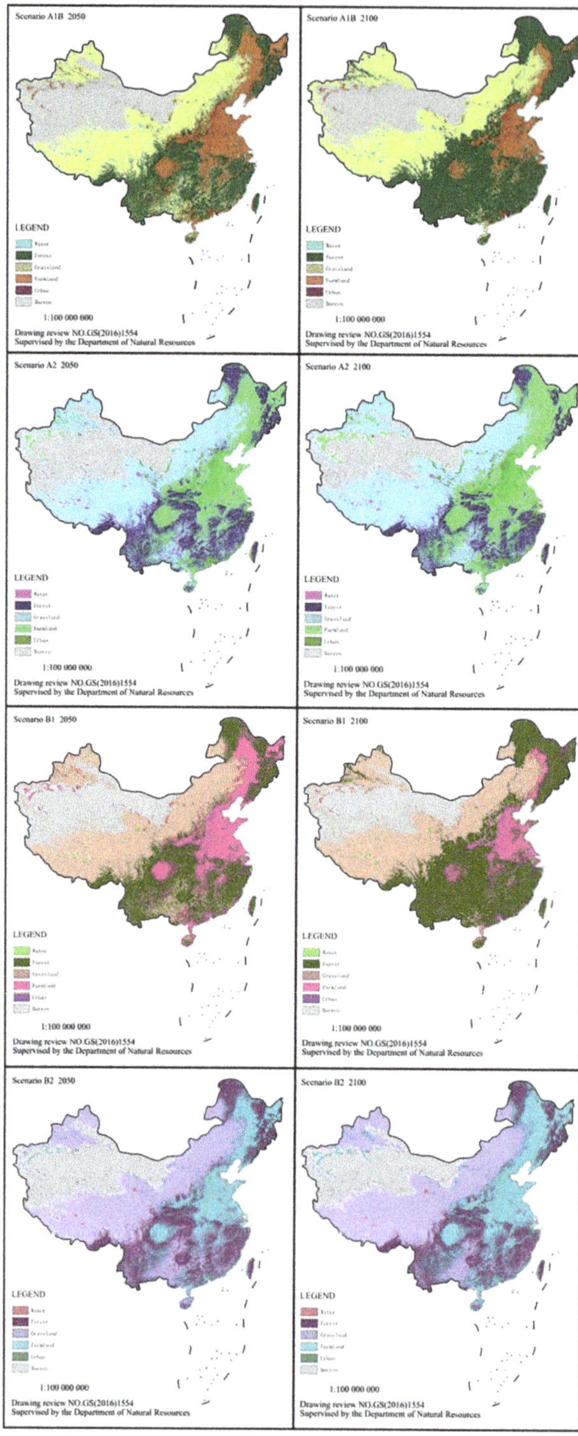

Figure 5. China's land use distribution in different scenarios (with no modification to base map).

Table 4. Change in farmland area and value in China under different scenarios (2010–2100).

Scenarios	Year	Area (10^8 ha)	Value (10^{12} CNY)
Baseline Scenario	2010	1.10	8.86
A1B Scenario	2050	2.29	14.86
	2100	0.96	7.92
A2 Scenario	2050	1.93	13.44
	2100	2.40	15.22
B1 Scenario	2050	1.50	11.31
	2100	0.74	6.32
B2 Scenario	2050	1.76	12.66
	2100	2.31	14.93

Note: A1B Scenario: low-speed population growth; sprawling city; super high-speed economic growth; rapid technological innovation; strong bio-fuel demand, balanced development of various energy. A2 Scenario: high-speed population growth; sprawling city; medium-speed economic growth; slow technological innovation; lower bio-fuel demand. B1 Scenario: low-speed population growth; compact city; high-speed economic growth; slower technological innovation; low overall energy consumption, low bio-fuel demand. B2 Scenario: medium-speed population growth; compact city; medium-speed economic growth; slower technological innovation; less overall energy consumption, low bio-fuel consumption.

4. Discussion

It is worth noting that meta-analysis is a method of performing quantitative comprehensive analysis and variation source analysis on existing results, so the effectiveness of meta-analysis-based value transfer method depends on the quantity and quality of existing empirical study to a great extent. But at present, there are a few research on farmland ecosystem in China, which may affect the research results. So, we suggest to further improve the quantity and quality of relevant empirical studies and attach importance to the technical perfection and index integrity in the process of ecosystem service value assessment. Meanwhile, it is also found from this study that whether urban development or national development in future, it is essential to pay attention to protecting arable land minimum and persist in permanent basic farmland policy, which could bring huge benefit for numerous natural ecosystems and social ecosystems.

5. Conclusions

China has always been a great agricultural country, in which farmland ecosystem development directly relates to national development. In this paper, MRA was made on the service value of farmland ecosystem in China, providing a novel contribution to the study of farmland ecosystem service value transfer. We discover that farmland characteristics, assessment method, and service type all affect the value of farmland. For example, farmland area negatively affects farmland value, and the farmland with cultivated land type as paddy field is more valuable than that as dry land. In addition, the farmland that provides crops and that has good soil conservation function is more valuable than that used for recreational tourism. This demonstrates that converting farmland into tourism development may bring down its value, but protecting it may increase its value. Another result is that the farmland value evaluated by using market valuation method is obviously superior to other value assessment methods, so the market valuation approach should be preferred in the farmland ecosystem service value assessment method in future. By sorting the value observations in ascending order, it is noted that with the increase of value observations, the model predicted value gradually transits from overestimation to underestimation, and transfer error also tends to decrease. This result is consistent with the research results of Brander et al. wetland ecosystem) and Salem et al. (mangrove forest ecosystem). Recently, there is still no reasonable explanation to such phenomenon, but it can be taken as a reference for value transfer process.

Regarding value transfer model construction, compared to the published meta-analysis studies, this paper used fewer explanatory variables, without considering such influential factors as the physical characteristics, environmental quality, climate change, and human activities of farmland ecosystem. The reason is the lack of locating observation and experimental research of ecosystem function in the study of farmland ecosystem in China, as well as the formation process of various ecosystem services and the mechanism study of human influence, so it is difficult to include relevant influential factors in this meta-regression analysis model. This is also a regret of this study. Regarding the results, we used the leave-one-out cross validation to obtain the error range of 88.87% and its median of 23.04%, both of which are higher than in-sample error by 36.74% and 16.59%. The FWLS error fluctuation results are significantly better than the WLS model. The studies on value transfer of farmland ecosystem in China between 2010 and 2100 indicate that the change trend in total values of farmland area and ecosystem is consistent, conforming to the opinions of Brander et al. and Woodward et al. However, the specific change conditions are somewhat varied under different scenarios, in which the total value increases evidently under A2 scenario, while the total value loss is the maximum under B1 scenario.

Author Contributions: Methodology, L.N., B.C., Y.L. (Yixin Luo) and J.M.; software, N.X. and Z.Y.; validation, T.Z. and P.L.; investigation, Y.L. (Yue Li) and T.Z.; data curation, Z.Y.; writing—original draft, L.N.; writing—review and editing, B.C. and J.M.; visualization, Y.L. (Yixin Luo), Y.L. (Yue Li) and P.L. All authors have read and agreed to the published version of the manuscript.

Funding: This research was funded by [Zhejiang Province Public Welfare Technology Application Research Project] grant number [No LGN21D010001] and [Scientific Research Fund of Zhejiang Provincial Education Department] grant number [No Y202147326].

Informed Consent Statement: Not applicable.

Data Availability Statement: Not applicable.

Conflicts of Interest: The authors declare no conflict of interest.

References

1. Liu, J.; Diamond, J. China's environment in a globalizing world. *Nature* **2005**, *435*, 1179–1186. [CrossRef] [PubMed]
2. Li, X.; Chen, G.; Liu, X.; Liang, X.; Wang, S.; Chen, Y.; Pei, F.; Xu, X. A New Global Land-Use and Land-Cover Change Product at a 1-km Resolution for 2010 to 2100 Based on Human–Environment Interactions. *Ann. Assoc. Am. Geogr.* **2017**, *107*, 1040–1059. [CrossRef]
3. Ye, Y.Q.; Zhang, J.E.; Qin, Z.; Li, Y.M.; Li, Y. Ecological profit and loss of farmland ecosystem in Foshan. *Acta Ecol. Sin.* **2012**, *32*, 4593–4604.
4. Xie, G.D.; Xiao, Y. Research progress of farmland ecosystem service and its value. *Chin. J. Eco-Agric.* **2013**, *21*, 645–651. [CrossRef]
5. Luo, Q.; Luo, L.; Zhou, Q.; Song, Y. Does China's Yangtze River Economic Belt policy impact on local ecosystem services? *Sci. Total Environ.* **2019**, *676*, 231–241. [CrossRef]
6. Sun, X.Z.; Zhou, H.L.; Xie, G.D. Service function and economic value of farmland ecosystem in China. *China Popul. Resour. Environ.* **2007**, *19*, 55–60.
7. Yin, F.; Mao, R.Z.; Fu, B.J.; Liu, G.H. Farmland ecosystem service function and its formation mechanism. *Chin. J. Appl. Ecol.* **2006**, *17*, 929–934.
8. Zhu, H.H.; Zhang, Y. Changes of farmland ecosystem service value in Xinjiang and its influential factor analysis. *J. Shihezi Univ. (Nat. Sci.)* **2020**, *38*, 340–346.
9. Yuan, Y.; Liu, J.T.; Jin, Z.Z. Comprehensive evaluation on positive and negative effect of farmland ecosystem service function in Luancheng county. *Chin. J. Ecol.* **2011**, *30*, 2809–2814.
10. Zhou, Z.M.; Zhang, L.P.; Cao, W.D.; Huang, Y.F. Assessment on ecosystem service function value of winter green manure-spring maize farmland. *Ecol. Environ. Sci.* **2016**, *25*, 597–604.
11. Sheng, J.; Chen, L.G.; Zhu, P.P. Assessment on ecosystem service function value of paddy-wheat rotation farmland. *Chin. J. Eco-Agric.* **2008**, *16*, 1541–1545. [CrossRef]
12. Zhang, L.; Li, X.J.; Zhou, D.M.; Zhang, Y.R. Study of China's lacustrine wetland ecosystem service value transfer based on Meta-analysis. *Acta Ecol. Sin.* **2015**, *35*, 5507–5517.
13. Xue, B.L.; Zhang, L.F.; Zhang, T.L.; Sun, W.C.; Li, Z.J. Paddy field ecosystem service value assessment—Take Hunan province as an example. *China Rural Water Hydropower* **2020**, *45*, 52–57.
14. Tang, H.; Zheng, Y.; Chen, F.; Yang, L.G.; Zhang, H.L.; Kong, Q.X. Assessment on ecosystem service value of different farmland types and cropping patterns in Beijing region. *Ecol. Econ.* **2008**, *24*, 56–59.
15. Xie, G.D.; Zhang, C.X.; Zhang, L.M.; Chen, W.H.; Li, S.M. Improvement to ecosystem service value method based on equivalent value factor per unit area. *J. Nat. Resour.* **2015**, *30*, 1243–1254.
16. Rao, L.Y.; Zhu, J.Z. Preliminary evaluation on ecosystem service function value of Simianshan forest in Chongqing. *J. Soil Water Conserv.* **2003**, *17*, 5–6.

17. Xiao, H.; Ouyang, Z.Y.; Zhao, J.Z.; Wang, X.K. Initial exploration of forest ecosystem service function and its ecological economic value assessment—Take Jianfengling tropical forests in Hainan Island as an example. *Chin. J. Appl. Ecol.* **2000**, *11*, 481–484.
18. Xue, D.Y.; Bao, H.; Li, W.H. Assessment on indirect economic value of forest ecosystem in Changbaishan Nature Reserve. *China Environ. Sci.* **1999**, *6*, 247–252.
19. Wu, L.L.; Lu, J.J.; Tong, C.F.; Liu, C.Q. Assessment on ecosystem service function value of wetland in Yangtze Estuary. *Resour. Environ. Yangtze Basin* **2003**, *12*, 411–416.
20. Wang, S.B.; Wang, P.J.; Hu, Z.Y.; Wang, X.R. Assessment on ecosystem landscape service value by contingent valuation method—Take Suzhou Creek in Shanghai as an example. *J. Fudan Univ. (Nat. Sci.)* **2003**, *42*, 463–467.
21. Zhang, Z.Q.; Xu, Z.M.; Cheng, G.D.; Su, Z.Y. Contingent valuation on ecosystem service recovery of Zhangye region in Heihe River Basin. *Acta Ecol. Sin.* **2002**, *22*, 885–893.
22. Lu, C.X.; Xie, G.D.; Cheng, S.K. Recreational entertainment function of river ecosystem and its value assessment. *Resour. Sci.* **2001**, *23*, 77–81.
23. Zhao, R.Q.; Huang, A.M.; Qin, M.Z.; Yang, H. Study on farmland ecosystem service function and evaluation method. *Syst. Sci. Compr. Stud. Agric.* **2003**, *19*, 267–270.
24. Zhang, D.; Li, X.S.; Chen, Y.H. Evaluation on farmland ecosystem service value classification in Huailai county. *Res. Soil Water Conserv.* **2016**, *23*, 234–239.
25. Zhang, H.F.; Ouyang, Z.Y.; Zheng, H.; Xiao, Y. Assessment on farmland ecosystem service function value in Manas River Basin. *Chin. J. Eco-Agric.* **2009**, *17*, 1259–1264. [CrossRef]
26. Moeltner, K.; Boyle, K.J.; Paterson, R.W. Meta-analysis and benefit transfer for resource valuation-addressing classical challenges with Bayesian modeling. *J. Environ. Econ. Manag.* **2007**, *53*, 250–269. [CrossRef]
27. Bergstrom, J.C.; Taylor, L.O. Using meta-analysis for benefits transfer: Theory and practice. *Ecol. Econ.* **2006**, *60*, 351–360. [CrossRef]
28. Wu, Z.J.; Zeng, H. Assessment of forest ecosystem service value in China based on Meta-analysis. *Acta Ecol. Sin.* **2021**, *41*, 5533–5545.
29. Miguel, M.F.; Butterfield, H.S.; Lortie, C.J. A meta-analysis contrasting active versus passive restoration practices in dryland agricultural ecosystems. *PEERJ* **2020**, *8*, e10428. [CrossRef]
30. Liu, H.; Li, C.B.; Han, X.L.; Li, M.Y. Land use changes and ecosystem service value transfer of Changbai Mountain region based on Meta-analysis. *Res. Soil Water Conserv.* **2020**, *27*, 293–300.
31. Ivezic, V.; Yu, Y.; van der Werf, W. Crop Yields in European Agroforestry Systems: A Meta-Analysis. *Front. Sustain. Food Syst.* **2021**, *5*, 193. [CrossRef]
32. Li, Q.B. Assessment on Ecological Protection Value of Wetland in Sanjiang Plain Based on Benefit Transfer Method. Ph.D. Thesis, Northeast Agricultural University, Harbin, China, 2018.
33. Costanza, R.; D'Arge, R.; de Groot, R.; Farber, S.; Grasso, M.; Hannon, B.; Limburg, K.; Naeem, S.; O'Neill, R.V.; Paruelo, J.; et al. The value of the world's ecosystem services and natural capital. *Nature* **1997**, *387*, 253–260. [CrossRef]
34. Raskin, P.D. Global Scenarios: Background Review for the Millennium Ecosystem Assessment. *Ecosystems* **2005**, *8*, 133–142. [CrossRef]
35. Xie, G.D.; Lu, C.X.; Leng, Y.F.; Zheng, D.; Li, S.C. Value assessment for ecological assets in Qinghai-Tibet Plateau. *J. Nat. Resour.* **2003**, *18*, 189–196.
36. Salem, M.E.; Mercer, D.E. The Economic Value of Mangroves: A Meta-Analysis. *Sustainability* **2012**, *4*, 359–383. [CrossRef]
37. De Groot, R.; Brander, L.; Van Der Ploeg, S.; Costanza, R.; Bernard, F.; Braat, L.; Christie, M.; Crossman, N.; Ghermandi, A.; Hein, L.; et al. Global estimates of the value of ecosystems and their services in monetary units. *Ecosyst. Serv.* **2012**, *1*, 50–61. [CrossRef]
38. Zhao, G.S.; Wen, Y.F.; Yu, F.W. Research progress, problem and trend of ecosystem service function value calculation. *Ecol. Econ.* **2008**, *18*, 100–103.
39. Chen, R.; Hu, X.Q. Applied research of farmland ecological value in new urban governance—Based on experience of Chongqing region. *Rural Econ. Sci.* **2020**, *31*, 10–14.
40. Zhou, R.J.; Wu, S.B.; Pi, X.P. Environmental value assessment of positive externality of agricultural production and its promotion study—Take Hunan province as example. *Res. Agric. Mod.* **2017**, *38*, 383–388.
41. Li, S.N. Study of Farmland Ecosystem Service Value Calculation in Qianxi County. Ph.D. Thesis, China University of Geosciences, Beijing, China, 2015.
42. Chen, Z.Z.; Zhang, Y.Q.; Wu, B.; Li, Z.P.; Geng, X.G.; Feng, J.Y.; Tian, S.Y.; Lei, N. Ecosystem service function value calculation of agricultural shelter forest in Shandong province. *Chin. J. Ecol.* **2012**, *31*, 59–65.
43. Li, S.; Ren, X.X. Farmland ecosystem service value assessment in Shanxi province. *Sci. Technol. Qinghai Agric. For.* **2019**, *49*, 54–59.
44. Zhu, Y.W.; Sang, B.Y.; Wang, Y.H.; Liu, K.; Chen, Q.M.; Chu, F.F. Calculation of farmland shelter forest's wind prevention and sand fixation service function in Xinjiang. *Chin. Agric. Sci. Bull.* **2015**, *31*, 7–12.
45. Kong, F.J.; Chen, Y.C.; Chen, Q.H.; Mou, Q.J.; Yan, J.Z. Spatial-temporal change characteristics of farmland ecological service value in Chongqing and its driving factor analysis. *Chin. J. Eco-Agric.* **2019**, *27*, 1637–1648.
46. Wang, J.; Li, Q.Y.; Guan, X. Value assessment of farmland ecosystem service function in Jinshi city in Changde. *Hunan Agric. Sci.* **2013**, *42*, 76–78.

47. Miao, J.Q.; Yang, W.T.; Yang, B.J.; Ma, Y.Q.; Huang, G.Q. Ecosystem service function and value assessment of the Hakkas terrace region in Chongyi county. *J. Nat. Resour.* **2016**, *31*, 1817–1831.
48. Zhang, W.W.; Li, J.; Liu, Y.X. Assessment on farmland ecosystem service value in Guanzhong-Tianshui Economic Zone. *Agric. Res. Arid Areas* **2012**, *30*, 201–205.
49. Sun, Y.Q.; Qiu, Z.Q.; Huo, H.C.; Yin, R.X.; Wen, Z.Z. Assessment on farmland ecosystem economic value of Guanshanhu District in Guiyang city. *Guizhou Agric. Sci.* **2019**, *47*, 155–158.
50. Liu, X.D.; Zhao, Z.B.; Li, K.G. Study on farmland ecosystem service function value calculation of Beidaihe District in Hebei Province. *J. Agric. Resour. Environ.* **2017**, *34*, 390–396.
51. Li, L.W.; Zhang, Y.; Min, G.F. Assessment on ecological benefits value of farmland shelter-belt in Henan Province. *For. Resour. Manag.* **2005**, *34*, 38–40.
52. Lei, N.; Zhang, Y.Q.; Wu, B.; Li, Z.P.; Feng, J.Y.; Chen, Z.Z. Agricultural shelter forest ecosystem service function value accounting in Henan Province. *Bull. Soil Water Conserv.* **2012**, *32*, 97–102.
53. Feng, W.; Zhang, H.N.; Sun, X.S.; Yue, Z.L. Assessment on comprehensive service function value of different crops in Heilonggang region. *Ecol. Econ.* **2017**, *33*, 146–150.
54. Wang, X.X. Study of Oilseed Rape-Paddy Rotation Farmland Ecosystem Service Value in Hunan. Ph.D. Thesis, Hunan Agricultural University, Changsha, China, 2016.
55. Wang, K.L.; Huang, G.Q.; Luo, Q.X.; Li, Z.Z. Assessment on ecosystem service value of paddy field multiple cropping system in hilly area of south Yangtze basin. *Acta Agric. Jiangxi* **2010**, *22*, 157–160.
56. Jiang, Z. Study on Economic Value Assessment of Paddy Field Ecosystem Service Function in Jingzhou City. Ph.D. Thesis, Yangtze University, Jingzhou, China, 2016.
57. Wang, M.; Zhang, J.; Gao, D.P.; Zhai, Y.L. Agricultural shelter forest ecosystem service function value accounting in Liaoning province. *J. Northeast For. Univ.* **2014**, *42*, 86–89.
58. Zhuang, S.H.; Chu, Y.; Bian, F.H.; Wang, Z.Q. Assessment of farmland ecosystem service function value in costal zone of Miaodao Archipelago. *J. Yantai Univ. (Nat. Sci. Eng. Ed.)* **2008**, *21*, 273–280.
59. Li, F.Y.; Wu, F.W. Assessment of farmland ecosystem service function value in Shanghai suburbs. *Shanghai Rural Econ.* **2006**, *13*, 22–25.
60. Wang, X.Y.; Tashpolat, T.; Zhang, F. Changes in farmland ecosystem service value in Ebinur Lake basin and the regression analysis of its impact factors—Take Jinghe county as an example. *China Rural Water Hydropower* **2016**, *41*, 103–107.
61. Nelson, J.P.; Kennedy, P.E. The Use (and Abuse) of Meta-Analysis in Environmental and Natural Resource Economics: An Assessment. *Environ. Resour. Econ.* **2009**, *42*, 345–377. [CrossRef]
62. Sundt, S.; Rehdanz, K. Consumers' willingness to pay for green electricity: A meta-analysis of the literature. *Energy Econ.* **2015**, *51*, 1–8. [CrossRef]
63. Chaikumbung, M.; Doucouliagos, H.; Scarborough, H. The economic value of wetlands in developing countries: A meta-regression analysis. *Ecol. Econ.* **2016**, *124*, 164–174. [CrossRef]
64. Kochi, I.; Hubbell, B.; Kramer, R. An Empirical Bayes Approach to Combining and Comparing Estimates of the Value of a Statistical Life for Environmental Policy Analysis. *Environ. Resour. Econ.* **2006**, *34*, 385–406. [CrossRef]
65. Johnston, R.J.; Besedin, E.Y.; Iovanna, R.; Miller, C.J.; Wardwell, R.F.; Ranson, M.H. Systematic Variation in Willingness to Pay for Aquatic Resource Improvements and Implications for Benefit Transfer: A Meta-Analysis. *Can. J. Agric. Econ.* **2005**, *53*, 221–248. [CrossRef]
66. Ghermandi, A.; Bergh, J.V.D.; Brander, L.; de Groot, H.L.; Nunes, P.A.L.D. Values of natural and human-made wetlands: A meta-analysis. *Water Resour. Res.* **2010**, *46*, W12516. [CrossRef]
67. Chen, Q. *Advanced Econometrics and Stata Application*; Higher Education Press: Beijing, China, 2014.
68. Brander, L.M.; Wagtendonk, A.J.; Hussain, S.; McVittie, A.; Verburg, P.H.; de Groot, R.S.; van der Ploeg, S. Ecosystem service values for mangroves in Southeast Asia: A meta-analysis and value transfer application. *Ecosyst. Serv.* **2012**, *1*, 62–69. [CrossRef]
69. Wu, N.; Song, X.Y.; Kang, W.H.; Deng, X.H.; Hu, X.Q.; Shi, P.J.; Liu, Y.Q. InVEST model based watershed ecological compensation standards under different perspectives—Take Gansu section of Weihe River as an example. *Acta Ecol. Sin.* **2018**, *38*, 2512–2522.
70. Dobbs, T.L.; Pretty, J. Case study of agri-environmental payments: The United Kingdom. *Ecol. Econ.* **2008**, *65*, 765–775. [CrossRef]
71. Brander, L.; Florax, R.J.G.M.; Vermaat, J. The Empirics of Wetland Valuation: A Comprehensive Summary and a Meta-Analysis of the Literature. *Environ. Resour. Econ.* **2006**, *33*, 223–250. [CrossRef]
72. Woodward, R.T.; Wui, Y.S. The economic value of wetland services: A meta-analysis. *Ecol. Econ.* **2001**, *37*, 257–270. [CrossRef]
73. Brander, L.M.; van Beukering, P.; Cesar, H.S. The recreational value of coral reefs: A meta-analysis. *Ecol. Econ.* **2007**, *63*, 209–218. [CrossRef]

74. Letourneau, A.; Verburg, P.H.; Stehfest, E. A land-use systems approach to represent land-use dynamics at continental and global scales. *Environ. Model. Softw.* **2012**, *33*, 61–79. [CrossRef]
75. Li, X.; Yeh, A.G.-O. Modelling sustainable urban development by the integration of constrained cellular automata and GIS. *Int. J. Geogr. Inf. Sci.* **2000**, *14*, 131–152. [CrossRef]

International Journal of
*Environmental Research
and Public Health*

MDPI

Article

Do Environmental Regulations Promote or Inhibit Cities' Innovation Capacity? Evidence from China

Xiaowen Zeng [1], Ming Jin [2] and Shuang Pan [3,*]

1 The Faculty of Economics, Guangdong University of Finance & Economics, Guangzhou 510320, China
2 School of Economics, Zhejiang University of Finance and Economics, Hangzhou 310018, China
3 School of Accountancy, Jiangxi University of Finance and Economics, Nanchang 330013, China
* Correspondence: pasan2016@163.com

Abstract: The "Porter Hypothesis" proposes that appropriate environmental regulations would promote firm innovation. This study aims to build a theoretical model for illustrating the impact and mechanism of environmental regulation on urban innovation through a panel of 281 Chinese prefecture-level cities during 2003–2016. The results indicated that an increase in environmental regulation markedly suppressed the innovative capacity of Chinese cities during the sample period. This inhibitory effect is primarily transmitted through two mediating variables: lower regional fiscal revenue and reduced manufacturing output. Moreover, improved regional economic development level helps generate positive incentives for environmental regulation and mitigate its inhibitions to innovation. Environmental regulation and urban innovation might have a non-linear U-shape relation, with the former helping improve urban innovation capacity upon reaching a particular level.

Keywords: environmental regulation; urban innovation; mediating effect

Citation: Zeng, X.; Jin, M.; Pan, S. Do Environmental Regulations Promote or Inhibit Cities' Innovation Capacity? Evidence from China. *Int. J. Environ. Res. Public Health* **2022**, *19*, 16993. https://doi.org/10.3390/ijerph192416993

Academic Editors: Hongxiao Liu, Tong Wu and Yuan Li

Received: 27 October 2022
Accepted: 13 December 2022
Published: 17 December 2022

1. Introduction

Recently, environmental problems have become more and more serious in the world. Global climate changes, acid rain, water pollution, air pollution and other types of environmental degradation are becoming increasingly common. China is not exempt from such environmental problems either. Relying on massive inputs of production factors from the reform and opening to foreign investment in 1978, the Chinese economy has achieved remarkable success, growing at over 9% per year from 1978 to 2021. It is undeniable that while the extensive growth approach has brought huge economic gains, it has also caused severe environmental pollution. Recently, China has been deepening sustainable development strategies and promoting a comprehensive green transformation of its development. This has led to the introduction of numerous environmental policies and regulations throughout China to prevent and control environmental risks tightly. Environmental quality keeps improving with the implementation of "closure, suspension, merger, or shifting to different line of production" of high-energy-consuming and high-polluting enterprises. However, this inevitably has a significant impact on regional economic development. When governments implement stringent environmental regulations and low-carbon policies, enterprises face various risks when transitioning to more environmental practices due to uncertainty about the future and increased operating costs. Moreover, local development may also be affected. Thus, the study of environmental regulation is essential to regional development.

Whether environmental protection and economic growth can be achieved concurrently has been extensively debated in both domestic and international academic circles. It is suggested that environmental regulation would aggravate the financial burden on firms and thus reduce their international competitiveness [1] and affect firms' total factor growth rate [2]. However, several researchers proposed that environmental regulation can have positive effects. For example, Porter argued that appropriate environmental regulation stimulated innovation and increased the competitiveness of firms, despite raising their costs, which is subsequently summarised as the "Porter hypothesis" [3]. However, there is still no clear consensus on whether this view is applicable to China. When firms face environmental regulation, a more effective solution is to reduce corporate pollution and increase revenue through innovative means. Some thus argued that incentive-based environmental regulation in China could enhance corporate innovation [4]. However, firms face transition risks such as lower revenues and higher costs under environmental regulations. Additionally, research and development (R&D) activities are inherently risky corporate behaviours, and enterprises may also lower R&D expenditures due to difficulty in providing sufficient funding or reducing corporate risk [5]. Moreover, poorer economic development can cause environmental regulations to inhibit innovation [6].

According to the existing studies, the change in environmental regulation intensity will have a significant impact on the innovation behaviour of enterprises, and the overall regional innovation level is closely related to the innovation ability of local enterprises. Therefore, it is not difficult to guess that environmental regulation is likely to have a corresponding impact on regional innovation. Although some scholars have analysed the micro-impact of environmental regulation from the perspective of enterprise behaviour, they mainly focus on listed companies or enterprises above a designated size, which will bring sample selection bias and make it difficult to identify the impact of environmental regulation. In addition, the relationship between urban innovation and environmental regulation will be more complex. On the one hand, local enterprises will be forced to transform due to stricter environmental regulations, which will significantly increase regional innovation. On the other hand, when the intensity of environmental regulations has been enhanced, polluting enterprises may shut down or reduce production in order to reduce the cost of pollution control, which will lead to the reduction of enterprise innovation activities and the decline of regional innovation ability. These changes are difficult to observe only through enterprise-level data. Therefore, this paper believes that the empirical study on the panel data of 281 prefecture-level cities is helpful to further explore the micro-impact mechanism of environmental regulation policies on regional innovation ability. Based on the city as an innovation carrier, this paper made several contributions: (1) Being different from the Porter hypothesis, this study finds that an increase in environmental regulation significantly suppressed cities' innovation capacity. At the same time, this paper adopts urban innovation data at the prefecture level, which can avoid sample selection bias. It can reflect the overall impact of environmental regulation policies on urban innovation ability and provide research support for investigating the spillover effect of environmental regulation policies. (2) This paper innovatively explores how Chinese environmental regulation inhibits urban innovation capacity. On the one hand, with the increase of environmental regulations, the decline in business efficiency of enterprises in the short term will bring a significant decline in local fiscal revenue, and the local government may be forced to reduce the subsidy support for enterprises' innovation, which will bring a restraining effect on urban innovation ability. On the other hand, since most Chinese manufacturing enterprises are still in the transition stage from extensive production to efficient production, blindly strengthening environmental regulation intensity is easy to lead to the decline of local manufacturing output and obstacles to technology research, thus restricting the improvement of urban innovation ability. (3) This paper conducts a variety of robustness tests on urban heterogeneity, environmental regulation policy categories, green technology innovation and other aspects and draws a series of new conclusions: in areas with poor economic development and less fixed investment, environmental regulation has a more significant inhibiting effect on urban innovation capability. In non-knowledge-intensive cities, the impact of environmental regulation is significantly negative, while in knowledge-intensive cities, the impact is not significant. Both market-oriented and command-and-control environmental regulation policies have a significant inhibitory effect on urban innovation ability in the sample period. (4) This study demonstrated the specific non-linear association between environmental regulation and urban innovation, which lends empirical proof for the Porter hypothesis at the meso-level. At the same time, it is clearly pointed out that China is still in a painful period of transition from extensive development to high-quality development, and the negative effects of environmental regulation policies will temporarily outweigh the positive effects. It is necessary to pay attention to the negative spillover effects of environmental regulation on technological innovation. This manuscript is organised below: Chapter 2 reviewes the relevant works. Chapter 3 introduces the theory model derivation. Chapter 4 provides the empirical test analysis and further research. The final section offers conclusions and policy advice.

2. Literature Review

2.1. Environmental Regulation and Urban Innovation

Environmental protection and economic growth are enduring topics of academic debate, and the loss of economic benefits due to environmental protection has long been thought-provoking for numerous scholars. Environmental regulation by governments can be a valuable motivator for firms to implement environmental initiatives [7]. Magat found that environmental regulation can influence firm innovation, but the effects vary depending on the environmental regulation type [8]. Scholars also argued that environmental regulation could lower the productivity level and growth rate of industries [1]. Porter countered that environmental regulation does not necessarily result in economic losses, suggesting that appropriate environmental regulation may spur 'innovation compensation' to enhance firm innovation [3]. Subsequently, Porter and Linde specified that innovation may occur when organisations attempt to increase the environmental efficacy in resource use, thus helping the production processes and product quality improvement [9].

2.2. Environmental Regulation and Economic Development

At present, China continues to implement low-carbon environmental policies, accompanied by increased environmental regulation in various regions. Many scholars have begun to discuss the link between environmental regulation and economic development. Several academics argued that environmental regulations caused the operating costs to increase and productivity to decline to some extent. For example, Guo discovered that environmental regulations fail to enhance green growth straightforwardly [10]. Yuan verified that environmental regulation reduces R&D investment over a long period utilizing panel data on Chinese manufacturing [5]. He utilised a Chinese water quality monitoring system and found that local governments implement more rigorous environmental criteria for enterprises upstream of monitoring sites [2]. Wu identified a significant U-shape relationship between environmental regulation and the green factor productivity of China's energy sector [11]. In contrast, Du argued that poor economic development can cause environmental regulation to stifle green technological innovation [6]. However, several academics hold the view that environmental regulation could prompt technological innovation and productivity improvement, especially in clean production industries [12]. Fu and Li found that environmental regulations promote innovation while increasing firms' costs, thereby improving their competitiveness [13]. Pan found that as market-based environmental regulations progressively enhance the energy efficiency, technological innovation will also be impacted by these regulations [4]. Additionally, market-based and voluntary environmental regulations possess a greater incentive effect on business innovation than that from command/control-based environmental regulations. The above thereby validated "Porter's hypothesis".

2.3. Environmental Regulation and Industries' Innovation Capacity

Concurrently, numerous researchers addressed the connection between environmental regulation and industries' innovation capacity in the Chinese manufacturing sector. Yuan conducted a study on Chinese manufacturing firms and determined that environmental regulation decreases their R&D investment [5]. Studies based on industry classification have apparent advantages: as performance measures tend to be consistent across industries, industry classification studies specify environmental regulation's impact on industries with varying development and pollution levels. Thus, their findings are more credible. However, there are limitations in that multiple industries exist in the same region. Thus, local governments need to consider multiple industries when making decisions, as different types of industries are affected by environmental regulations to different degrees. Hence, the regional role of environmental regulation also requires investigation. For companies of the same region environmentally regulated with similar degrees, an examination of regional environmental regulation has greater potential to highlight its regional role and be more informative for the implementation of local governmental policy.

Additionally, other scholars have conducted region-based studies about environmental regulation and firm innovation. Nie found that environmental regulation fostered innovation among less developed regions of western China and demonstrated the applicability of Porter's hypothesis in less developed regions of developing countries [14]. Li revealed that environmental regulations did not significantly affect the efficiency of urban science and technology innovation in Xi'an, suggesting the inapplicability of the Porter hypothesis in Xi'an [15].

In summary, focusing on Porter's hypothesis, the established research mainly examined the influence of environmental regulations on innovation at a micro level. However, research at the prefecture level has been insufficient and thus needs to be expanded. Concurrently, studies at the regional level centred on the correlation of environmental regulation to regional economic development but have been unable to identify and validate the mechanisms of its impact. Thus, this study examined the effects and transmission mechanisms of environmental regulation towards urban innovation using meso-level data to serve as a reference for setting regional environmental regulation policies.

3. Mathematical Model Analysis

The ability of cities to innovate depends on economic support and talent development. Many firms are at the forefront of urban innovation. "Porter's hypothesis" suggests that there exists a possible non-linear relationship between environmental regulation and firm innovation, i.e., environmental regulation may inhibit firm innovation in the short term but increase firm innovation and competitiveness in the long run. Similarly, this study argues that environmental regulation may ultimately have a corresponding impact on the innovation capacity of urban areas by influencing firms' production and business behaviour. Thus, the following mathematical model was derived by drawing on the research method of Zhang et al. (2011) [16]. We mainly extended Zhang's theoretical model to the city level and shifted the research perspective to the field of urban innovation ability.

Assumptions: Manufacturers conduct production activities in a perfectly competitive product and factor market; as production expands, the pollution generated by the manufacturers increases accordingly.

Let the vendor's revenue function be:

$$R = P * A(K_A)f(K_P)$$

where P denotes the price of products, K_A denotes the capital input used in production for technological innovation, and K_P represents the capital input to the daily production of the firm. $A(K_A)$ represents the level of technological innovation in production, and $f(K_P)$ represents the level of output at the given level of innovation.

Then, the output function of the manufacturer can be expressed as $F = A(K_A)f(K_P) = Af$. Here, it is assumed that the level of innovation in production with technological innovation is Hicks neutral.

As manufacturers produce emissions in the production process and pollution has negative externalities, the government will regulate manufacturers' pollution behaviour by specifying a level of pollution, i.e., environmental regulation (ERS). Existing research suggests that, under government environmental regulation, manufacturers may first reduce the level of pollution emissions from their production process through technological innovation in their production processes to achieve a lower level of pollution; then, manufacturers may increase their R&D investment to increase their output level. Although this will lead to more pollution emissions, firms can increase their pollution control expenditure in response to environmental regulation, benefiting from the increased scale of output and profits. It can be deduced that manufacturers' technological innovation is related to their own production technology A and pollution control technology E. Based on this, the present study argues that urban innovation in the context of environmental regulation can also be composed of two parts: regional productive technological innovation (CI_A) and the pollution control technology innovation induced by environmental regulation (CI_E). This study further assumes that the urban innovation function is separable; thus, $CI = CI_A + CI_E$, which satisfies $CI' = (\mathrm{A}, \cdot) > 0, CI' = (\cdot, E) > 0$.

Additionally, the manufacturer's emission function is assumed to be $W = (F, E)$, which is a function of the level of output and the pollution control technology. This function has the following properties:

First, pollution emissions increase with the increase in the scale of output, i.e., $W' = (F, \cdot) > 0$. Second, pollution emissions decrease as pollution control technology improves, i.e., $W' = (\cdot, E) > 0$. Apparently, E is positively correlated with the intensity of environmental regulation; it is believed that an increase in the intensity of environmental regulation will be followed by an increase in technological innovation in pollution treatment.

Assuming that the portion of firms' total output devoted to pollution control α denotes the intensity coefficient of environmental regulation, where α represents a real number between 0 and 1, then $\alpha A(K_A)f(K_P) = E$. Thus, the final profit function of the manufacturer is as follows: $\pi = P[A(K_A)f(K_P) - \alpha A(K_A)f(K_P)]$. The constraint under which the manufacturer produces is then as follows:

$ERS = W[A(K_A)f(K_P),\ \alpha A(K_A)f(K_P)]$, i.e., the pollution emissions are equal to the environmental regulation.

Constructing Lagrangian functions:

$$L = P[A(K_A)f(K_P) - \alpha A(K_A)f(K_P)] + \lambda\{W[A(K_A)f(K_P),\ \alpha A(K_A)f(K_P)] - ERS\}$$

The first-order optimality condition for the manufacturer is solved by the Lagrangian function as:

$$P(1-\alpha)A'f + \lambda \frac{\partial W[A(K_A)f(K_P),\ \alpha A(K_A)f(K_P)}{\partial K_A} = 0 \tag{1}$$

$$P(1-\alpha)Af' + \lambda \frac{\partial W[A(K_A)f(K_P),\ \alpha A(K_A)f(K_P)}{\partial K_P} = 0 \tag{2}$$

$$-PAf + \lambda \frac{\partial W[A(K_A)f(K_P),\ \alpha A(K_A)f(K_P)}{\partial \alpha} = 0 \tag{3}$$

From Equation (3), we obtained

$$P = \lambda \cdot \partial W / \partial E \tag{4}$$

Bringing Equation (4) into Equation (1) yielded the following:

$$\partial W / \partial E = -\partial W / \partial F \tag{5}$$

This equation demonstrates that the optimal option for a manufacturer facing environmental regulations is to make the increase in marginal pollution in production equal to the decrease in marginal pollution from pollution control inputs, i.e., the level of emissions of the manufacturer decreases as the intensity of environmental regulation increases.

From Equations (2), (3) and (5), we derived the following: $\partial W/\ \partial K_A > 0$, and since $P(1-\alpha)A'f > 0$, it is introduced that $\lambda < 0$.

Substituting this into Equation (3), we obtained $W/\partial\alpha < 0$. This implies that the manufacturer's pollution emissions will keep decreasing as its investment in pollution control keeps increasing during the production process due to the effects of environmental regulations.

Next, the impact of environmental regulation on urban innovation was examined from a technological innovation perspective.

According to $CI' = (A, \cdot) > 0$, then $\partial CI/\partial A > 0$, and it can be deduced that:

$$\frac{\partial CI}{\partial A} = \frac{\partial CI}{\partial W} \cdot \frac{\partial W}{\partial A} + \frac{\partial CI}{\partial W} \cdot \frac{\partial W}{\partial E} \cdot \frac{\partial E}{\partial A} > 0 \quad (6)$$

and because $\frac{\partial W}{\partial A} = \frac{\partial W}{\partial E} \cdot f + \frac{\partial W}{\partial E} \cdot \alpha f$, and $CI = CI_A + CI_E$, we eventually derived that:

$$\frac{\partial CI}{\partial A} = (\frac{\partial CI_E}{\partial W} + \frac{\partial CI_A}{\partial W}) \cdot [\frac{\partial W}{\partial F} f(1 - 2\alpha)] > 0 \quad (7)$$

where α denotes the intensity of the environmental regulation. From $1 - 2\alpha > 0$, $\partial W/\partial F > 0$, $\partial W/\partial E < 0$, we obtained $\partial W/\partial\alpha < 0$.

From Equation (7), it can be deduced that when $0.5 > \alpha > 0$, then $\partial CI_A/\partial W > 0$. This means that when the level of environmental regulation faced by enterprises is low, with the increase in the intensity of environmental regulation, the emissions of enterprises will decline. However, this will also lead to a decline in technological innovation in enterprise production, which is ultimately detrimental to the improvement of the level of urban innovation. When $\alpha > 0.5$ and tends to 1, $\frac{\partial W}{\partial F} f(1 - 2\alpha) < 0$ and $\partial CI_E/\partial W$ tends to 0, we obtain $\partial CI_A/\partial W < 0$. At this time, the improvement of technological innovation in enterprises is negatively correlated with pollution emissions. This suggests that the higher the intensity of environmental regulation, the lower the emissions of enterprises, which will increase the technological innovation in enterprise production, leading to a further rise in the level of urban innovation. Therefore, it can be deduced that the effect/impact of the level of environmental regulation on urban innovation is not unique, and further empirical test analysis is required.

4. Empirical Design and Analysis

4.1. Data Sources and Variable Descriptions

The main sources of panel data for the 281 prefecture-level cities for the period from 2003–2016 are the China Urban Statistical Yearbook, the China City and Industry Innovation Report 2017, the CSMAR database, the WIND database, and www.zhuanli.com (accessed on 24 March 2020).

The explained variable, the urban innovation index (innovation), was measured using the urban innovation index, which is currently a more standardised indicator for measuring innovation capacity at the city level, in addition to patent data. This study also used the patent grant numbers data from www.zhuanli.com (accessed on 24 March 2020) for robustness testing.

The core explanatory variable, i.e., environmental regulation intensity (ERS), was obtained by measuring five indicators using the entropy method with reference to Wang [17]. These five indicators include the sulphur dioxide removal rate, soot removal rate, comprehensive utilisation rate of industrial solid waste, domestic wastewater treatment rate, and domestic waste harmless treatment rate. The specific treatment methods are listed below:

(1) Raw data standardisation.

Positive indicator: $x'_{ij} = (x_{ij} - \overline{x})/s_j$ Reverse indicator: $x'_{ij} = (\overline{x} - x_{ij})/s_j$

where x_{ij} indicates the raw data of the jth indicator of the ith city, x'_{ij} represents the standardised indicator values and $\overline{x}$ and s_j denote the mean and standard deviation of the jth indicators, respectively. As there were negative values in the standardised data and the entropy method requires logarithmic processing, the standardised data were thus converted to positive values by adding the following constants: $Z_{ij} = x'_{ij} + A$

(2) Isomorphism of the indicators and calculating the proportion (p_{ij}) of the ith city in the jth indicator (p_{ij})

$$p_{ij} = \frac{Z_{ij}}{\sum_{i=1}^{n} Z_{ij}} (i = 1,2,\ldots,281; j = 1,2,\ldots,5)$$

(3) Calculation of the entropy value (e_j) of the jth indicator:

$$e_j = -k \sum_{i=1}^{n} p_{ij} \ln(p_{ij}), \text{ where } k = \frac{1}{\ln(n)}, \; e_j \geq 0$$

(4) Calculation of the differentiation factor (g_j) of the jth indicator: $g_j = 1 - e_j$

(5) Normalising the coefficient of variation and calculating the weights (w_j) of the jth indicator: $w_j = g_j / \sum_{j=1}^{m} g_j \; (j = 1,2,\ldots,m)$

(6) Calculation of the environmental regulation intensity (ERS_j) of the ith city: $ERS_j = \sum_{j=1}^{m} w_j \, p_{ij}$

Referring to related studies for other control variables, the ratio of secondary and tertiary industries in cities was selected for measuring industrial structure (Industry). The proportion of loan balance in the gross domestic product (GDP) was used to measure financial development (Finde). The natural rate of population growth was used to measure the population growth rate (Growth). The natural logarithm of population size was used to measure the population structure (Lnpeosize). The proportion of financial spending on science and education was used for measuring regional government behaviour (Sciedu), and the Log GDP for measuring regional economic development (LnGDP), among others. These were all control variables in this study. To reduce the estimation bias caused by heteroskedasticity, the standard errors of clustering to the city level were used in this study.

The explained, explanatory, and control variables mentioned above were set up as listed in Table 1. Descriptive statistics of variables are shown in Table 2.

Table 1. Setting of the main variables.

<table>
<tr><th></th><th>Name</th><th>Definition</th><th>Properties</th></tr>
<tr><td>Explained variables</td><td>Innovation</td><td>Innovation index, number of patents</td><td>Innovation indicators</td></tr>
<tr><td>Explanatory variables</td><td>ERS</td><td>Environmental regulation intensity score</td><td>Environmental regulation variables</td></tr>
<tr><td rowspan="6">Control variables</td><td>Industry</td><td>Ratio of the secondary and tertiary industries</td><td rowspan="6">Urban Characteristics</td></tr>
<tr><td>LnGDP</td><td>Logarithm of the gross national product</td></tr>
<tr><td>Finde</td><td>Financial development</td></tr>
<tr><td>Growth</td><td>Natural population growth rate</td></tr>
<tr><td>Lnpeosize</td><td>Population size in logarithms</td></tr>
<tr><td>Sciedu</td><td>Proportion of financial spending on science and education</td></tr>
</table>

Table 2. Variables description.

Variable	N	Mean	Sd	Min	Max
Innovation	3929	7.053	39.03	0	1100
ERS	3929	0.652	0.157	0.169	0.978
Industry	3926	1.467	0.783	0.106	10.60
Finde	3646	0.824	1.557	0.0753	90.16
Growth	3911	5.952	4.843	−8.900	40.78
Lnpeosize	3929	5.856	0.693	2.796	8.129
Sciedu	3926	0.198	0.047	0.0158	0.497
LnGDP	3927	6.719	1.063	3.459	10.25

4.2. Empirical Regression Results

(1) Baseline regression model

To examine how environmental regulation intensity affects urban innovation capacity, a benchmark regression model was set in this study as follows:

$$Innovation_{it} = \alpha + \beta_1 ERS_{it} + \sum_j \beta_j \, control_{it} + \mu_i + \gamma_t + \varepsilon_{it} \tag{8}$$

where $Innovation$ denotes the explanatory variables, ERS as the environmental regulation intensity, μ_i denotes the city fixed effect, γ_t is the year fixed effect and controlit denotes the control variable and was set as Table 1.

Table 3 presents the baseline regression results. Column (1) denotes results without control variables and fixed effects, Column (2) denotes results without fixed effects and Column (3) represents results having control variables and city fixed effects, while Column (4) denotes results with control variables and city, as well as time fixed effects. The regression results indicate the negative coefficients of environmental regulation intensity and the negative correlation between environmental regulation intensity and China's urban innovation capacity over the sample period. This result suggests that while intensive environmental regulation may be beneficial to energy conservation and emission reduction at present, but is detrimental to the enhancement of urban innovation capacity. This, in turn, may undermine the ability of cities to achieve long-term energy conservation and emission reduction through innovation over time. Moreover, when two-way city and time fixed effects are added, the coefficient signs of control variables like industrial structure and economic development change significantly. This suggests that heterogeneity may still exist in the way environmental regulation affects the innovation capacity of regional cities and that further mechanism analysis is needed. Additionally, the Porter hypothesis suggests that environmental regulation may enable firms to avoid the cost effects of environmental regulation by promoting innovation for high-quality development.

Table 3. Baseline regression results.

	(1) Innovation	(2) Innovation	(3) Innovation	(4) Innovation
ERS	−15.0208 **	−25.7736 ***	−24.8920 ***	−18.5348 ***
	(−2.54)	(−4.36)	(−3.97)	(−3.08)
Industry		−7.1888 ***	−7.2840 ***	2.3303 **
		(−6.43)	(−6.74)	(2.41)
Finde		1.8014 **	1.8023 **	−0.6837 ***
		(1.99)	(2.26)	(−2.94)
Growth		0.3309	0.4262 *	0.3173 ***
		(1.59)	(1.90)	(2.89)
Lnpeosize		−8.3218 ***	−11.1270 ***	97.8779 ***
		(−4.70)	(−5.19)	(2.65)
Sciedu		−2.6257	9.7068	121.5947 ***
		(−0.27)	(0.99)	(5.35)
LnGDP		20.1563 ***	23.2931 ***	−15.9533 ***
		(7.95)	(7.87)	(−3.59)
_cons	343.8323 ***	56.0059 ***	−52.5999 ***	-2.1×10^2
	(3.99)	(−7.83)	(−7.59)	(−0.84)
City	YES	NO	YES	YES
Year	YES	NO	NO	YES
N	3929	3624	3624	3624
R-sq	0.540	0.186	0.199	0.595

Note: ***, ** and * represent the significance at the 1%, 5% and 10% levels, respectively. The *t*-values are in parentheses.

(2) Robustness test

① The impact of price fluctuations

In this paper, we used the GDP deflator index to convert cities' GDP data into the constant price based on 2003, which can eliminate the influence of price factors on the regression results and then test the robustness of the baseline regression result. The result report is shown in Table 4. It can be seen that the result after price treatment is still consistent with the baseline regression result, indicating that our baseline regression result is still robust.

Table 4. Robustness test results of GDP constant price.

	(1) Innovation	(2) Innovation	(3) Innovation	(4) Innovation
ERS	−15.0208 **	−18.2479 ***	−22.3618 ***	−17.3725 ***
	(−2.54)	(−3.61)	(−3.88)	(−3.53)
Industry		−7.9571 ***	−6.9628 ***	1.9263 **
		(−6.71)	(−6.72)	(2.43)
Finde		1.8835 **	1.8521 **	−0.5264 ***
		(2.20)	(2.43)	(−3.12)
Growth		0.4635 **	0.7811 *	0.4281 ***
		(2.13)	(1.92)	(2.99)
Lnpeosize		−11.0072 ***	−10.2381 ***	96.2313 ***
		(−5.29)	(−4.92)	(3.78)
Sciedu		3.3873	9.1273	123.234 ***
		(0.34)	(0.12)	(4.72)
LnGDP		22.7693 ***	24.2341 ***	−16.1926 ***
		(7.98)	(6.92)	(−4.21)
_cons	343.8323 ***	−54.4965 ***	−52.2381 ***	-2.1×10^2
	(3.99)	(−7.77)	(−6.46)	(−0.75)
City	YES	NO	YES	YES
Year	YES	NO	NO	YES
N	3929	3624	3624	3624
R-sq	0.540	0.192	0.199	0.595

Note: ***, ** and * represent the significance at the 1%, 5% and 10% levels, respectively. The *t*-values are in parentheses.

② Regional heterogeneity regression test

China's urban innovation capacity is characterised by a clear regional development imbalance. In particular, provincial capitals and municipalities boast salient advantages in terms of innovation, as they are able to quickly circumvent the adverse effects of environmental regulation through their innovation activities involving innovation funds and talents. However, other small and medium-sized cities may not have such conditions. Thus, the inclusion of provincial capitals and municipalities in the full-sample regression may raise a sample selectivity bias. Hence, the baseline model was reassessed after excluding provincial capitals and municipalities, with results presented in Table 5, and revealed that the environmental regulation intensity still negatively correlated to the innovation capacity of cities, indicating a still robust baseline regression result.

Table 5. Robustness test results excluding provincial capitals and municipalities.

	(1) Innovation	(2) Innovation	(3) Innovation	(4) Innovation
ERS	−12.0634 **	−12.3951 **	−10.3517 *	−13.4969 **
	(−2.25)	(−2.31)	(−1.95)	(−2.40)
Industry		−3.7860 ***	−4.3018 ***	0.3077
		(−4.41)	(−4.21)	(0.52)
Finde		0.9207 ***	1.0316 ***	−0.2241 *
		(2.65)	(2.86)	(−1.72)

Table 5. *Cont.*

	(1) Innovation	(2) Innovation	(3) Innovation	(4) Innovation
Growth		0.4800 **	0.5767 **	0.1030
		(2.21)	(2.34)	(1.31)
Lnpeosize		−7.5658 ***	−9.5432 ***	96.5647 **
		(−4.08)	(−4.07)	(2.23)
Sciedu		2.7756	3.5675	55.9606 ***
		(0.34)	(0.41)	(3.10)
LnGDP		11.0440 ***	13.6805 ***	−7.9811 ***
		(4.75)	(4.58)	(−3.07)
_cons	7.5912 **	−16.3936 ***	−15.6138 ***	-5.8×10^2 **
	(2.31)	(−8.71)	(−8.12)	(−2.15)
City	YES	NO	YES	YES
Year	YES	NO	NO	YES
N	3509	3235	3235	3235
R-sq	0.494	0.140	0.160	0.576

Note: ***, ** and * represent the significance at the 1%, 5% and 10% levels, respectively. The *t*-values are in parentheses.

In addition, considering the regional differences in environmental regulations, this paper added more regional heterogeneity regression tests. We made a supplementary investigation on the heterogeneity of economic development and fixed investment in 281 prefecture-level cities by taking the annual mean as the dividing standard. The regression results are shown in Table 6. According to the results in Columns (2) and (4) of Table 6, there is a significant negative correlation between environmental regulation and urban innovation capability in areas with poor economic development and less fixed investment, which indicates that the enhancement of environmental regulation has a more inhibitory effect on the innovation capability of backward areas, while the inhibitory effect on the innovation capability of developed areas is not significant. The possible reasons for this result are as follows: due to the lack of sufficient innovation methods in less developed areas, with the increase of environmental regulation, it is more likely to increase the business pressure of enterprises and reduce the innovation input of enterprises, thus inhibiting the improvement of urban innovation ability. On the contrary, areas with better economic development have a variety of means to avoid environmental regulations, which is more apt to delay the negative impact of environmental regulations on cities' innovation ability. Moreover, in order to investigate the heterogeneity of urban innovation, we conducted a grouping regression to investigate the difference between knowledge-intensive cities and non-knowledge-intensive cities. Since China began to issue the national innovative City construction list in 2008, the cities on the construction list have had significant advantages in human capital, technological innovation and other aspects. Therefore, this paper took the cities in the list as knowledge-intensive cities, while those not in the list as non-knowledge-intensive cities, andmade regression estimations respectively. The regression results are reported in Columns (5) and (6) of Table 6. It can be found that in non-knowledge-intensive cities, the impact of environmental regulation is significantly negative, while in knowledge-intensive cities, the impact is not significant. This indicated that the increase in environmental regulation intensity has a more significant inhibitory effect on non-knowledge-intensive cities. In addition, since non-knowledge-intensive cities still occupy most of the samples, environmental regulation policies overall still have an inhibiting effect on urban innovation ability.

Table 6. Robustness test results of regional economic development, fixed investment and knowledge-intensive city.

	(1) High_GDP	(2) Low_GDP	(3) High_FIX	(4) Low_FIX	(5) Knowledge-Intensive City	(6) Non-Knowledge-Intensive City
ERS	−23.1933	−1.0307 ***	−25.7282	−0.6652 **	−50.8068	−2.1908 ***
	(−0.72)	(−3.85)	(−0.86)	(−2.52)	(−0.58)	(−3.37)
Industry	17.0594 ***	0.1807 **	17.6586 ***	0.2892 ***	122.8785 **	0.0423
	(3.17)	(2.32)	(2.80)	(3.87)	(2.15)	(0.27)
Finde	−8.9773	0.3993 *	−3.0338 ***	0.3195 *	−35.3804	0.0725
	(−0.99)	(1.90)	(−3.51)	(1.84)	(−1.40)	(1.63)
Growth	1.0362 *	0.0217 ***	0.9156 *	0.0212 ***	1.9645	0.0614 ***
	(1.68)	(3.11)	(1.71)	(3.09)	(1.54)	(3.72)
Lnpeosize	190.0496 *	9.2486 ***	165.6567 **	5.2244 ***	365.4422 **	6.2423
	(1.92)	(4.15)	(2.03)	(4.34)	(2.59)	(1.52)
Sciedu	248.5177 ***	10.0942 ***	256.0448 ***	8.5292 ***	-8.1×10^2 **	30.9878 ***
	(2.88)	(6.48)	(2.91)	(6.67)	(−1.97)	(7.25)
LnGDP	−43.3764 **	−0.0782	−58.8607 ***	−1.1771 ***	−17.9567	−1.9222 ***
	(−2.16)	(−0.24)	(−3.39)	(−3.67)	(−0.38)	(−2.86)
_cons	-6.5×10^2	−51.3693 ***	-3.6×10^2	−22.1732 ***	-1.9×10^3 *	71.7138 **
	(−0.97)	(−3.93)	(−0.69)	(−3.34)	(−1.73)	(2.11)
city	YES	YES	YES	YES	YES	YES
year	YES	YES	YES	YES	YES	YES
N	941	2683	1055	2569	298	3326
R-sq	0.640	0.643	0.634	0.812	0.864	0.736

Note: ***, ** and * represent the significance at the 1%, 5% and 10% levels, respectively. The *t*-values are in parentheses.

③ Replacement of urban innovation index

To avoid regression estimation bias due to the indicator measure, the core explained variable in the benchmark regression in the previous section, the urban innovation index (Innovation), was replaced. Patents are an effective indicator of innovation levels and can reflect the level of innovation output of a city. Thus, this study conducted robustness testing by replacing the urban innovation index with the number of patents granted (Patent) as obtained from www.zhuanli.com (accessed on 24 March 2020) for each city as the explained variable (Table 7), and proved the consistency with the baseline regression result, demonstrating the robust regression results in this study.

Table 7. Results of robustness tests after replacing the urban innovation index.

	(1) Patent	(2) Patent	(3) Patent	(4) Patent
ERS	−4.2646 ***	−3.8763 ***	−3.0519 ***	−3.8006 ***
	(−5.80)	(−4.80)	(−4.19)	(−5.40)
Industry		−1.7383 ***	−1.8409 ***	−0.2981
		(−8.73)	(−8.76)	(−1.30)
Finde		0.5098 ***	0.5309 ***	0.0923 **
		(7.05)	(8.42)	(2.01)
Growth		0.1099 **	0.1401 ***	0.0949 ***
		(2.56)	(2.97)	(4.47)
Lnpeosize		−3.3814 ***	−4.0289 ***	11.6817
		(−7.23)	(−7.43)	(1.55)
Sciedu		9.8662 ***	11.8644 ***	32.5693 ***
		(3.80)	(4.36)	(5.53)
LnGDP		5.8164 ***	6.5426 ***	−2.2969 ***
		(11.78)	(11.33)	(−2.65)
_cons	5.7103 ***	−14.9866 ***	−14.3915 ***	−65.7841
	(7.56)	(−13.31)	(−13.22)	(−1.37)
City	YES	NO	YES	YES
Year	YES	NO	NO	YES
N	3047	3024	3024	3024
R-sq	0.734	0.340	0.358	0.757

Note: *** and ** represent the significance at the 1% and 5% levels, respectively. The *t*-values are in parentheses.

According to the green patent list and the international classification code provided by the World Intellectual Property Organization (WIPO), we summed up green patent data at the city level and divided them into two parts: green patent application and green patent authorisation. According to the regression results in Table 8, the impact of environmental regulation on green patents is significantly negative at the level of 1% for both green patent applications and green patent grants. This means that environmental regulation policies will lead to the decline of green innovation patents. This is consistent with the original regression results, indicating that the original conclusion has good robustness.

Table 8. Results of robustness tests after dividing green patent into two parts: green patent application and green patent authorisation.

	(1)	(2)	(3)	(4)	(5)	(6)	(7)	(8)
	Green Patent Application				Green Patent Authorisation			
	GreenIno1	GreenIno1	GreenIno1	GreenIno1	GreenIno2	GreenIno2	GreenIno2	GreenIno2
ERS	−6.4662 ***	−0.9222 ***	−0.7430 ***	−0.5525 ***	−5.5415 ***	−0.3178 ***	−0.4506 ***	−0.3098 ***
	(−43.33)	(−8.27)	(−6.44)	(−4.80)	(−39.34)	(−2.86)	(−3.92)	(−2.75)
Industry		−0.2646 ***	−0.2368 ***	−0.2940 ***		−0.2368 ***	−0.2480 ***	−0.3111 ***
		(−12.52)	(−11.24)	(−14.23)		(−11.01)	(−11.57)	(−15.19)
Finde		0.1210 ***	0.1175 ***	0.1104 ***		0.1254 ***	0.1245 ***	0.1136 ***
		(3.23)	(3.22)	(4.73)		(2.92)	(3.06)	(4.55)
Growth		−0.0010	−0.0043	0.0013		−0.0037	−0.0026	0.0029
		(−0.37)	(−1.45)	(0.40)		(−1.40)	(−0.93)	(0.91)
Lnpeosize		−0.3069 ***	−0.2883 ***	−0.4833 ***		−0.2969 ***	−0.3588 ***	−0.5278 ***
		(−10.62)	(−9.41)	(−14.13)		(−9.93)	(−11.38)	(−15.32)
Sciedu		−1.1194 ***	−0.8162 ***	−0.5861		−1.1134 ***	−0.7788 **	−0.9079 **
		(−3.71)	(−2.62)	(−1.60)		(−3.59)	(−2.46)	(−2.47)
LnGDP		1.5265 ***	1.5159 ***	1.7063 ***		1.4555 ***	1.5170 ***	1.7105 ***
		(69.17)	(61.23)	(66.06)		(62.78)	(59.30)	(64.71)
_cons	−1.4687 ***	−5.8220 ***	−5.7905 ***	−4.8744 ***	−1.6625 ***	−5.8496 ***	−5.7679 ***	−4.7252 ***
	(−15.82)	(−40.03)	(−38.57)	(−23.66)	(−19.42)	(−39.16)	(−37.49)	(−22.65)
City	YES	NO	YES	YES	YES	NO	YES	YES
Year	YES	NO	NO	YES	YES	NO	NO	YES
N	3901	3599	3599	3599	3901	3599	3599	3599
R-sq	0.799	0.319	0.803	0.843	0.765	0.170	0.771	0.814

Note: *** and ** represent the significance at the 1% and 5% levels, respectively. The *t*-values are in parentheses.

④ Endogenous problem

Additionally, there may exist endogeneity between environmental regulation and urban innovation, as the more innovative a city is, the more likely it is to reduce pollution emissions through innovation, thus rendering it unnecessary for the city to reduce pollution emissions through environmental regulation policies. Thus, a certain reciprocal causality exists. With this in mind, a two-stage least squares analysis of the baseline model was conducted, with air circulation coefficients as instrumental variables. The main considerations for selecting instrumental variables were as follows. First, there exists no theoretical relationship between the air circulation coefficient and the innovation capacity of cities, which satisfies the exogeneity requirement of the instrumental variable. Second, air circulation coefficients correlated to environmental regulation by directly determining the level of air pollution dissipation. Specifically, when at a uniform level of pollution, cities with better air circulation assume that as pollution dissipates quickly, they do not need to adopt stricter environmental regulation policies. However, cities with lower air mobility need to increase their level of environmental regulation to control pollution emissions. Hence, when the air circulation coefficient is higher, the regional environmental regulation policy is theoretically weaker, thus satisfying the exogeneity requirement of instrumental variables. Table 9 presented the endogeneity test with the inclusion of the instrumental variable of air circulation coefficient. The findings of this study remained robust. Moreover, this study conducted a weak instrumental variables test, considering the issue of possible weak instrumental variables between the instrumental variables and environmental regulation. As deduced from the weak instrumental variables test, the F-values (F-weak) in the first stage were all beyond the empirical value of 10, suggesting the absence of weak instrumental variables. Additionally, the regression results for the instrumental variables indicated the conclusions of the current study remained valid following the endogeneity issue addressed.

Table 9. Endogeneity test with the inclusion of the instrumental variable (air circulation coefficient).

	(1) Innovation	(2) Innovation	(3) Innovation	(4) Innovation
ERS	−17.6497 *	−29.3929 ***	−26.3977 ***	−19.9488 *
	(−1.65)	(−4.38)	(−3.83)	(−1.89)
Industry		−7.2118 ***	−7.2766 ***	2.3639 *
		(−8.80)	(−8.72)	(1.77)
Finde		1.8223 ***	1.8091 ***	−0.6815 *
		(4.62)	(4.61)	(−1.86)
Growth		0.3404 ***	0.4280 ***	0.3239 **
		(2.65)	(3.26)	(2.27)
Lnpeosize		−8.5855 ***	−11.1972 ***	98.1400 ***
		(−6.70)	(−8.41)	(10.28)
Sciedu		−1.8387	9.8361	121.6783 ***
		(−0.13)	(0.67)	(6.90)
LnGDP		20.5667 ***	23.4263 ***	−15.8856 ***
		(20.04)	(21.71)	(−3.69)
_cons	396.2816 ***	−55.0671 ***	−62.2522 ***	-1.7×10^2 **
	(32.27)	(−8.67)	(−8.74)	(−2.27)
City	YES	NO	YES	YES
Year	YES	NO	NO	YES
N	3643	3619	3619	3619
R-sq	0.573	0.186	0.199	0.595
F-weak	27.84	25.21	36.23	26.35

Note: ***, ** and * represent the significance at the 1%, 5% and 10% levels, respectively. The *t*-values are in parentheses.

⑤ Different kinds of environmental regulation policies

In order to reflect the heterogeneity of different environmental regulation policies, we considered two pilot environmental policies as representatives of market-oriented and command-and-control environmental regulations and adopted the difference-in-differences (DID) model to study this problem.

We use carbon emission trading pilot cities as a proxy for market-oriented environmental regulation. This is a Chinese policy launched in 2011, which is similar to the European Union Emissions Trading System (EU-ETS) and aims to achieve optimal economic output with minimal environmental costs. We took the pilot city as the experimental group and the non-pilot city as the control group and designed a DID model as model (9), in which the POL variable was a dummy variable. When the city was included in the list of pilot cities, it was set as 1; otherwise, it was set as 0, and the remaining variables remained unchanged. We focused on the regression coefficient and significance of POL. The regression results are reported in Table 10. Column (1) is listed as the two-way fixed effect regression results without control variables, and Columns (2)–(4) are listed as the regression results of the year fixed effect and city fixed effect gradually added after the addition of control variables. It can be seen that the coefficient of POL has been significantly negative at the level of 1%, which is consistent with the baseline regression results of this paper. It shows that market-oriented environmental regulation can also harm the improvement of urban innovation levels.

$$Innovation_{it} = \alpha + \beta_1 POL_{it} + \sum_j \beta_j\, control_{it} + \mu_i + \gamma_t + \varepsilon_{it} \quad (9)$$

Table 10. DID regression results of market-oriented environmental regulation.

Market-Oriented	(1) Innovation	(2) Innovation	(3) Innovation	(4) Innovation
POL	−48.9229 ***	−33.6805 ***	−35.9997 ***	−35.4096 ***
	(−4.91)	(−4.20)	(−4.36)	(−5.64)
Industry		−5.9974 ***	−6.4375 ***	0.9284
		(−6.79)	(−7.19)	(1.18)
Finde		1.7430 *	1.8286 **	−0.4375 **
		(1.74)	(2.01)	(−2.20)
Growth		0.0565	0.1869	0.0858
		(0.31)	(0.95)	(0.88)
Lnpeosize		−4.6313 ***	−8.5713 ***	94.1931 ***
		(−3.50)	(−4.87)	(2.79)
Sciedu		−25.2315 **	−11.0340	112.8472 ***
		(−2.16)	(−1.00)	(5.24)
LnGDP		15.0302 ***	19.3583 ***	−12.3454 ***
		(10.23)	(9.28)	(−3.32)
_cons	4.3632 ***	−57.4117 ***	−54.1978 ***	-2.4×10^2
	(17.39)	(−8.64)	(−8.18)	(−1.02)
City	YES	NO	YES	YES
Year	YES	NO	NO	YES
N	3929	3624	3624	3624
R-sq	0.582	0.215	0.233	0.614

Note: ***, ** and * represent the significance at the 1%, 5% and 10% levels, respectively. The *t*-values are in parentheses.

Secondly, we adopt the Air pollution Prevention and Control Action Plan policy as the representative of command-controlled environmental regulation. The Air Pollution Prevention and Control Action Plan is a policy initiated by China in 2013. It is issued by the State Council on the air pollution prevention and control Action Plan, which makes mandatory requirements for the reduction of the concentration of inhalable particulate matter in different regions of the country. Therefore, it can be used as a representative of command-controlled environmental regulation policy. We took the pilot city as the experimental group and the non-pilot city as the control group and designed a DID model as model (10), in which the variable of AIR was a dummy variable. When the city was included in the list of pilot cities, it was set as 1; otherwise, it was set as 0, and the remaining variables remained unchanged. We focused on the regression coefficient and significance of AIR. The regression results are reported in Table 11. Column (1) is listed as the two-way fixed effect regression result without control variables, Columns (2)–(4) are listed as the regression results of the year fixed effect and city fixed effect gradually added after the addition of control variables. It can be found that the coefficient of AIR is always significantly negative at the level of 1%, which is consistent with the benchmark regression result of this paper. It shows that command-and-control environmental regulation will also damage the improvement of urban innovation levels.

$$Innovation_{it} = \alpha + \beta_1 AIR_{it} + \sum_j \beta_j \, control_{it} + \mu_i + \gamma_t + \varepsilon_{it} \tag{10}$$

Table 11. DID regression results of command-controlled environmental regulation.

Command-Controlled	(1) Innovation	(2) Innovation	(3) Innovation	(4) Innovation
AIR	−16.2614 ***	−14.9303 ***	−7.0667 ***	−7.7409 ***
	(−6.19)	(−4.57)	(−4.89)	(−4.19)

Table 11. *Cont.*

Command–Controlled	(1) Innovation	(2) Innovation	(3) Innovation	(4) Innovation
Industry		−7.0636 ***	−7.3537 ***	2.2818 **
		(−6.44)	(−6.71)	(2.39)
Finde		1.6783 *	1.6931 **	−0.6838 ***
		(1.90)	(2.18)	(−2.94)
Growth		0.2704	0.3965 *	0.3241 ***
		(1.30)	(1.80)	(2.93)
Lnpeosize		−6.7099 ***	−9.9680 ***	97.6943 ***
		(−4.14)	(−5.08)	(2.65)
Sciedu		−9.8392	4.1007	121.1801 ***
		(−1.02)	(0.43)	(5.33)
LnGDP		17.5897 ***	21.1076 ***	−16.1490 ***
		(8.67)	(8.37)	(−3.63)
_cons	3.5886 ***	−63.1212 ***	−58.9005 ***	-2.2×10^2
	(10.78)	(−7.88)	(−7.66)	(−0.87)
City	YES	NO	YES	YES
Year	YES	NO	NO	YES
N	3929	3624	3624	3624
R-sq	0.529	0.180	0.195	0.595

Note: ***, ** and * represent the significance at the 1%, 5% and 10% levels, respectively. The *t*-values are in parentheses.

4.3. Analysis of the Impact Mechanisms

Regarding the previous baseline regression findings, we concluded that the environmental regulation intensity was strongly negatively related to the innovation capacity of cities during the sample period. We believe that there are two potential mechanisms: government subsidy and enterprise operation. Firstly, with the increase of environmental regulations, the decline in business efficiency of enterprises in the short term will bring a significant decline in local fiscal revenue, and the local government may be forced to reduce the subsidy support for enterprises' innovation, which will bring a restraining effect on urban innovation ability. Secondly, since most Chinese manufacturing enterprises are still in the transition stage from extensive production to efficient production, blindly strengthening environmental regulation intensity is easy to lead to the decline of local manufacturing output and obstacles to technology research, thus restricting the improvement of urban innovation ability.

Thus, we then clarified the specific mechanism of environmental regulations in affecting urban innovation capacity by constructing the corresponding mediating effect model with two mediating variables: regional fiscal revenue and regional manufacturing output. Drawing on the practice of Xu and Liu [18], the mediating effect was verified by investigating the regression coefficients of a recursive simultaneous equation using a stepwise approach. Taking regional revenue (REV) as a mediating variable, the following test model was constructed.

$$Innovation_{it} = \alpha + \beta_1 ERS_{it} + \sum_j \beta_j\, control_{it} + \mu_i + \gamma_t + \varepsilon_{it} \tag{11}$$

$$REV_{it} = \alpha_2 + \delta_1 ERS_{it} + \sum_j \delta_j\, control_{it} + \mu_i + \gamma_t + \varepsilon_{it} \tag{12}$$

$$Innovation_{it} = \alpha_3 + \omega_1 ERS_{it} + \omega_2 REV_{it} + \sum_j \omega_j\, control_{it} + \mu_i + \gamma_t + \varepsilon_{it} \tag{13}$$

We tested, in turn, the coefficients β_1 of the stepwise model (11), δ_1 of model (12) and ω_2 of model (13). If all three coefficients were significant, the mediating effect of REV was significant. This indicated that environmental regulation would influence the explained variable urban innovation capacity through the mediating variable REV, with a mediating effect size of $\delta_1 \times \omega_2$. The regression results of regional fiscal revenue are presented in Columns (1)–(4) of Table 12. The mediating effect of regional fiscal revenue was $\delta_1 \times \omega_2$, which was significantly negative. This indicated that environmental regulation negatively influenced urban innovation capacity through the mediating effect of regional fiscal revenue. Specifically, Column (2) in Table 12 proved the negative relationship between environmental regulation and regional fiscal revenue. Concurrently, Column (3) of Table 12 showed that an increase in fiscal revenue could enhance the city's innovation capacity. Ultimately, enhanced environmental regulation will bring reduced regional fiscal revenue, which in turn will lead to a decrease in local government support for enterprise innovation policies and, ultimately, a decrease in the urban innovation capacity. These findings are consistent with previous theoretical explanations.

Table 12. Mechanisms by which environmental regulation affects the innovation capacity of Chinese cities.

	(1) Innovation	(2) REV	(3) Innovation	(4) Innovation	(5) IND	(6) Innovation	(7) Innovation
ERS	−18.5348 ***	−1.0947 ***		−3.6768 **	−2.0583 ***		−14.6193 **
	(−3.08)	(−3.60)		(−2.29)	(−3.61)		(−2.47)
REV			0.1358 ***	0.1357 ***			
			(10.03)	(10.03)			
IND						1.9322 ***	−1.9023 ***
						(6.55)	(−6.46)
Industry	2.3303 **	−0.0910	3.5331 ***	3.5658 ***	7.4853 ***	16.6603 ***	16.5699 ***
	(2.41)	(−1.43)	(5.18)	(5.21)	(12.47)	(5.36)	(5.33)
Finde	−0.6837 ***	0.0133	−0.8683 ***	−0.8641 ***	0.3236 ***	−0.0745	−0.0680
	(−2.94)	(0.83)	(−6.58)	(−6.56)	(4.13)	(−0.38)	(−0.35)
Growth	0.3173 ***	0.0246 ***	−0.0176	−0.0166	−0.0108	0.2937 ***	0.2969 ***
	(2.89)	(4.14)	(−0.34)	(−0.32)	(−0.77)	(2.82)	(2.85)
Lnpeosize	97.8779 ***	3.9291 ***	44.5504 ***	44.5507 ***	−5.1356 **	88.1182 **	88.1082 **
	(2.65)	(2.66)	(2.68)	(2.68)	(−2.28)	(2.57)	(2.57)
Sciedu	121.5947 ***	10.3820 ***	−19.4134	−19.3114	−15.3751 ***	91.9168 ***	92.3459 ***
	(5.35)	(7.23)	(−1.14)	(−1.13)	(−6.01)	(4.71)	(4.74)
LnGDP	−15.9533 ***	0.1233	−17.7765 ***	−17.6274 ***	8.0326 ***	−1.0194	−0.6725
	(−3.59)	(0.41)	(−6.95)	(−6.91)	(5.46)	(−0.27)	(−0.18)
_cons	-2.1×10^2	−5.0168	-1.5×10^2	-1.4×10^2	−11.0899	-2.4×10^2	-2.3×10^2
	(−0.84)	(−0.48)	(−1.22)	(−1.21)	(−0.67)	(−1.00)	(−0.98)
city	YES	YES	YES	YES	YES	YES	YES
year	YES	YES	YES	YES	YES	YES	YES
N	3624	3624	3624	3624	3624	3624	3624
R-sq	0.595	0.793	0.875	0.875	0.945	0.609	0.610

Note: *** and ** represent the significance at the 1% and 5% levels, respectively. The t-values are in parentheses.

Similarly, regional manufacturing output was also adopted as the mediating variable investigating how environmental regulation contributes to the innovation capacity of cities. We found that the mediating effect of regional manufacturing output was also significant and had a negative coefficient. This indicated that environmental regulations markedly dampened urban innovation capacity through the mediating effect of manufacturing output. Specifically, as can be deduced from Column (5) of Table 12, environmental regulation is significantly negatively related to regional manufacturing output, while Column (6) of Table 12 shows that manufacturing output is positively related to urban innovation capacity. In summary, increased environmental regulation leads to a certain degree of decline in regional manufacturing output, which may constrain firms' R&D and innovation behaviour, and ultimately lead to a decline in regional urban innovation capacity, validating the relevant explanations presented in the previous section.

To further examine the effect of mediating variables, we have added the interaction term between fiscal revenue and environmental regulations, as well as the interaction term between manufacturing output value and environmental regulations, to model Equation (8), which will be used as a supplement for the robustness test. After adding interaction items into model Equation (8) separately, the new regression results are shown in Table 13. It can be found that no matter the interaction term between fiscal revenue and environmental regulation, or the interaction term between industrial output and environmental regulation, are all significantly negative. This means that environmental regulation policies really have an impact on urban innovation ability through the two factors of regional fiscal revenue and industrial output value. Meanwhile, the higher the regional fiscal revenue and industrial output value, the more significant the inhibition effect of environmental regulation on regional innovation ability. For local governments in China, the negative effects brought by environmental regulation policies will temporarily outweigh the positive effects, which is also an issue that government departments need to consider further.

Table 13. Regression results of interaction terms between intermediate variable and environmental regulations.

	(1) Innovation	(2) Innovation
REV × ERS	−6.7466 *** (−3.88)	
REV	19.7336 *** (2.73)	
IND×ERS		−2.4905 *** (−4.61)
IND		−0.3617 * (−1.75)
ERS	1.7487 (0.34)	109.0976 *** (3.89)
Industry	3.6497 *** (5.20)	15.0186 *** (5.26)
Finde	−0.9281 *** (−6.05)	0.0758 (0.41)
Growth	−0.0265 (−0.49)	0.2791 *** (2.69)
Lnpeosize	43.7520 *** (2.80)	87.6656 ** (2.56)
Sciedu	−21.4370 (−1.27)	100.4303 *** (5.12)
LnGDP	−18.9218 *** (−6.27)	2.7377 (0.75)
_cons	-1.5×10^2 (−1.33)	-3.5×10^2 (−1.44)
city	YES	YES
year	YES	YES
N	3624	3624
R-sq	0.877	0.615

Note: ***, ** and * represent the significance at the 1%, 5% and 10% levels, respectively. The t-values are in parentheses.

However, although environmental regulation may inhibit innovation in the short term, there may exist a compensatory effect of innovation; with appropriate environmental regulation incentives, firms may also increase their R&D and innovation, boost energy efficiency and emission reduction and improve their innovation and competitiveness. This finding may also apply to the association between environmental regulation and urban innovation; if regions are able to provide attractive incentives during the environmental regulation process, they may drive business involvement in technological innovation, thus promoting the city's ability to innovate. Thus, this study further examined the moderating effect of regional economic development and environmental investment on the relationship between environmental regulation and urban innovation capacity by cross-multiplying regional gross national product per capita (GDPPER) and regional environmental investment in pollution control (ENIVES) with the environmental regulation intensity, respectively (Table 14). The coefficients of both the cross multiplier of environmental investment and that of GDP per capita were distinctly positive. Concurrently, the environmental regulation's coefficient remained apparently negative, but its magnitude was significantly reduced. This indicated that as the regional economy expanded, the inhibitory influence of environmental regulation regulating urban innovation capacity could be effectively mitigated by increasing the positive regulation incentive. Moreover, when the government increases the environmental investment in pollution control, it can somewhat reduce the detrimental influence of environmental regulation on innovation.

Table 14. Moderating effects of regional economic development and investment.

	(1) Innovation	(2) Innovation
ENIVES × ERS	3.2339 ***	
	(3.59)	
ENIVES	0.0019	
	(0.52)	
GDPPER × ERS		6.9643 ***
		(3.06)
GDPPER		3.7167 ***
		(3.58)
ERS	−2.0510 **	−2.0686 **
	(2.44)	(−2.40)
Industry	−0.7466 **	10.7979 ***
	(−2.58)	(5.93)
Finde	−2.9813	−1.1922 ***
	(−1.51)	(−3.51)
Growth	0.0167	0.1701 **
	(0.69)	(2.10)
Lnpeosize	21.6479 **	101.6441 ***
	(1.98)	(3.32)
Sciedu	9.7721 **	−40.1351
	(2.15)	(−1.59)
LnGDP	−2.1484 *	−44.3047 ***
	(−1.77)	(−4.47)
_cons	−54.7029	−44.8829
	(−0.70)	(−0.24)
City	YES	YES
Year	YES	YES
N	1055	3624
R-sq	0.619	0.690

Note: ***, ** and * represent the significance at the 1%, 5% and 10% levels, respectively. The t-values are in parentheses.

4.4. Further Discussion

As mentioned earlier, environmental regulation and innovation have a non-linear connection. At lower environmental regulation intensity, it is unnecessary to innovate for environmental protection due to the low expenditure on circumventing environmental regulation. However, the adoption of other methods to circumvent environmental regulation will crowd out investment in innovation to a certain extent. In such cases, environmental regulation can further reduce innovation, whereas as its intensity surpasses a certain level, it becomes challenging to circumvent environmental regulation by other means. Thus, firms may be forced to get around the negative consequences of environmental regulation via innovation. Hence, there may exist a non-linear U-shape relationship between environmental regulation and urban innovation. The current study analysed this issue deeply.

To examine this issue, the squared environmental regulation was introduced in the model, which was designed as follows:

$$innovation_{it} = \alpha + \beta_1 ERS_{it} + \beta_2 ERS_{it}^2 + \sum_j \beta_j\, control_{it} + \mu_i + \gamma_t + \varepsilon_{it} \tag{14}$$

If non-linear characteristics existed, then the inflection point at which environmental regulation affects urban innovation was calculated as follows:

$$Inflection\ Point = -\frac{\beta_1}{2\beta_2} \tag{15}$$

Table 15 reports the regression results for model (14). In the full sample, the squared environmental regulation was remarkably positive, while the first-order term coefficient was obviously negative, indicating the validation of the non-linear characteristic of the U-shape. The calculated inflection points at approximately 0.7 could turn the impact of environmental regulation from negative to positive. Heterogeneity existed in the results for the above three regions. The second-order term coefficient for environmental regulation was insignificant in the eastern region. Conversely, the central and western regions exhibited a non-linear U-shaped characteristic, with an inflection point at approximately 0.6. We think there may be several reasons for this phenomenon. Firstly, the eastern region, as an economically developed region, has been attaching more and more importance to the development of a green economy in recent years, and the intensity of environmental regulations has been increasing year by year. However, local enterprises are generally still in the transition of technological innovation and cannot meet the requirements of environmental regulations in the short term. As a result, environmental regulation has led to the increase of pollution treatment costs and the deterioration of the business environment; innovation behaviour will be significantly inhibited accordingly. Secondly, due to the underdeveloped economy, the central and western regions often enjoy various industrial support policies and capital subsidies from different level governments. This can effectively reduce the problem of rising costs caused by increasing R&D investment. Due to a higher degree of marketisation and fewer government subsidies for production in the eastern region, based on cost-benefit analysis, local enterprises' operation strategies will be more cautious. Especially in the face of increased environmental regulation, enterprises are more inclined to maintain the stability of daily operation funds, thus reducing R&D investment.

Table 15. Non-linear relationship between environmental regulation and the innovation capacity of cities.

	Full Sample	Eastern Region	Central Region	Western Region
ERS	−96.4268 ***	−92.3252 *	−62.3718 ***	−60.8416 ***
	(−4.05)	(−1.65)	(−5.90)	(−4.72)
ERS^2	64.8753 ***	44.3182	52.3886 ***	53.2727 ***
	(3.36)	(1.01)	(5.99)	(4.98)
Industry	2.6925 *	6.3187 *	0.3550	0.2932
	(1.93)	(1.85)	(0.35)	(0.51)
Finde	−0.6514 *	−1.1470 *	0.8292	8.6236 ***
	(−1.71)	(−1.82)	(1.26)	(6.34)
Growth	0.3052 **	0.3501	0.1127	0.0837
	(2.05)	(1.11)	(1.45)	(1.08)
Lnpeosize	95.2399 ***	113.7912 ***	−6.4220	25.2700 ***
	(9.55)	(7.15)	(−0.80)	(3.09)

Table 15. *Cont.*

	Full Sample	Eastern Region	Central Region	Western Region
Sciedu	115.6751 ***	183.1929 ***	54.7605 ***	45.7342 ***
	(6.27)	(5.16)	(5.18)	(4.43)
LnGDP	−15.2994 ***	−22.4736 **	10.0248 ***	8.8039 ***
	(−3.42)	(−2.39)	(4.03)	(3.55)
_cons	-1.8×10^2 **	-2.6×10^2 **	−10.6416	-2.3×10^2 ***
	(−2.36)	(−2.03)	(−0.23)	(−4.27)
City	Yes	yes	yes	yes
Year	Yes	yes	yes	yes
N	3624	1765	934	925
R-sq	0.596	0.607	0.628	0.611
Inflection point	0.7		0.6	0.6

Note: ***, ** and * represent the significance at the 1%, 5% and 10% levels, respectively. The *t*-values are in parentheses.

5. Conclusions

Porter's hypothesis has prompted scholars to address the micro effects of environmental regulations from a firm's perspective. The current research examined the micro effects and mechanisms affecting the innovation capacity of cities by constructing theoretical models and empirical tests. We discovered that environmental regulation substantially negatively correlated to China's urban innovation capacity during the sample period, and its increasing intensity could dampen China's urban innovation capacity. Concurrently, this inhibitory effect was mainly transmitted through two mediating variables: lower regional fiscal revenue and reduced manufacturing output. Increased regional economic development helps to bring positive incentives for environmental regulation, thus somewhat mitigating the inhibiting effect of increased environmental regulation on urban innovation capacity. Additionally, there may exist a non-linear U-shape relationship between environmental regulation and urban innovation; the sufficiently high intensity of environmental regulation will force firms to innovate and circumvent the drawbacks of environmental regulation.

Thus, the following issues should be considered during the policy development of environmental regulation in China. Firstly, as China is in the transition from extensive development to high-quality development, the negative effects of environmental regulation policies will temporarily outweigh the positive effects. Therefore, it is necessary to be alert to the negative spillover effects of environmental regulation on technological innovation. Secondly, local governments should be cautious about the adverse effects brought about by environmental regulation at the initial stage and implement reasonable emission policies to enable enterprises to survive the period of loss resulting from environmental regulation to help enhance innovation in companies and cities. Thirdly, local governments must rationally judge the characteristics of environmental regulation inflection points on their own circumstances and formulate corresponding environmental regulation policies. Fourthly, in more economically developed regions, local governments may consider supporting innovative firms with financial subsidies or tax concessions to enter the innovation dividend period of environmental regulation. Fifthly, for economically less-developed regions, due to the lack of sufficient innovative means, the intensity of environmental regulation by the local government should not be too high, which can avoid directly increasing the pressure on business operations. It will be helpful to mitigate the negative impact of environmental regulation on local economic development. Future studies may focus on a comparison between developing and developed regions in China to find the differences in innovation levels and remedies to boost innovation through environmental regulations in the deprived areas.

Author Contributions: Conceptualization, X.Z., M.J. and S.P.; methodology, X.Z.; software, S.P.; validation, X.Z. and M.J.; formal analysis, X.Z.; investigation, X.Z. and S.P.; resources, X.Z.; data curation, X.Z. and S.P.; writing—original draft preparation, X.Z. and S.P.; writing—review and editing, M.J.; visualization M.J.; project administration, X.Z.; funding access, X.Z. All authors have read and agreed to the published version of the manuscript.

Funding: This work is partially sponsored by the Youth Innovative Project of Guangdong Universities (2021WQNCX022).

Institutional Review Board Statement: These studies did not involve human participants and were reviewed and approved by the ethics committee of the Guangdong University of Finance and Economics.

Informed Consent Statement: Not applicable.

Data Availability Statement: No datasets were generated or analyzed in this study.

Conflicts of Interest: The authors declare no conflict of interest.

References

1. Gray, W.B. The cost of regulation: OSHA, EPA and the productivity slowdown. *Am. Econ. Rev.* **1987**, *77*, 998–1006.
2. He, G.; Wang, S.; Zhang, B. Watering down environmental regulation in China. *Q. J. Econ.* **2020**, *135*, 2135–2185. [CrossRef]
3. Porter, M.E. Towards a dynamic theory of strategy. *Strateg. Manag. J.* **1991**, *12*, 95–117. [CrossRef]
4. Pan, X.; Ai, B.; Li, C.; Pan, X.; Yan, Y. Dynamic relationship among environmental regulation, technological innovation and energy efficiency based on large scale provincial panel data in China. *Technol. Forecast. Soc. Chang.* **2019**, *144*, 428–435. [CrossRef]
5. Yuan, B.; Xiang, Q. Environmental regulation, industrial innovation and green development of Chinese manufacturing: Based on an extended CDM model. *J. Clean. Prod.* **2018**, *176*, 895–908. [CrossRef]
6. Du, K.; Cheng, Y.; Yao, X. Environmental regulation, green technology innovation, and industrial structure upgrading: The road to the green transformation of Chinese cities. *Energy Econ.* **2021**, *98*, 105247. [CrossRef]
7. Henriques, I.; Sadorsky, P. The relationship between environmental commitment and managerial perceptions of stakeholder importance. *Acad. Manag. J.* **1999**, *42*, 87–99. [CrossRef]
8. Magat, W.A. The effects of environmental regulation on innovation. *Law Contemp. Probl.* **1979**, *43*, 4–25.
9. Porter, M.E.; Van der Linde, C. Toward a new conception of the environment-competitiveness relationship. *J. Econ. Perspect.* **1995**, *9*, 97–118. [CrossRef]
10. Guo, L.L.; Qu, Y.; Tseng, M.L. The interaction effects of environmental regulation and technological innovation on regional green growth performance. *J. Clean. Prod.* **2017**, *162*, 894–902. [CrossRef]
11. Wu, H.; Hao, Y.; Ren, S. How do environmental regulation and environmental decentralization affect green total factor energy efficiency: Evidence from China. *Energy Econ.* **2020**, *91*, 104880. [CrossRef]
12. Wang, Y.; Shen, N. Environmental regulation and environmental productivity: The case of China. *Renew. Sustain. Energy Rev.* **2016**, *62*, 758–766. [CrossRef]
13. Fu, J.Y.; Li, L.S. A Case Study on the Environmental Regulation, the Factor Endowment and the International Competitiveness in Industries. *Manag. World* **2010**, *10*, 87–98.
14. Nie, X.; Wu, J.; Zhang, W.; Zhang, J.; Wang, W.; Wang, Y.; Luo, Y.; Wang, H. Can environmental regulation promote urban innovation in the underdeveloped coastal regions of western China? *Mar. Policy* **2021**, *133*, 104709. [CrossRef]
15. Li, H.; Zhang, J.; Wang, C.; Wang, Y.; Coffey, V. An evaluation of the impact of environmental regulation on the efficiency of technology innovation using the combined DEA model: A case study of Xi'an, China. *Sustain. Cities Soc.* **2018**, *42*, 355–369. [CrossRef]
16. Zhang, C.; Lu, Y.; Guo, L.; Yu, T.S. The Intensity of Environmental Regulation and Technological Progress of Production. *Econ. Res. J.* **2011**, *2*, 113–124.
17. Wang, F.X.; Mao, A.H.; Li, H.L. Quality Measurement and Regional Difference of Urbanization in Shandong Province Based on the Entropy Method. *Geogr. Sci.* **2013**, *11*, 42–48.
18. Xu, M.; Liu, J.S. How inter-provincial counterpart support affects the economic performance of recipient regions–and the multiple mediating effects of economic growth and urban-rural income convergence. *Econ. Sci.* **2018**, *4*, 75–88.

International Journal of
Environmental Research and Public Health

MDPI

Article

The Impact of Managers' Environmental Cognition on Urban Public Service Innovation from the Perspective of Green Ecology

Ling Liu [1] and Yuanyuan Zhou [2,3,*]

[1] School of Tourism, Xinyang Normal University, Xinyang 464000, China
[2] School of Business Administration, Zhongnan University of Economics and Law, Wuhan 430073, China
[3] School of Business, Xinyang Normal University, Xinyang 464000, China
* Correspondence: zhouyy@stu.zuel.edu.cn

Abstract: With the continuous improvement in urban managers' cognition of green ecology, it is a hot issue in current research to explore the role of managers' cognition on urban public service innovation from the perspective of green ecology. The main purpose of this study was to explore ways to build green ecological cities and provide high-quality urban public services based on managers' environmental cognition. Through sorting out and discriminating the concepts related to green ecology, this research improves the current theoretical system related to green ecological city services. A theoretical model of a green ecological city public service system was constructed, and its influence path and effect on green ecological city public service innovation were analyzed in detail. This research provides a good tool and method for follow-up research to better understand the composition and innovation of green ecological city public service systems.

Keywords: green ecology; environmental cognition; urban public service system; system building

Citation: Liu, L.; Zhou, Y. The Impact of Managers' Environmental Cognition on Urban Public Service Innovation from the Perspective of Green Ecology. *Int. J. Environ. Res. Public Health* **2022**, *19*, 15945. https://doi.org/10.3390/ijerph192315945

Academic Editors: Hongxiao Liu, Tong Wu and Yuan Li

Received: 2 November 2022
Accepted: 22 November 2022
Published: 29 November 2022

1. Introduction

A city is the carrier of social development nowadays. With the acceleration of global urbanization, there are increasing requirements for urban public services. Green ecological urban construction and public services have become a current development trend. In this context, whether city managers have formed effective environmental cognition and whether they can provide green ecological public services for the city have not yet been concluded. Starting from the construction of a green ecological city public service system, this paper explores the effect path of city managers' cognition on it through improving the system.

For city managers, the formation of their environmental cognition is accompanied by the deepening of the concept of green ecology. Regarding how to manage a city and provide public services, one of the earliest studies conducted in the 19th century gathered a large amount of information on urban population and industry convergence for the living environment of urban residents. This produced more serious challenges in order to explore effective, appropriate, and sustainable urban development. In many cities, the management development concept arose at a historic moment [1]. For example, the concept of a "garden city" in the early period and the concepts of an "ecological city", a "healthy city", and a "green city" have been put forward by subsequent scholars around the world. In recent years, with the rise of the green concept, the global urbanization wave has once again pushed the concepts of a "low-carbon city", a "low-carbon eco-city", and a "green eco-city" to a new research frontier [2]. Regarding urban management and public service provision under the concept of a "green ecological city", most scholars believe that it is a sociological construct that includes both natural ecology and human ecology, requiring managers to pay attention to the protection of natural resources and the environment in the process of urban management [3]. At the same time, attention should be paid to adding more practical content such as "green life" and "green culture" in the process of providing public services for urban residents [4].

The main contribution of this paper lies in the in-depth analysis of managers' environmental cognition on the management planning of an urban spatial layout, including the infrastructure facilities, housing, transportation, ecology, green space, and other public services. At the same time, by building a green ecological city public service system model, the current green ecological context connotation can be enriched and the extension of urban public service supplies—namely, in the city of the carrier—through a public service supply can achieve a political, economic, cultural, social, and ecological civilization construction of "Five-one project", promote harmony between humans and nature as well as society, the economy, and building a community of life to realize the green development practice of "whole elements" of nature, society, and economy; the "whole space" of towns, agriculture, and ecology; and the "whole process" of past, present, and future developments.

The purpose of this study was to build a theoretical model of a green eco-city public service system from the three aspects of the construction criteria, influencing factors, and action mechanisms by applying the methodology of urbanology, ecology, and systems engineering to analyze its influence path and effect on green eco-city public service innovation by combining the environmental cognition of urban managers. The specific application of the research has two aspects. On the one hand, it can better construct an urban public service system under the background of current green and low-carbon developments and provide more green and healthy public services for urban residents. On the other hand, it can also help urban managers with green ecological environment cognition to discover the key and difficult problems in the provision of urban public services and to make targeted improvements.

2. Materials and Methods

2.1. Characteristics of Managers' Environmental Cognition from the Perspective of Green Ecology

From the perspective of green ecology, city managers form different environmental cognitive characteristics based on different goal directions and management objectives, which will have different impacts on urban public service innovation. This study summarizes the composition characteristics, target characteristics, evolution characteristics, measures, and value characteristics of urban managers' environmental cognition and explains the possible impact on urban public services (Table 1).

Table 1. Environmental cognitive characteristics of urban managers.

Characteristics of Categories	Specific Characteristics	Explanation
Constituent Characteristics	System	The composition of managers' environmental cognition is systematic. As city management involves many elements, their environmental cognition must seek the balance of the whole system in multi-dimensions.
	Open	The process of urban management is a process of exchange and replacement with the outside world, so its environmental cognition must ensure that the city always maintains an open circulation and circulation with the outside world.
	Complexity	Environmental cognition is complicated because the core purpose of both urban management and public services is to serve urban residents. The complexity of human beings results in the complexity of environmental cognition.
Target Characteristics	Diversity	The management of a green ecological city should show diversity in levels, elements, and associations to better guarantee the diversity and vitality of the city.
	High Efficiency	From the perspective of green ecology, urban management should be efficient to achieve sustainable urban development.
	Compound	In urban management, the green concept and ecological concept are fully combined to realize the coordination and unity of ecological, social, production, cultural, and other urban functions.

Table 1. *Cont.*

Characteristics of Categories	Specific Characteristics	Explanation
	Security	Safety is an important prerequisite for management. From the perspective of green ecology, managers' environmental cognition goals should also pay attention to the security guarantee of urban social stability, economic development, natural resources, the environment, public safety, and health.
	Health	Urban management should contribute to the health of urban residents, their living environment, and their future development.
Evolution Characteristics	Dynamic	As things develop, the environmental cognition formed by managers should also have a certain dynamic and it should dynamically adjust with the process of urban development.
	Adaptive	The urban system responds to the external environment and eventually feeds back into the human and social systems. Therefore, managers' environmental cognition also needs to adapt to this feedback and achieve the dynamic adaptation of the internal and external circulation of management through continuous adjustment, feedback, re-adjustment, and re-feedback.
Features of the Measures	Cyclical	From the perspective of green ecology, the measures taken by managers based on environmental cognition should reflect the concept of circular development, which is a virtuous cycle of all systems and levels of the city, to improve the overall urban public service efficiency and resource utilization rate.
	Compactness	Compactness refers to the specific measures formulated by managers to pay attention to the appropriate and compact use of injected land, public infrastructure, and public services rather than ignoring costs in order to achieve green ecological goals.
Value Characteristics	Human Nature	The value of urban managers' environmental cognition is primarily people-oriented. Whether it is the construction of green ecological cities or the provision and innovation of public services, the fundamental goal and the method of value realization are to realize the comprehensive and free development of human beings on the basis of a harmonious coexistence between man and nature.
	Symbiotic	Symbiosis emphasizes that managers should pay attention to the city rather than a single individual in their environmental cognition. In the process of providing green and ecological public services, they should pay attention to mutual assistance and mutualism outside the city, between cities and regions, and between cities.
	Harmony	It is emphasized that managers' environmental cognition should be based on harmony. It is not necessary to sacrifice the interests of a small number of groups to achieve a certain goal, but all urban residents should do their best and give full play to their strengths whilst enjoying the wellbeing brought about by green ecological cities so that harmony and consistency can help urban development.

After analyzing and summarizing the environmental cognitive characteristics of city managers from the perspective of green ecology, this paper further analyzes and explains the structure of the house chart of environmental cognitive characteristics (Figure 1). As can be seen from the figure, complexity, systematicness, and openness constitute the basis for the formation of managers' environmental cognition [5]. Dynamic and adaptive characteristics are necessary for managers' environmental cognition [6]. Diversity, efficiency, complexity, safety, and health are the main goals for managers' environmental cognition to be applied to green ecological city public services [7]. Circularity and compactness are the main measures for managers to provide green ecological city public services based on environmental cognition. Finally, symbiosis, human nature, and harmony are the ultimate value pursuits of urban management and urban public services provided by managers based on environmental cognition [8].

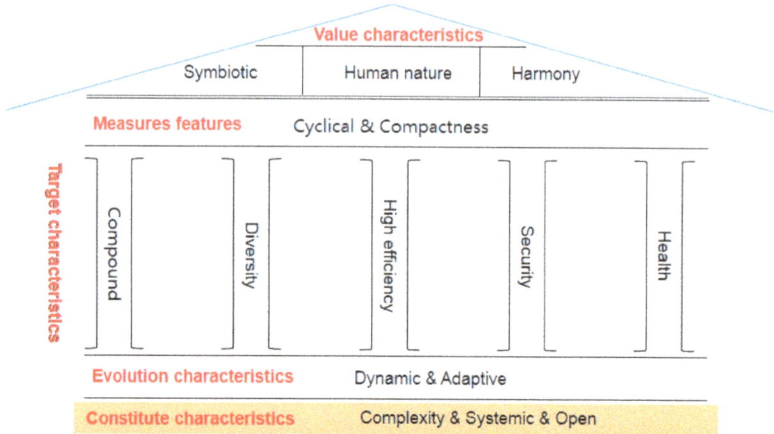

Figure 1. House diagram of environmental cognition characteristics of urban managers.

2.2. Connotations of Green Ecological City Public Services

Under the promotion of the concept of sustainable development, urban public services have also begun to advocate sustainable development. Meanwhile, sustainable urban public services pay more attention to improvement in the inclusiveness and equality of urban public services. However, in the sustainable development of urban public services, the climate, biology, energy, and economy are rarely involved. Therefore, on this basis, current city managers move out of the single urban public service category and consider a more global and multi-factor perspective, propose low-carbon city and low-carbon ecological city public services, plan the targets of urban public services more specifically, and add more factors for harmony between people and nature [9]. This has formed the basic prototype of a green city and ecological city public service in the process of urban management. Urban public service from a perspective of green ecology is the integration of a green city and an ecological city public service. No matter what the type of public service delivery method is, the ultimate goal is to achieve a perfect combination of an artificial environment, a natural environment, and a human environment in urban public services, ensuring the sustainable development of the city [10]. Based on the above discrimination, Figure 2 depicts the subordinate relationship of various urban public services from the perspective of sustainable development and green ecology.

As can be seen from the figure, both "green" and "ecological" in the public service of a green ecological city have been endowed with broader connotations. "Green" no longer reflects environmental factors, but is deeply combined with the connotation of sustainable development. Specific to the provision of urban public services, it means that urban public services are clean, safe, stable, and comfortable, with natural harmony and human health. "Ecological" is beyond the scope of the original ecology of a natural ecological system with traditional boundaries. From the phylogeny to the urban public services that provide an urban public service, there is an open working of the body that has complex and diverse characteristics. Urban public services provided by the target should have efficient coordination to achieve the city's economic development and social wellbeing as well as an excellent ecological environment system.

2.3. The Theoretical Connection between Managers' Environmental Cognition and Urban Public Services

After explaining the connotation of a green ecological city public service and the cognitive characteristics of managers' environments from the perspective of green ecology, the theoretical basis of their influence and connection mainly comes from three theoretical pillars. One is the urban theory related to urban planning, including urban development theory, urban space theory, regional coordination theory, and urban economics theory. Second is the ecological sustainable development theory related to environmental ecology, including circular economy theory, compound ecology theory, ecological sustainable theory, and carrying capacity theory. The third is system theory, which mainly includes cybernetics, operations research, complex adaptive system theory, and system dynamics [11]. The specific theoretical basis is shown in Figure 3.

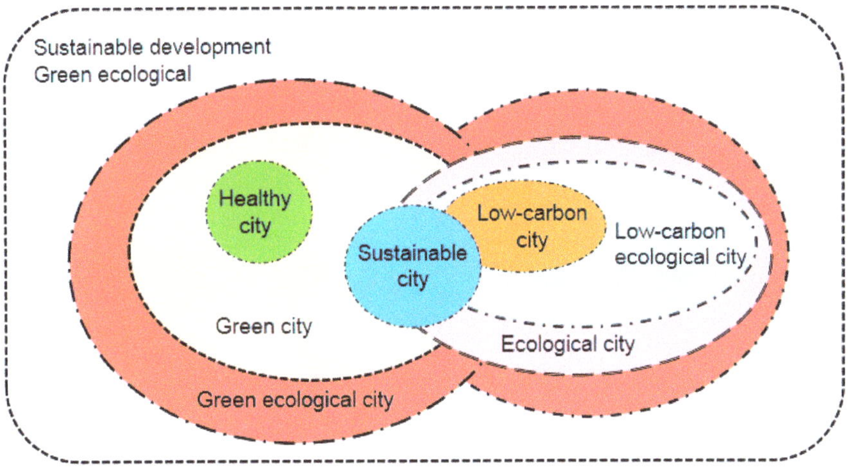

Figure 2. The subordination relationship of various urban public services from the perspective of green ecology.

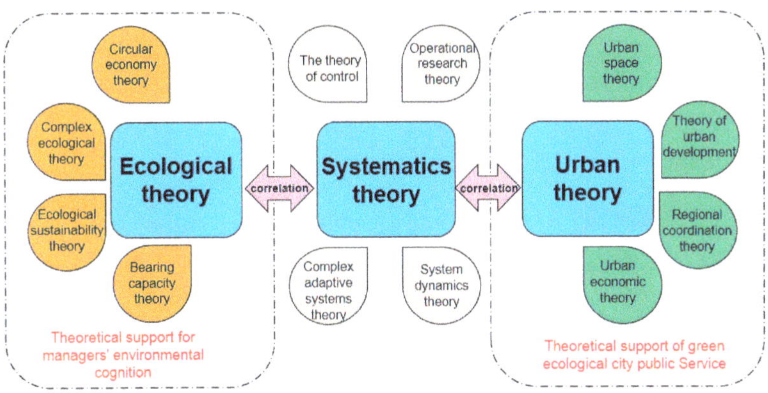

Figure 3. Composition of theoretical connection.

3. Analysis of the Results and Discussion

The public service innovation activity of a green ecological city is systems engineering. This chapter intends to simplify the expression of this complex system by constructing a theoretical model of a green ecological city public service system. Guided by the value objectives and development criteria of a green ecological city, and taking urban structure composition and organizational elements as objects supported by the logic of the urban development model and operation mechanism, it applies the principles, viewpoints, and methods of systems engineering to the research and solution of urban public service problems.

It provides a conceptual model tool to analyze the impact of managers' environmental cognition on urban public service provision; urban planning; management, construction and operation; and long-term sustainable development. The model constructed in this study also more clearly analyzed the specific aspects of the environmental cognition of city managers on green ecological city public service facilities under the background of green ecology, and through what methods.

3.1. Model Construction Criteria

A rule of model building is the goal of this research institute to build a model needed to achieve the results based on an analysis for managers handling the cognitive environment. The building green ecological perspective of the criterion is based on environmental cognition through city planning, construction, and operation management; the influence of these makes the city provide the public service that helps to promote a green ecological city construction [12]. Specifically, the criteria of this model include the general criteria and implementation criteria of a target set constructed at a general and sublevel. The construction criteria of this analysis model are shown in Figure 4.

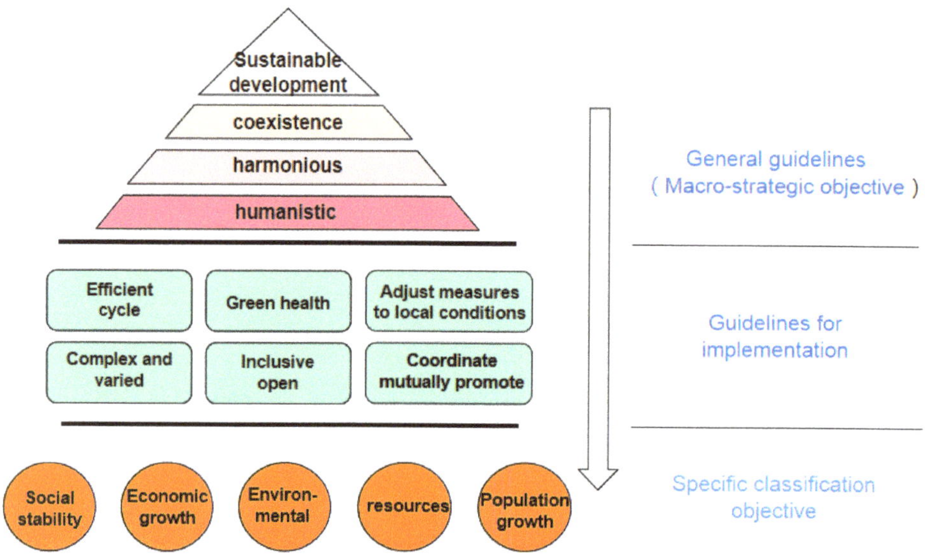

Figure 4. Partition of model construction criteria.

It can be seen from the above figure that, from the perspective of green ecology, the general criteria of green ecological city public service model construction—namely, the macro-strategic goal—are sustainable development, coexistence, harmony, and being people-oriented. Managers must make their own management decisions by providing public services in the city, which achieve the economic constructions, political constructions, cultural constructions, social constructions and ecological civilization constructions of a "five one" all-round development through the mobilizing of all elements of nature, society, and economy in rural towns as well as the ecological whole space, taking into account the whole process of the past, present, and future and realizing the sustainable development of a harmonious coexistence between man and nature, society, and economy [13].

Furthermore, the model was mainly based on six specific implementation criteria to achieve specific classification objectives. Efficient circulation: In the implementation process of a green ecological city public service, there should be efficient circulation among the various subsystems. Various resources and energies can be fully utilized, and coordination between the various systems can be formed to achieve the most effective and reasonable utilization. Green health: Green ecological city public services should provide a safe and healthy urban environment, which can give full play to the creativity and productivity of people and maximize the protection of residents' physical and mental health and living environment quality. The measures should be adjusted to local conditions; in a green ecological city construction of public services, obvious regional differences and characteristics of city managers will inevitably exist in the provision of public services and should fully consider the management of the city itself in terms of the geographical environment, cultural environment, economic environment, technology environment, and business and social environment as well as the differences and features [14]. The green ecological public service system constructed in different cities should be different rather than prevailing in different places and lacking differentiated public service systems. On the one hand, it is difficult to give full play to the fundamental goal of public service systems to serve a local green ecological city construction and the residents; on the other hand, it does not conform to the specific ecological conditions of local cities and other characteristics [15]. Compound diversity: The innovative development of green ecological city public services also needs to follow the principle of compound diversity. On the one hand, the diversification of urban public service forms to achieve the spatial integrity, continuity, and richness of urban public services. On the other hand, the concept of green ecology should be fully upheld to ensure the diversity of humans and nature in urban life and accommodate diverse people and natural creatures [16]. Coordination and mutual promotion: Green ecological city public services should also contribute to coordinated and mutual development among urban residents, the surrounding natural environment, and the local economy as well as society and the livelihood of people. Especially under the environmental cognition of green compound ecology, urban management needs to pay more attention to the benign interaction and development of economic systems, social systems, and natural systems. Inclusive and open: In the complex system of a green ecological city, which is filled with various elements such as the capital, resources, and information, public services from the perspective of green ecology cannot be analyzed from a closed and narrow perspective. The analysis should be carried out with an open attitude, and a diversified and composite open system should be created, which coexists with the natural environment where organisms exist equally and mutually, activities are convenient, and the values of greenness and inclusivity should be advocated.

3.2. Main Influencing Factors

After analyzing the construction criteria of the green ecological city public service system theoretical model combined with the classification of urban managers' environmental cognitive characteristics, we further analyzed the influencing factors on green ecological city public services. The relevant influencing factors were selected and analyzed from the perspective of urban management, centering on the development requirements of urban social, economic, and environmental systems and under the construction principle of a green ecological city public service system model.

For the study of the influencing factors, there were three main factors for different research purposes. One was to divide cities into natural, economic, and social aspects according to the management objectives and discuss the role of managers' environmental cognition on their public services from three aspects [17]. The first category was classified from the characteristics of the public services themselves; the differential influence of managers' environmental cognition was studied according to different types of public services [18]. The other one took the city as an ecosystem and analyzed the role of managers' cognition from the aspects of the system structure, function, and coordination [19].

This study draws on the analysis experience of previous research viewpoints and discusses the impact of managers' environmental cognition on urban public services from five aspects, including social status, economic status, natural environment, resources, and population whilst taking the city as a whole ecosystem. In other words, from the perspective of green ecology, managers' environmental cognition mainly affects urban public services through the above five aspects.

According to the above analysis and the theoretical basis of the connotation of the public service system of a green ecological city, we considered which specific variables would make further responses or changes to the above factors. In a public service system, basic public service industry, ecological protection public service, or public space or site, the public cultural facilities, transportation, social security public service, green science, technology public service, and urbanization construction public service will be affected. According to the function hierarchy and order of the above five factors, the main influencing factors of managers' environmental cognition on urban public services from the perspective of green ecology were constituted [20]. The composition relationship of the specific influencing factors is shown in Figure 5.

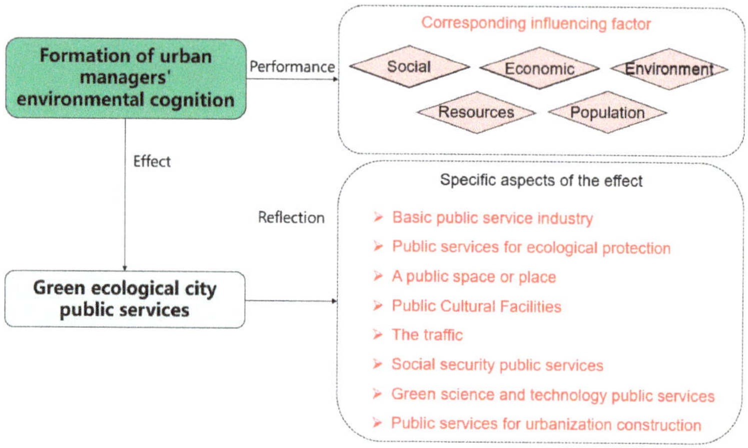

Figure 5. Composition of major influencing factors of managers' environmental cognition on public service innovation.

3.3. Mechanism of the Model

After obtaining the main influencing factors, to build a green ecological urban public service system theory model managers need, through the analysis of the cognitive environment, the specific role of the green ecological city public service innovation mechanism through to the internal organization structure and interaction between the influence factors and the method of analysis. This enables the connected factors to form a final analysis model. Figure 6 depicts the main path of action.

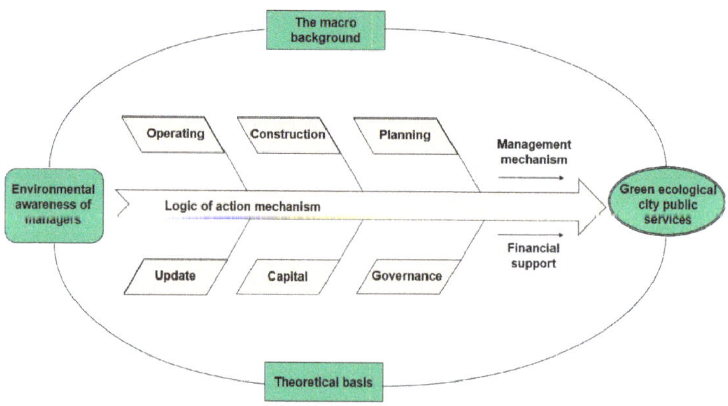

Figure 6. Managers' environment cognition and the mechanism of action of the public service of a green ecological city.

As can be seen from the figure, the mechanism of the impact of managers' environmental cognition on green eco-city public services is a model where the logic of the whole process of the action mechanism is formed by taking the macro-background and theoretical support as the external framework and the operation, construction, planning, renewal, capital, and governance as the internal components. The driving forces, management mechanisms, and financial support are the auxiliary components.

External framework: In terms of the macro-background, the process of urban management includes the economic conditions, cultural structure, national characteristics, geographical environment, historical tradition, social form, and other time characteristics of the urban residents in a specific macro-background. The research on the impact of green ecological city public services also needs to be carried out under a specific background and macro-environment. Theoretical support: The theory technology level also affects and restricts the city development; for the urban planning design and construction of public services management, with enough theory as the guidance, the city builders and managers' cognitive connotation of city development can decide their ability to solve the problem of urban development for different countries and regions with different natures and scales as well as the different stages and cultural cities. The theoretical basis supporting and guiding urban development is also completely different, which requires reasonable screening and selection [21].

Internal composition and driving force: the sustainable development of a city is based on the results of the joint action of various driving forces, which are mutually conditioned and coordinated. Various dynamic factors not only reflect the quality of urban development, but also enrich and transform the development of urban politics, economy, society, and humanity. The rapid development and innovation of urban public services is also inseparable from the creation and renewal of internal compositions and driving forces, so governments need to manage the city from an operational, construction, planning, renewal, capital, and governance point of view along with other aspects. At the same time, from the perspective of green ecology and based on the concept of development, the urban management of city managers should be carried out based on the value goals of environmental cognition. The main body of these urban management activities is not completely the government, nor is it completely dependent on administrative means and other coercive forces. Urban management is the cooperation between governments and urban people, government and non-governmental organizations, public and private institutions, and compulsory and voluntary cooperation. The provision of urban public services is also a process of top-down interactions. The government, non-governmental organizations, and various private institutions deal with public affairs through common goals based on diversified means such as cooperation and consultation, so their power dimensions are diversified rather than pure top-down. Therefore, the development and innovation of public services in a green ecological city is a process in which the government, together with society and other forces, participate in and act together as the driving force for development [22].

Auxiliary means: Managers' environmental cognition also needs the existence of auxiliary mechanisms to realize the mechanism of urban public services, which involves two aspects. On the one hand is the management system. The provision of urban public services and policy operation cannot be separated from the institutional guarantee, especially the urban administrative system, which is based on the economic system and adapts to the current economic system, determining the functional development and positioning of the city. With an economic system, it is important to note the other side of auxiliary means; namely, urban financial support. Urban financial support is the guarantee for a successful implementation of the various measures and means. The system and scale of urban financial revenue and expenditure directly restrict and influence the speed and quality of urban public service development. The construction of a green ecological city, the guarantee of a balanced, equal, reliable, and adaptive provision of public services, the adjustment of urban income distribution and urban economic development, and the improvement of the urban governance capacity are inseparable from the support of urban finance.

The logic mode of the mechanism of action: First of all, managers' environmental cognition affects urban public services in a whole process and managers' influence on green ecological city public services is a whole process practice. Meanwhile, in the practice process, managers should fully combine their own local characteristics and various adjustment differences to obtain a reasonable layout. Therefore, its function is the fusion of a variety of factors through a long period of effect evolution and formation. It is also a balanced path and self-sustaining system to observe the whole process of urban management and public service.

Time effectiveness: Managers also differ in their cognition of green ecology at a time level because the provision of public services is a whole process practice that includes planning, construction, and governance. According to the different stage requirements, managers should give play in an orderly, successive, and phased manner. They should recognize the effects of policy guidance, resource support, and cultural dominance in the early stage; capital drive, market construction, and industrial development in the middle stage; and green ecological location drive and value concept dissemination in the late stage to give full play to the improvement of intergenerational distribution efficiency based on policy recognition.

In general, through the analysis of each part of the model's mechanism, the paper analyzed the operation logic and rules of manager cognition in urban public service innovation, which has a certain stage and reference significance. At present, the values of urban management in the world are changing from the direct pursuit of economic efficiency maximization to the pursuit of social fairness and justice; from materialism to the protection and production of cultural and ecological values. It can be said that the research on the innovation and development of green ecological city public services in this paper came into being with the progress of urban management and environmental changes, which could provide a reference for the development and management of public services in other cities.

3.4. Model

Based on the analysis and construction of the construction criteria, main influencing factors, and mechanism mentioned above, this study comprehensively established a system model describing the impact of managers' environmental cognition on urban public service innovation and development, as shown in Figure 7. Based on this model, feasible concrete measures and influencing factors can be analyzed through the orientation of city managers for green ecological cities in terms of objectives, functions, and operation mechanisms. This paper provides a three-dimensional and comprehensive overview of the green eco-city public service, a complex whole system, and provides a good method and tool for green eco-city management.

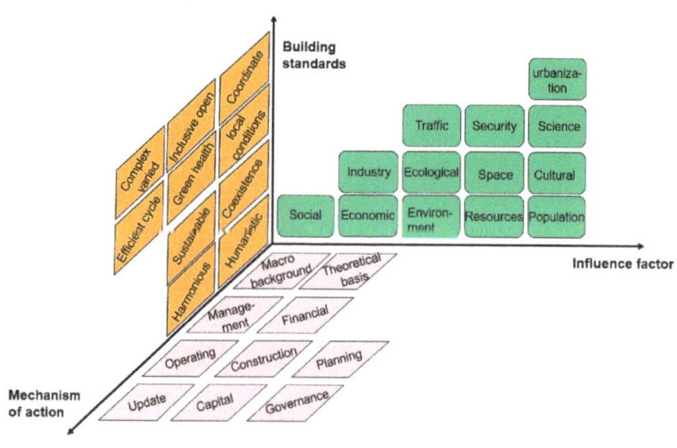

Figure 7. Green ecological urban public service system theory model.

4. Conclusions

Through the research, it was found that the theoretical basis of green ecological city public services mainly came from the three pillars of urbanology, ecology, and systematics. By using the above theories and a practical analysis, the following understandings were obtained. Managers' environmental cognition from the perspective of green ecology is characterized by systematization, openness, and complexity. The main goal of management is diversification, efficiency, compounds, safety, and health. In the management process, there are the development laws of being dynamic and adaptive. The management measures taken are mainly circular and compact, and the value goals of symbiosis, being people-oriented, and harmony are realized by acting on urban public services. A green ecological city public service is a comprehensive and complex system. The extension of "green" and "ecological" concepts requires more scientific, more systematic, and more natural management methods to provide public services. The environmental cognition that urban managers should have include being systematic, open, and complex. Urban management should pursue diversification, efficiency, safety, and health. The theoretical model of a green ecological city public service system constructed in this study analyzed and explained the relationship between managers' behavior and urban public service innovation and development in a relatively three-dimensional and comprehensive way, providing a reference for a better understanding of the system composition as well as the innovation and development laws of green ecological city public services.

Author Contributions: Conceptualization, L.L. and YY.Z.; Formal analysis, YY.Z.; Funding acquisition, L.L.; Methodology, L.L.; Resources, YY.Z.; Writing—original draft, L.L. All authors have read and agreed to the published version of the manuscript.

Funding: National Social Science Foundation of China, Study on the mechanism of government empowerment for improving villagers' livelihood ability in rural tourism destinations. Grant No. 20CJY053.

Data Availability Statement: The labeled dataset used to support the findings of this study is available from the corresponding author upon request.

Conflicts of Interest: The authors declare that there are no conflict of interest.

References

1. Pan, H.; Tang, Z.; Han, J.; Sun, H. From Green to Healthy, Analysis of Humanistic Strategy under Ecological City Space. *IOP Conf. Ser. Earth Environ. Sci.* **2020**, *615*, 012025. [CrossRef]
2. Yang, Q.; Gao, D.; Song, D.; Li, Y. Environmental regulation, pollution reduction and green innovation: The case of the Chinese Water Ecological Civilization City Pilot policy. *Econ. Syst.* **2021**, *45*, 100911. [CrossRef]
3. Su, T. Research on the Allocation of Urban and Rural Public Service Facilities Based on Life Circle Theory—A Case Study of Honghu City, Hubei Province. *Open Access Libr. J.* **2021**, *8*, 7. [CrossRef]
4. Jaung, W.; Carrasco, L.R.; Richards, D.R.; Shaikh, S.F.E.A.; Tan, P.Y. The role of urban nature experiences in sustainable consumption: A transboundary urban ecosystem service. *Environ. Dev. Sustain.* **2022**, 1–21. [CrossRef]
5. Vinci, S.; Egidi, G.; López Gay, A.; Salvati, L. Population Growth and Urban Management in Metropolitan Regions: The Contribution of Natural Balance and Migration to Polycentric Development in Barcelona. *Appl. Spat. Anal. Policy* **2021**, *15*, 71–94. [CrossRef]
6. Simone, C.; Iandolo, F.; Fulco, I.; Loia, F. Rome was not built in a day. Resilience and the eternal city: Insights for urban management. *Cities* **2021**, *110*, 103070. [CrossRef]
7. Suleimany, M. Smart Urban Management and IoT; Paradigm of E-Governance and Technologies in Developing Communities. In Proceedings of the 2021 5th International Conference on Internet of Things and Applications, Isfahan, Iran, 19–20 May 2021.
8. Muafeq, H.; Ghalib, G.; Gaefer, A. Activating the ecological role of urban management tasks in the city environment. *IOP Conf. Ser. Earth Environ. Sci.* **2021**, *790*, 012050. [CrossRef]
9. Xi, X.; Han, F.; Xie, Y.; Yang, L.; Yan, H.; Luo, C.; Sun, L. The Key Factors Influencing the Decline of Carbon Emission Intensity in Low-Carbon Cities and Countermeasure Research—A Case of Fuzhou, Jiangxi. *IOP Conf. Ser. Earth Environ. Sci.* **2021**, *769*, 022040. [CrossRef]
10. Jin, Z. Green city economic efficiency based on cloud computing and machine learning. *Arab. J. Geosci.* **2021**, *14*, 2599. [CrossRef]
11. Yan, J.; Yang, J.; Zhu, F.; Teng, Z. Green city and government ecological environment management based on ZigBee technology. *Environ. Technol. Innov.* **2021**, *23*, 101711. [CrossRef]

12. Kimpouni, V.; Nzila, J.D.D.; Watha-Ndoudy, N.; Bilongo, E.C.K.; Mouhamed, S.Y.; Kampe, J.-P.; Louembe, D. Sociocultural and Ecological Dynamics of Green Spaces in Brazzaville (Congo). *Int. J. Ecol.* **2020**, *2020*, 3719267. [CrossRef]
13. Haider, S.; Hafeez, I. A step toward smart city and green transportation: Eco-friendly waste PET management to enhance adhesion properties of asphalt mixture. *Constr. Build. Mater.* **2021**, *304*, 124702. [CrossRef]
14. Wang, R.; Wu, H.; Chiles, R. Ecosystem Benefits Provision of Green Stormwater Infrastructure in Chinese Sponge Cities. *Environ. Manag.* **2022**, *69*, 558–575. [CrossRef]
15. Shen, F.; Liu, B.; Luo, F.; Wu, C.; Chen, H.; Wei, W. The effect of economic growth target constraints on green technology innovation. *J. Environ. Manag.* **2021**, *292*, 112765. [CrossRef] [PubMed]
16. Babar, M.; Khattak, A.S.; Jan, M.A.; Tariq, M.U. Energy Aware Smart City Management System using Data Analytics and Internet of Things. *Sustain. Energy Technol. Assess.* **2021**, *44*, 100992. [CrossRef]
17. Cai, M.; Shi, Y.; Ren, C. Developing a High-resolution Emission Inventory Tool for Low-Carbon City Management Using Hybrid Method—A Pilot Test in High-density Hong Kong. *Energy Build.* **2020**, *226*, 110376. [CrossRef]
18. Sun, X. Green city and regional environmental economic evaluation based on entropy method and GIS. *Environ. Technol. Innov.* **2021**, *44*, 101667. [CrossRef]
19. Laurini, R. A primer of knowledge management for smart city governance. *Land Use Policy* **2021**, *111*, 104832. [CrossRef]
20. Abhirami, K.S.; Kunnel, A.P.; Archa, R.; Latheef, A.; Bindiya Hari, P. Assessment of Municipal Solid Waste Management in Kochi City. *IJERT—Int. J. Eng. Res. Technol.* **2021**, *9*, 64–68.
21. Huiyi, Z.; Vignesh, C.C.; Daniel, J.A. Influence of Social and Environmental Responsibility in Energy Efficiency Management for Smart City. *J. Interconnect. Netw.* **2021**, *22*, 2141002.
22. Wu, Y. Ecological Smart City Construction Based on Ecological Economy and Network Governance. *Comput. Intell. Neurosci.* **2022**, *2022*, 5682965. [CrossRef] [PubMed]

International Journal of
Environmental Research and Public Health

MDPI

Article

Investigating the Usage Patterns of Park Visitors and Their Driving Factors to Improve Urban Community Parks in China: Taking Jinan City as an Example

Deyi Kong [1], Zujian Chen [1,*], Cheng Li [2] and Xinhui Fei [1]

[1] College of Landscape Architecture and Art, Fujian Agriculture and Forestry University, Cangshan District, Fuzhou 350002, China
[2] College of Arts, Shandong Jianzhu University, Jinan 250101, China
* Correspondence: zujianchen@fafu.edu.cn

Abstract: Urban community parks have significant benefits for city residents, both physical and spiritual. This is especially true in developing countries, such as China. The purpose of our study is to describe the current situation of the community parks in five main districts of Jinan City while recognizing features of the community parks that influence usage patterns. Our study also means to determine the desired improvements of visitors that promote access to and use of community parks on the basis of the Chinese context. We conducted a survey among 542 community park visitors and obtained valid responses. The findings of respondents show that community parks are mostly used by people over 55 years (34.7%) and children under 10 years (23.6%). The main motives for using community parks are for exercise (24.2%) and to socialize with others (21.6%). The majority of respondents (65.7%) rated the community park as satisfactory and considered only a few improvements needed. Regarding the desired improvements, numerous respondents mentioned adding more physical training facilities (13.3%) and activity areas (7.6%), as well as emergency call buttons in areas frequented by children and older people (7.6%). Furthermore, most of the respondents (79.9%) indicated that they would like to use the community parks more frequently if there is additional progress to make the parks more attractive, cleaner, and friendlier. These results can help park designers, government agencies, and community groups to provide the planning and design strategies for community parks to promote their upgrading in China.

Keywords: community parks; urban green space; mixed methods; park usage; driving factors

Citation: Kong, D.; Chen, Z.; Li, C.; Fei, X. Investigating the Usage Patterns of Park Visitors and Their Driving Factors to Improve Urban Community Parks in China: Taking Jinan City as an Example. *Int. J. Environ. Res. Public Health* **2022**, *19*, 15504. https://doi.org/10.3390/ijerph192315504

Academic Editors: Yuan Li, Hongxiao Liu and Tong Wu

Received: 28 October 2022
Accepted: 17 November 2022
Published: 23 November 2022

Publisher's Note: MDPI stays neutral with regard to jurisdictional claims in published maps and institutional affiliations.

1. Introduction

Urbanization is advancing worldwide, and by 2050, 68% of the world's population is projected to live in urban areas [1]. This rapid urbanization has led to an accelerating decline in urban green space [2,3]. Therefore, the planning and design of urban green infrastructure (GI) is attracting increasing global attention, which is a network in the city of interconnected green spaces consisting of patches, corridors, and matrixes to protect the diverse interests of human beings, biodiversity, and habitats of various species [4,5]. In the current situation of accelerated urbanization and surging urban population, plenty of researchers emphasize the introduction of community parks or numerous minor urban green spaces as a critical "patch" of the urban green infrastructure, which can serve the interests of numerous communities [6–8].

On 1 June 2018, the Chinese Ministry of Housing and Urban-Rural Development issued a standard for urban green space classification [9], defining "community park" as an independent area with basic recreation and service facilities and its urban green spaces developed mainly for residents of a particular community to conduct daily recreational activities on 1–10 hectares of land. Community parks have myriad benefits for residents and the city's environment. In terms of residents, they provide more opportunities for people to connect with nature, provide a space for social engagement and exercise, reduce obesity, depression, and anxiety, and positively impact people's bodies and minds [10–12]. In terms of the environment, they have been called the "lungs of the city" because they help provide fresh air, increase biodiversity in the area, improve the microclimate, and mitigate the heat island effect [13–15].

Over the past decade, urban green infrastructure (GI), such as rainwater gardens, linear green spaces, and wetlands, have attracted the attention of numerous researchers, with large public parks being the most studied [16–18]. However, studies on community parks remain limited [19]. Most of them have targeted specific people and content, such as older adults and children [20,21], and several studies have provided valuable insights into residents' preferences for community parks in different thematic styles and how landscape features influence people's access to, interaction with, and satisfaction with the parks [22–24]. Dunton et al. [25] concluded that although a community park area is small, a large number of them are in a city, and they are the most underrated but potentially valuable ecological resources of urban GI. Furthermore, in recent years, urban densification, particularly in developed countries (such as Germany and The United States) and in developing countries in transition (such as China and Malaysia), has strained limited land resources and reduced the opportunity to create larger green spaces in cities [26]. The loss of green spaces has reduced residents' access to nature. This affects the frequency of use, where longer distances reduce the frequency of use, harming public health [27–29]. As a result, numerous cities need to pay more attention and establish more small urban green spaces, such as community parks located around living and working environments, to improve urban ecology, enhance urban quality and improve the quality of life for residents [30,31]. This is vital as community parks are the most exposed green infrastructures in urban areas, with a well-designed and managed community park increasing access to nature and opportunities for health and fitness for diverse communities [32,33].

Even though community parks have been created to benefit urban populations, especially within developing countries such as China, research on their usage and perception is lacking. Currently, in urban construction in China, community park construction is mainly contracted to the design company, which simply passes the designer's landscape awareness to the visitors, and the visitors can only be given local or forced changes after obtaining the right to use the park [16,18,34–37]. Shinew et al. [38] concluded in his study of American urban parks that the subjective initiative of the visitors serves as the constructing and operating logic of community parks. However, the current Chinese community parks violate this logic. Research on visitor usage patterns of and desired improvements to urban community parks is vital, as those small urban green spaces such as community parks that do not meet most visitors' expectations could be easily replaced with other land uses [39,40]. More research on community parks is needed to determine their importance to city residents and to encourage their construction. China is in the midst of an urban renewal movement. The urgency of community park research stems from the fact that it can effectively help residents meet their demands for a better life. It is necessary to study the usage patterns of park visitors and their driving factors to improve urban community parks to guide the revitalization and construction of community parks [41,42].

In addition, research on community parks is mainly focused on North American or European contexts currently [43–45]. Sparse research has been performed on the usage patterns of community parks, especially in the Chinese context. It is essential to diversify this research because people's outdoor activities are often determined by their cultural backgrounds, and there are wide differences among countries in beliefs, politics, and lifestyles [46–48]. The findings and conclusions of studies conducted in North American or European contexts do not apply to China, which is relatively integrated in terms of ethnicity, culture, and religion. Therefore, an understanding of the usage and perception of community parks in China context is necessary [49,50]. In this study, we focus on community parks following the definition given by the Chinese Ministry of Housing and Urban-Rural Development [9]. To the best of the author's knowledge, there is no study of visitor usage patterns and desired improvements in community parks in Jinan City or in any other region of China. Our study is formulated to fill the gap of knowledge in this area through the following questions: (a) Who uses community parks? (b) How do park visitors use the community parks? (c) What are the visitors' desired improvements to the community parks in Jinan City?

2. Methodology

2.1. Study Area

Located in central Shandong Province, Jinan City is the provincial capital and spans a total area of 10,244.45 km^2 (Figure 1). It is geographically located between 36°02′~37°54′ north latitude and 116°21′~117°93′ east longitude. It belongs to the warm temperate continental monsoon climate zone, with abundant sunlight and four distinct seasons. The average temperature is 14.2 °C, and the annual precipitation is 548.7 mm. According to the Seventh National Census (Department of Statistics Jinan City, 2020), there are 9,202,400 permanent residents in Jinan, with 50.13% males and 49.87% females. The urban population was 6,760,000, or 73.46%, and the rural population was 2,442,425, or 26.54%.

Figure 1. Location map of Jinan City (Form Standard map service: http://bzdt.ch.mnr.gov.cn, accessed on 22 February 2022).

As shown in Figure 2, Jinan is divided into nine strategic zones, including seven urban districts (Jiyang, Zhangqiu, Shizhong, Lixia, Huaiyin, Tianqiao, and Licheng) and two counties (Shanghe and Pingyin). As the urban population is mainly concentrated in Shizhong, Lixia, Huaiyin, Tianqiao, and Licheng districts, the field research for our study focuses on these five urban districts.

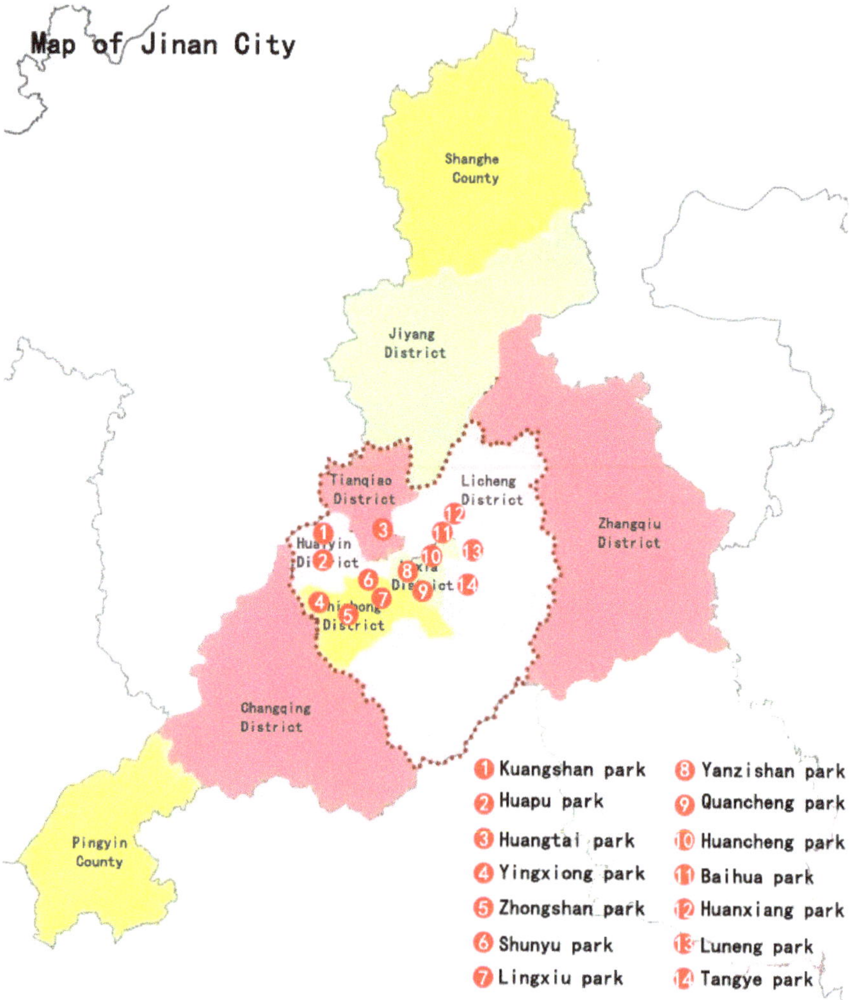

Figure 2. Location map of Jinan community parks.

2.2. Study Site

There are about 16 community parks identified in the five urban districts in Jinan City. Fourteen representative community parks of similar size were selected, excluding two parks under repair and closed. The remaining 14 community parks and their details are shown in Table 1. The sample community parks are free public parks and open to the public 24 h a day. All the community parks have plants, benches, viewpoints, trails, and activity facilities, and both of them could perform the basic functions of beautiful landscapes and entertainment for residents' health.

Table 1. The community parks in Jinan City.

Number	Community Park Real Photos	Community Park Details
1		Name: Kuangshan park Size: 6.9 hm^2 District: Huaiyin district Street name: Jiqing Street GPS location: E 116.954878° N 36.691597°
2		Name: Huapu park Size: 5.8 hm^2 District: Huaiyin district Street name: Jingshi Street GPS location: E 116.949529° N 36.659509°
3		Name: Huangtai park Size: 4.7 hm^2 District: Tianqiao district Street name: Hangyun Street GPS location: E 117.055937° N 36.711579°
4		Name: Yingxiong park Size: 7.2 hm^2 District: Shizhong district Street name: Yingxiongshan Street GPS location: E 117.010969° N 36.645278°

Table 1. *Cont.*

Number	Community Park Real Photos	Community Park Details
5		Name: Zhongshan park Size: 6.3 hm^2 District: Shizhong district Street name: Jingsan street GPS location: E 116.995771° N 36.667601°
6		Name: Shunyun park Size: 4.5 hm^2 District: Shizhong district Street name: Shunyu Street GPS location: E 117.022598° N 36.633229°
7		Name: Lingxiu park Size: 4.1 hm^2 District: Shizhong district Street name: Jianxiu Street GPS location: E 117.009673° N 36.597034°
8		Name: Yanzishan park Size: 4.2 hm^2 District:Lixia district Street name: Peace Street GPS location: E 117.06713° N 36.663794°

Table 1. *Cont.*

Number	Community Park Real Photos	Community Park Details
9		Name: Quancheng park Size: 7.1 hm^2 District: Lixia district Street name: Jingshi Street GPS location: E 117.026361° N 36.651235°
10		Name: Huancheng park Size: 6.4 hm^2 District: Lixia district Street name: Jiefang Street GPS location: E 117.038033° N 36.668754°
11		Name: Baihua park Size: 6.9 hm^2 District: Licheng district Street name: Minziqian stret GPS location: E 117.07713° N 36.681782°
12		Name: Huanxiang park Size:4.5 hm^2 District: Licheng district Street name: Huanxiangdian Street GPS location: E 117.083152° N 36.719796°

Table 1. *Cont.*

Number	Community Park Real Photos	Community Park Details
13		Name: Luneng park Size: 5.3 hm^2 District: Licheng district Street name: Shiji Street GPS location: E 117.21927° N 36.698993°
14		Name: Tangye park Size: 6.5 hm^2 District: Licheng district Street name: Tangye East street GPS location: E 117.235363° N 36.686717°

2.3. Sample Selection and Design

To gauge the visitor usage patterns and desired improvements to community parks in Jinan City, the respondents were randomly selected in 14 community parks and comprised of workers, retirees, homemakers, children, and so on. Chinese community park visitors of all ages (mainly of Han nationality), obtained through an opportunistic sampling strategy, were included as respondents in this study (M = 42 years old, SD > 0.9). Respondents volunteered for the survey and were not offered compensation. Because of the low-risk nature of the study, including this not being focused on patients or people with specific health conditions, no ethical approval process was required. However, we informed the respondents of the purpose of the study before they agreed to be interviewed, and their verbal consent was obtained. They all agreed that their interview results would be published anonymously.

2.4. Survey Instruments and Procedures

This study adopted a mixed research method combining complementary quantitative and qualitative data for collection and analysis. The quantitative data collection component was conducted using interviewer-completed questionnaires, which consisted of four parts: (i) demographic data; (ii) visitors' usage patterns; (iii) motivations for visiting; and (iv) desired improvements in community parks in Jinan. Meanwhile, qualitative semi-structured interviews provided more detailed data on perceptions and desired improvements among visitors to community parks.

The survey was conducted in December 2019 (winter), March 2020 (spring), July 2020 (summer), and October 2020 (autumn) in the 14 community parks. According to Chinese people's physiological habits, excluding lunch and dinner time, the surveys were conducted on both weekdays and weekends, in the early morning (6:00–9:00 a.m.), morning (9:00–12:00 a.m.), afternoon (2:00–5:00 p.m.), and evening (7:00–9:00 p.m.) to obtain a sufficiently representative study sample of community park visitors in Jinan City. A total of 550 questionnaires were handed out and eight completed questionnaires were excluded because of missing information. The remaining 542 questionnaires were available. Meanwhile, we divided eight respondents into a group, and the fourth respondent in each group will be interviewed in depth. If the interviewee is not willing to undergo an in-depth interview, the in-depth interview will be postponed to the fifth or sixth place. Based on this, we also conducted 66 in-depth interviews with specific visitors who were willing to engage in detailed discussions. During the interview, the field researcher provided positive feedback to the participants through nods and acknowledgments, and shared the researcher's personal history, identity, and biases to reassure the respondents.

2.5. Analysis

This study used Excel 2210 and SPSS 2.0 software for descriptive analyses of the collected data. The findings are presented and explained quantitatively and qualitatively to determine visitor usage patterns of and desired improvements to urban community parks in Jinan City, China.

3. Results and Discussion

Compared to other cultures, our study found several similarities and differences in Chinese perceptions and attitudes about community parks and the improvements desired for their use. To the best of the author's knowledge, our study is the first to examine the attitudes of Chinese people, especially residents of Jinan City, toward community parks.

3.1. Who Uses Community Parks in Jinan City?

Demographic Data

As can be seen from Table 2, a total of 542 respondents participated in the survey, including children, students, office workers and retirees, and the majority of them were of Han nationality, which is consistent with the demographic characteristics of Jinan [51]. There were slightly more female visitors than male visitors, which is consistent with the findings of Jasmani et al. [52], Cohen et al. [44] and others. Because female visitors value safety more, community parks meet the conditions of safety, close to home and within their familiar social range, thus they visit them more frequently [12,45]. At the same time, we found in-depth interviews that in addition to safety concerns, additional aspects limited women's ability to visit parks further away, especially the need for domestic labor, resulting in insufficient time. Nordh and Ostby [39] and Cohen et al. [44] also confirmed that women have highly limited "leisure-time" or "self-time", whether they are working women, housewives or even retirees.

People over 55 years old accounted for 34.7% and children below 10 years old accounted for 23.6%; these age groups were the main visitors of the community parks. People aged 20–40 made up only 10.3% and they were the least to visit the parks. Because Chinese urban residents over 55 years mainly are retired people who generally fall into two categories. One group is taking care of their grandchildren, and the other is taking care of their fitness, both of which are reasons for older adults to go to the park [12,53,54]. Conversely, young people aged 20–40 are generally under pressure from work, and lacking time is the main factor limiting their access to parks [22,36]. What is noteworthy in our study is that middle-aged people between 40 and 55 years of age also visited the community parks and accounted for 18.8% of the sample. This is novel but understandable, and Veitch et al. [55] in his research of the relationship between physical activity and cognitive function in middle-aged adults, concluded that most middle-aged people have good family and social relationships and relatively stable employment, allowing them to explore their interests, develop and maintain interpersonal relationships, and shift their focus from career to family. The community park can provide them with an open space for family activities and social communication.

Table 2. Demographic information of the respondents.

Respondent Information	Total	
	Number (n)	Percentage (%)
	Sex?	
Male	264	48.7
Female	278	51.3
	Race?	
Han nationality	510	94.1
Hui nationality	32	5.9
	Age group?	
Under 10 years	128	23.6
10–20 years	68	12.5
20–40 years	56	10.3
40–55 years	102	18.8
Over 55 years	188	34.7
	Occupation?	
Children (≤6 years)	101	18.6
Student	66	12.2
Public sector	64	11.8
Private sector employees	44	8.1
Self-employed	29	5.4
Pensioner	212	39.1
Unemployed	26	4.8

3.2. *How Are the Community Parks Used?*

3.2.1. Visitor Usage Patterns

Table 3 shows that 35.4% of respondents go to community parks more than once a day, and 23.6% visit them 3–4 times a week. The most common travel time is 10–20 min (44.6%) and 5–10 min (33.2%), respectively. According to Dunton et al. [25], a community park is more likely to be visited when it is close to one's residence, and for every 100 m reduction in the distance between home/office and the nearest community park, the probability of visiting the park increases four-fold. The greater the commute from the home/office to a community park, the less frequently it is visited [56,57]. In Shunyu Park, a 58-year-old retired woman also expressed the same opinion:

> *"My grandson went to primary school last year, and now I don't have to look after him all day. So, I go to Shunyu Park, which is close to home and convenient every day, play chess, Tai Chi, and chat with my friends to enjoy time."*

Furthermore, a total of 52.2% of respondents walk to the park, while 31.0% ride bicycle/e-bike. Few people come to community parks by public transport (14.0%), car (2.2%), or taxi (0.6%). This is understandable as community parks are generally located downtown, where there is much traffic during rush hour, and taking cars or buses and other means of transportation can be time-consuming. Furthermore, a majority of community parks in Jinan are close to residential or office areas and can be easily reached by simple walks. This was also observed in the use of community parks among Danes, where 61% of visits are on foot [45], and in Los Angeles, where about 81% of visits are on foot [44]. According to Neuvonen et al. [27], people prefer to drive or take public transportation to large parks, such as botanical gardens, zoological parks, and amusement parks, on weekends or holidays. This is also confirmed in our study. Most respondents (50.4%) show that they would go to community parks on weekdays, while only a small number mentioned they would go to community parks on public holidays (13.7%) and special events (4.7%). In our in-depth interview, a 28-year-old housewife said,

"My husband usually drives the family to the zoo or botanical garden outside the city on weekends, but during the week I simply walk with my son to Lingxiu Park for him to socialize and experience nature. This is really convenient."

Table 3. Analysis of visitor usage pattern.

The Usage Pattern of Visitor	Total	
	Number (n)	Percentage (%)
Would you like to visit this community park alone or in a group?		
Alone	198	36.5
In a group	344	63.5
If in a group, with whom?		
Friends	358	66.1
Family members	176	32.5
Others	8	1.4
How often do you visit this community park?		
≥1 a day	192	35.4
3–4 times a week	128	23.6
1–2 times a week	92	17.0
1–2 times a month	66	12.2
1–2 times a year	64	11.8
How did you get to the park?		
Walk	283	52.2
Bicycle/E-bike	168	31.0
Public transport (bus, subway)	76	14.0
Car	12	2.2
Taxi	3	0.6
How long is the road travel time?		
≤5 min	48	8.9
5–10 min	180	33.2
10–20 min	242	44.6
20–30 min	56	10.3
≥30 min	16	3.0
When would you like to come to this community park?		
Weekdays	273	50.4
Weekends	169	31.2
Public holiday	74	13.7
Special event	26	4.7
What time would you like to visit this community park?		
6:00–9:00 am	89	16.4
9:00–12:00 am	125	23.1
2:00–6:00 pm	223	41.1
7:00–9:00 pm	105	19.4
How long do you normally stay at the park?		
≤30 min	51	9.4
30 min–1 h	217	40.0
1–2 h	254	46.9
≥2 h	20	3.7

What is remarkable in this study is that 12.2% of respondents visited the park once or twice a month and 11.8% once a year, while most of them were young people aged 20–40. This phenomenon is similar to that of Denver, USA, where young people tend to visit community parks less frequently [58]. According to Chen et al. [36], there are generally two reasons. First, lacking time (being overly busy) is the main reason that constraints young people to visit community parks. The other is that the Internet is more attractive than parks. They rarely go to parks unless they feel they are in suboptimal health [59]. This is true, and the findings of our in-depth interviews confirm this view. According to a 33-year-old male programmer in Tangye park:

> *"I prefer to lie down and play the mobile phone or computer games with my friends during my rest time. But since I was diagnosed with moderate fatty liver disease in my physical examination last week, I have decided to run in Tangye Park for an hour after work from now on."*

Staying at home for prolonged periods has negative effects on both physical (e.g., obesity, high blood pressure, and spinal damage) and mental (e.g., depression and social skills reduction) health [60–62]. Attracting people, especially young people, into parks has become vital for both Eastern and Western countries. Many researchers have carried out research and concluded that holding cultural activities in community parks, such as performances and exhibitions, has an extremely positive effect on attracting people to parks [59,63,64]. In addition, Rathore [65] concluded that interesting landscape viewpoints, such as a sea of flowers, exotic art sculptures, and interactive entertainment facilities, are all elements that will attract young people to the park and encourage them to recommend it to their peers.

46.9% of respondents would stay in the community park for 1–2 h, and 40.0% would stay for 30 min–1 h. On the contrary, 3.7% of respondents spent more than 2 h in the community park. This is reasonable, according to the global recommendations on physical activity for the health of the World Health Organization [66], an hour of exercise a day is not only in line with people's physiological habits but also extremely beneficial to physical health. Our survey confirmed that Chinese residents have a good health concept of "exercising one hour a day, reaping happiness a life", and the average time the respondents stayed in community parks was 1 h. In addition, our study found that a small number of respondents would stay in community parks for 2 h or more, mostly because of park activities being held, such as charity sales and performances. In line with the ideas of Veitch [55,67] park activities affect the frequency and length of park visits. It can be concluded that positive correlations between "socializing" and "park activities" indicate that these events also provide more opportunities for people to meet and socialize. In addition, there is one thing worth noting in the study: about 9.4% of respondents spend less than 30 min in community parks, as they usually use community parks as a shortcut to their destination. In the in-depth interview, a 32-year-old male office worker said:

> *"I prefer to pass through Luneng No. 7 Park on the way home from the bus station. Compared with the street sidewalk, the park road is quieter and safer during rush hours, without noise and cars."*

Moreover, a great number of respondents (63.5%) are inclined to visit the community parks in a group, where 66.1% are with friends, and 32.5% tend to visit with family members. This is true regardless of sex, with only a minority tending to visit community parks alone (36.5%). This is similar to previous studies of other Asian cultures, especially Malaysians [40], Turks [68,69] and Italians [70], where the majority of people tend to visit parks accompanied by spouses, friends, children or dogs in a large group. White U.S. visitors, however, were a different story, with about 88% preferring to visit the park alone or with one person [71,72]. This has also been observed in the Netherlands, where they prefer to go to parks in small groups, as couples, or alone [68,73]. In addition, according to Peschardt [45], visitors who want to "socialize" generally go to community parks in a large group, while those who need "rest and recuperation" generally visit alone. At the same time as studying the importance of being alone or with company for the psychological recovery of appreciating different environments, Staats and Hartig [74] found that people enjoy the company of friends in urban environments for a variety of reasons, including safety, but when safety is not an issue; in the absence of a companion, recovery is accelerated. This is also supported by several studies, showing that women tend to visit public parks in the company of family or friends rather than alone because their fear levels are significantly higher than men [48,75,76]. For all that, Gu et al. [37] explored the factors that prevent

Chinese people from visiting urban parks, founding that the lack of a companion is one of the factors hindering Chinese from visiting the park. In addition, Chinese females also prefer to visit parks in a group, as they feel safer when accompanied by family members or acquaintances [77].

According to the survey, residents in Jinan prefer to visit community parks at 2:00–5:00 p.m. (41.1%), followed by 9:00–12:00 a.m. (23.1%) and 7:00–9:00 p.m. (19.4%). There is no doubt about it because Jinan City has a moderate monsoon climate, and the temperatures in the early morning and night are lower; fewer people prefer to go out, especially in winter. This is consistent with our field observations. Community parks were least visited in the early morning (6:00–9:00 a.m.), especially in winter, and the main visitors were young people doing morning jogging activities. In the morning (9:00–12:00 a.m.), community parks ushered in a small peak of people to visit. This is because the old people entered the park doing tai chi, singing, and other recreational activities. In the afternoon (2:00–6:00 p.m.), community parks had the most visitors, including children, young people, middle-aged people, and old people, and all kinds of activities, such as chatting, sunbathing, and exercise, were taking place. However, the most common activity is interaction, including adult-adult interaction, adult-child interaction, and child–child interaction. In the evening (7:00–9:00 p.m.), the flow of people in the community park dropped, mostly middle-aged women gathering in brightly lit areas in the park for square dancing. This is somewhat different from Western or other cultures, such as India [12], Japan [78], and the United States [79]. The vast majority of women do not prefer going to a park at night; the lack of safety is the main limiting factor. On the contrary, 68% of women prefer to go to urban parks in the evening with a companion in China [80]. In our in-depth interview study, a 56-year-old retired female confirmed this view. She said:

> *"I go to Huangtai Park with five or six neighbors every night to dance regularly. The park is near my home, convenient, safe and enjoyable. Now our dancing team has grown to about 20 people, the dynamic music attracting many youthful people to join us in square dancing."*

In China, community parks generally can bring Chinese security even in the evening. The main reason is that in 1994, the Ministry of Public Security of the People's Republic of China issued the Urban People's Police Patrol Regulations [81], stipulating that community police must patrol the area under their control every four hours, which ensures urban and community security to a large extent. Therefore, Chinese residents are willing to go to community parks for recreation in the evening. The findings are directly in line with previous findings that both males and females are more likely to visit parks when they feel comfortable and safe [82]. Crenshaw [83] also concluded in her study that many women praised the lack of harassment and adequate safety as positive attributes that promoted their visits to parks.

3.2.2. Motives for Visiting

As shown in Figure 3, Chinese urban residents visit community parks for a variety of motives, among which the biggest motivation is "to exercise" (24.2%). This is similar to the situation in other countries [84,85]. This demonstrates the significance of these minor green spaces in the city, where they provide residents with free places to exercise conveniently [31,43,45]. It is not surprising that people use community parks mostly for "exercise" in China. That is largely because community parks, as a form of public welfare for urban residents, do not charge entrance fees and are equipped with free fitness equipment (e.g., spacewalk machines, hip twisters, and tai chi hand presses) sponsored by the Chinese Sports Lottery. These conditions encourage residents to use community parks for exercise. Respondents also visited community parks to "socialize with others" (21.6%), and some visited to "play with children" (12.4%). Others use the park to "take a shortcut" (9.6%), to relax and relieve pressure (8.9%), to get close to nature (7.7%), to enjoy beautiful scenery (7.2%), as well as for aimless (6.3%). This shows the benefits of community parks for people. Ward-Thompson et al. [86] also concluded in their study that urban green spaces such as community parks have far-reaching health benefits for residents, including increased sociability and reduced stress. Moreover, it was further stressed that being in a green environment has a positive effect on people of all ages. Lin et al. [87] and Jasmani et al. [52] also reported that natural features in green spaces have some highly beneficial effects, including reducing stress and improving moods. They found that residents prefer to live surrounded by community parks that can play an active role for them. This was echoed in our in-depth interview according to a 26-year-old office worker:

> *"I used to play in the mall with friends, but now we also choose to go to a nearby community park to enjoy nature and take a fresh breath, which makes us cheerful; after all, we always stay inside day and night."*

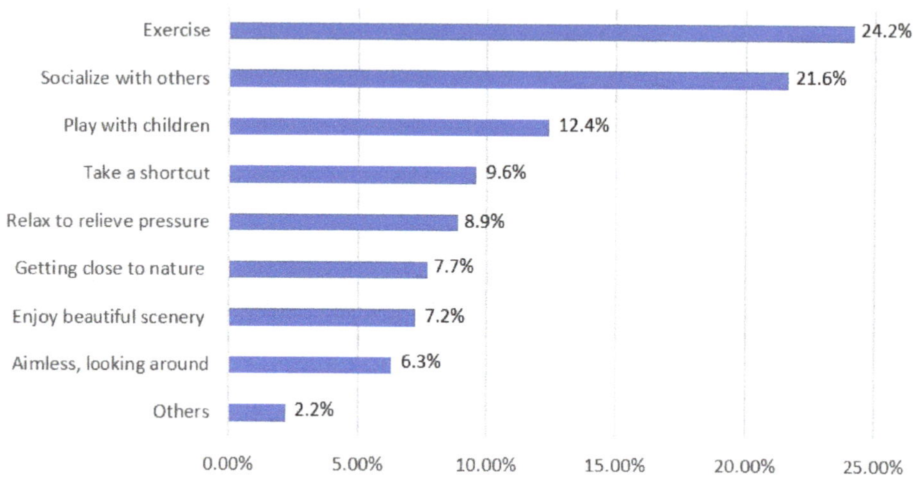

Figure 3. Motives for visiting community parks.

3.3. *What Are the Visitors' Desired Improvements to the Community Parks in Jinan?*

Desired Improvements

As shown in Table 4, most respondents (65.7%) were satisfied with community parks in Jinan City and thought only a small amount of improvements were needed. Some respondents (23.8%) were not satisfied and thought many improvements were needed. For the rest, 8.3% of respondents were very satisfied, and 2.2% were very dissatisfied.

In terms of the desired improvements of community parks, most respondents (13.3%) mentioned the addition of physical training facilities, activity areas (7.6%), activity facilities for children (6.8%), dog/pet activity area (4.8%), and bathroom (4.2%). The findings indicate that there is a short supply relationship between urban residents and available resources in community parks, which is in line with the findings in other countries such as America [88], Denmark [89], Indonesia [90] and Thailand [91]. This is real, and we also observed this in the field, e.g., people lining up to use exercise equipment, square dancers interacting with chess players in the same space, and only one sandpit in the park for children to play. A big reason for this result is community parks' poor planning and design. Vierikko et al. [92] expressed that the suitable space scale of community parks used to guide and control plays a crucial role; residents' activities are also affected by environmental constraints, and how to provide high-quality activity guide users to use the park, healthy, comfortable environment is a designer should be considered. Kruize et al. [85] and Lin et al. [87] also concluded that the function of the community park layout needs to meet the age people use at the same time, should not only provide an open space but also provide a certain space for a particular activity, ensure security and relatively independent activities, make the community park can meet the demand of many groups, multi-type leisure activities.

Williams et al. [93] added that the road system is the core of community parks. Smooth and reasonable park roads can not only serve as the basis for activities such as walking and running but also connect each activity node and organically connect each space. Basu and Nagendra [12] believe that the solution to the problem is to analyze and summarize the core functional needs of a high frequency of daily leisure activities of visitors by studying their usage patterns and desired improvements to urban community parks so as to guide the setting of leisure functions of community parks in the practice process. This is consistent with our research in Jinan City, which emphasizes that the designer of the park should fully understand the surrounding community environment and demographic composition, consider the setting of the park project from the perspective of users, arrange targeted leisure activities and functional sites, and then carry out the overall layout of community park space.

Table 4. Analysis of visitors' desired improvements.

Park Desired Improvements	Total	
	Number (n)	Percentage (%)
Are you satisfied with the community park?		
Very satisfied	45	8.3
Satisfactory, fewer improvements are needed	356	65.7
Unsatisfactory, more improvements are needed	129	23.8
Very unsatisfactory	12	2.2
What is the main desired improvement of you to the community park?		
No changes	32	5.9
Landscape viewpoints	18	3.3
Art sculptures	21	3.9
Vegetation	45	8.3
Unfavorable visitor behaviors	13	2.4
Park cleanliness	24	4.4
Add recycling bins	8	1.5
Add bathroom	23	4.2
Add concession stands	15	2.8
Lighting	35	6.5
Emergency buttons	41	7.6
Information/interpretive signs	17	3.1
Level off the road	13	2.4
Wheelchair accessibility	27	5.0
Add rest seats	34	6.3
Add activity area	41	7.6
Add activity facilities for children	37	6.8
Add physical training facilities	72	13.3
Add dog/pet activity area	26	4.8
Would you like to use community parks more often if changes were implemented?		
Yes	433	79.9
Maybe	64	11.8
No	23	4.2
No response	22	4.1

Furthermore, some respondents mentioned increasing landscape viewpoints (3.3%), art sculptures (3.9%), and vegetation (8.3%), which are related to the beauty of the park. Others mentioned keeping park cleanliness (4.4%), concession stands (2.8%), preventing unfavorable visitor behaviors (2.4%), and adding recycling bins (1.5%), which relates to the management and maintenance of community parks. According to the research of Nordh et al. [39,43] and van den Berg [94], the beauty degree of community parks is positively correlated with the participation of visitors. The research found that park "softscape" was positively correlated with access, while park visitors like parks with abundant "softscapes", such as trees, shrubs, flowers, and water features, but do not want parks with too numerous "hardscapes", such as stairs, roads, and buildings.

Based on the survey findings, the desired improvement, including emergency buttons (7.6%), lighting (6.5%), rest seats (6.3%), wheelchair accessibility (5.0%), information/interpretive signs (3.1%), and level off the road (2.3%). These are related to community park friendliness. In the in-depth interview, a 74-year-old retired professor from Shunyu Park said:

> *"Community parks are usually crowded with elderly people and children, who are more likely to have accidents, thus, the friendliness of the park is very important. It would be better if emergency equipment were installed in the park's activity areas."*

Previous studies have shown that people with poor physical and/or mental status are less inclined to visit community parks due to various limitations, but when parks become more friendly and accessible, they are more willing to visit them, which significantly improves their physical and/or mental status [95,96]. Indeed, taking the findings together, our study concludes that individuals with wheelchairs and/or other special needs need greater friendliness and accessibility to community parks. Finally, respondents were also asked if they would use the parks more if changes were made to make community parks more attractive, cleaner, and friendlier. A majority of respondents (79.9%) responded positively.

4. Conclusions

An investigation into urban community parks is a relevant topic today, especially for developing countries such as China that are growing rapidly. Studies on visitor usage patterns and desired improvements to urban community parks in the context of Chinese society are imperative in diversifying and growing the field of local knowledge. Whether from the view of the effect of planning a community park itself or the development needs of the current urban renewal, it is necessary to experience and study the community park from the perspective of visitors' life cognition and habit preference. Our study uses a combination of quantitative and qualitative analysis to clarify who uses the urban community parks and how, and what their desired improvements toward 14 community parks in Jinan City, China, are. This is helpful for the development of community parks. On the one hand, this highlights the importance of community parks in the daily lives of urban residents. On the other hand, it could be helpful to develop strategies to meet the needs of visitors when planning and designing or redesigning and upgrading community parks. This will ensure that community parks are not displaced by urbanization.

However, this study also has some limitations. Our study was conducted during the COVID-19 pandemic. According to China's epidemic prevention policy on access to public spaces, people should provide a 48 h negative nucleic acid certificate and take their temperature when entering the park. These controls somewhat reduced the frequency with which residents visited community parks, resulting in an incomplete sample size. In addition, it is not a complete study; it represents only selected respondents in community parks and cannot represent all city residents in Jinan City. Furthermore, the field researcher may have had personal biases in selecting respondents, such as unconsciously selecting the same number of men and women and favoring interviews with retirees, which is a potential source of error. We, therefore, suggest that further research should involve more researchers and a larger number of interviews with urban residents (including both visitors and non-visitors to community parks) and expand the sample size to obtain more comprehensive data on community parks. Such research could better prepare for the construction, enhancement, and redevelopment of future urban community parks.

Author Contributions: Conceptualization, D.K. and Z.C.; methodology, D.K.; validation, D.K. and C.L.; formal analysis, D.K.; investigation, D.K.; data curation, D.K. and X.F.; writing—original draft preparation, D.K.; writing—review and editing, Z.C. and C.L.; visualization, D.K. and X.F.; supervision, Z.C. and C.L. All authors have read and agreed to the published version of the manuscript.

Funding: This research received no external funding.

Institutional Review Board Statement: Not applicable.

Informed Consent Statement: Not applicable.

Data Availability Statement: Not applicable.

Acknowledgments: The authors gratefully acknowledge the 542 anonymous respondents in our experiments.

Conflicts of Interest: The authors declare no conflict of interest.

References

1. United Nations. Revision of World Urbanization Prospects 2018. Population Division of the United Nations Department of Economic and Social Affairs (UNDESA). Available online: https://desapublications.un.org/publications (accessed on 28 February 2022).
2. Imam, A.U.K.; Banerjee, U.K. Urbanisation and greening of Indian cities: Problems, practices, and policies. *Ambio* **2016**, *45*, 442–457. [CrossRef] [PubMed]
3. Quaas, M.F.; Smulders, S. Brown growth, green growth, and the efficiency of urbanization. *Environ. Resour. Econ.* **2018**, *71*, 529–549. [CrossRef]
4. Lee, H.; Song, K.; Kim, G.; Chon, J. Flood-adaptive green infrastructure planning for urban resilience. *Landsc. Ecol. Eng.* **2021**, *17*, 427–437. [CrossRef]
5. Olic, P.; Stober, D. Urban green infrastructure for shrinking city: Case study—City of Osijek. In Proceedings of the 3rd World Multidisciplinary Civil Engineering, Architecture, Urban Planning Symposium (WMCAUS), Prague, Czech Republic, 18–22 June 2018.
6. Houston, D.; Zuñiga, M.E. Put a park on it: How freeway caps are reconnecting and greening divided cities. *Cities* **2019**, *85*, 98–109. [CrossRef]
7. Myungshik, C.; Kim, H.W. Identifying priority management area for unexecuted urban parks in long-term by considering the pattern of green infrastructure. *J. Environ. Manag.* **2019**, *27*, 243–265.
8. Peschardt, K.K.; Stigsdotter, U.K. Evidence for designing health promoting pocket parks. *Int. J. Archit. Res. Archnet-Ijar* **2014**, *8*, 149–164. [CrossRef]
9. Chinese Ministry of Housing and Urban-Rural Development Standard for Classification of Urban Green Space, Industry Standard–Urban Construction. Available online: https://www.mohurd.gov.cn (accessed on 22 November 2021).
10. Carlson, S.A.; Brooks, J.D.; Brown, D.R.; Buchner, D.M. Racial/ethnic differences in perceived access, environmental barriers to use, and use of community parks. *Prev. Chronic Dis.* **2010**, *7*, A49.
11. Tzoulas, K.; Korpela, K.; Venn, S.; Yli-Pelkonen, V.; Kaźmierczak, A.; Niemela, J.; James, P. Promoting ecosystem and human health in urban areas using Green Infrastructure: A literature review. *Landsc. Urban Plan.* **2007**, *81*, 167–178. [CrossRef]
12. Basu, S.; Nagendra, H. Perceptions of park visitors on access to urban parks and benefits of green spaces. *Urban For. Urban Green.* **2021**, *57*, 126959. [CrossRef]
13. Sharifi, F.; Nygaard, A.; Stone, W.M.; Levin, I. Green gentrification or gentrified greening: Metropolitan Melbourne. *Land Use Policy* **2021**, *108*, 105577. [CrossRef]
14. Zou, H.; Wang, X.J. Progress and gaps in research on urban green space morphology: A review. *Sustainability* **2021**, *13*, 1202. [CrossRef]
15. Heilmayr, D.; Friedman, H.S. Cultivating healthy trajectories: An experimental study of community gardening and health. *J. Health Psychol.* **2020**, *25*, 2418–2427. [CrossRef] [PubMed]
16. Grahn, P.; Stigsdotter, U.A. Landscape planning and stress. *Urban For. Urban Green.* **2003**, *2*, 001–018. [CrossRef]
17. Zube, H.E.; Moore, G.T. (Eds.) *Advances in Environment, Behavior, Design Volume 1*; Springer Science + Business Media: New York, NY, USA, 1987; p. 77.
18. Chang, Q.; Liu, X.; Wu, J.; He, P. MSPA-based urban green infrastructure planning and management approach for urban sustainability: Case study of Longgang in China. *J. Urban Plan. Dev.* **2015**, *141*, A5014006. [CrossRef]
19. Taylor, R.B.; Haberman, C.P.; Groff, E.R. Urban park crime: Neighborhood context and park features. *J. Crim. Justice* **2019**, *64*, 13–22. [CrossRef]
20. Veitch, J.; Denniss, E.; Ball, K.; Koorts, H.; Deforche, B.; Timperio, A. Increasing translation of research evidence for optimal park design: A qualitative study with stakeholders. *Int. J. Behav. Nutr. Phys. Act.* **2020**, *17*, 49. [CrossRef]
21. Duan, Y.P.; Wagner, P.; Zhang, R.; Wulff, H.; Brehm, W. Physical activity areas in urban parks and their use by the elderly from two cities in China and Germany. *Landsc. Urban Plan.* **2018**, *178*, 261–269. [CrossRef]

22. Kim, G.; Miller, P.A. The impact of green infrastructure on human health and well-being: The example of the huckleberry Trail and the Heritage Community Park and Natural Area in Blacksburg, Virginia. *Sustain. Cities Soc.* **2019**, *48*, 101562. [CrossRef]
23. Li, Y.L.; Fan, S.; Li, K.; Zhang, Y.; Kong, L.; Xie, Y.; Dong, L. Large urban parks summertime cool and wet island intensity and its influencing factors in Beijing, China. *Urban For. Urban Green.* **2021**, *65*, 127375. [CrossRef]
24. Hsu, K.W.; Chao, J.C. Study on the value model of urban green infrastructure development—A case study of the central district of Taichung city. *Sustainability* **2021**, *13*, 7402. [CrossRef]
25. Dunton, G.F.; Almanza, E.; Jerrett, M.; Wolch, J.; Pentz, M.A. Neighborhood park use by children: Use of accelerometry and global positioning systems. *Am. J. Prev. Med.* **2014**, *46*, 136–142. [CrossRef] [PubMed]
26. Anguluri, R.N.; Narayanan, P. Role of green space in urban planning: Outlook towards smart cities. *Urban For. Urban Green.* **2017**, *25*, 58–65. [CrossRef]
27. Neuvonen, M.; Sievänen, T.; Tönnes, S.; Koskela, T. Access to green areas and the frequency of visits—A case study in Helsinki. *Urban For. Urban Green.* **2007**, *6*, 235–247. [CrossRef]
28. Amano, T.; Butt, I.; Peh, K.S.H. The importance of green spaces to public health: A multi-continental analysis. *Ecol. Appl.* **2018**, *28*, 1473–1480. [CrossRef] [PubMed]
29. Choumert, J. An empirical investigation of public choices for green spaces. *Land Use Policy* **2010**, *27*, 1123–1131. [CrossRef]
30. Wu, Z.; Chen, R.; Meadows, M.E.; Sen Gupta, D.; Xu, D. Changing urban green spaces in Shanghai: Trends, drivers and policy implications. *Land Use Policy* **2019**, *87*, 104080. [CrossRef]
31. Liu, H.X.; Li, F.; Li, J.; Zhang, Y. The relationships between urban parks, residents' physical activity, and mental health benefits: A case study from Beijing, China. *J. Environ. Manag.* **2017**, *190*, 223–230. [CrossRef]
32. Curtis, D.S.; Rigolon, A.; Schmalz, D.L.; Brown, B.B. Policy and environmental predictors of park visits during the first months of the COVID-19 pandemic: Getting out while staying in. *Environ. Behav.* **2022**, *54*, 487–515. [CrossRef]
33. Geng, D.C.; Innes, J.; Wu, W.; Wang, G. Impacts of COVID-19 pandemic on urban park visitation: A global analysis. *J. For. Res.* **2021**, *32*, 553–567. [CrossRef]
34. Huang, S.C.L. Park user preferences for establishing a sustainable forest park in Taipei, Taiwan. *Urban For. Urban Green.* **2014**, *13*, 839–845. [CrossRef]
35. Lin, B.S.; Lin, C.T. Preliminary study of the influence of the spatial arrangement of urban parks on local temperature reduction. *Urban For. Urban Green.* **2016**, *20*, 348–357. [CrossRef]
36. Chen, C.; Luo, W.; Li, H.; Zhang, D.; Kang, N.; Yang, X.; Xia, Y. Impact of perception of green space for health promotion on willingness to use parks and actual use among young urban residents. *Int. J. Environ. Res. Public Health* **2020**, *17*, 5560. [CrossRef] [PubMed]
37. Gu, X.; Li, Q.; Chand, S. Factors influencing residents' access to and use of country parks in Shanghai, China. *Cities* **2020**, *97*, 102501. [CrossRef]
38. Shinew, K.J.; Stodolska, M.; Floyd, M.; Hibbler, D.; Allison, M.; Johnson, C.; Santos, C. Race and ethnicity in leisure behavior: Where have we been and where do we need to go? *Leis. Sci.* **2006**, *28*, 403–408. [CrossRef]
39. Nordh, H.; Ostby, K. Pocket parks for people—A study of park design and use. *Urban For. Urban Green.* **2013**, *12*, 12–17. [CrossRef]
40. Balai Kerishnan, P.B.; Maruthaveeran, S.; Maulan, S. Investigating the usability pattern and constraints of pocket parks in Kuala Lumpur, Malaysia. *Urban For. Urban Green.* **2020**, *50*, 126647. [CrossRef]
41. Baran, P.K.; Smith, W.R.; Moore, R.C.; Floyd, M.F.; Bocarro, J.N.; Cosco, N.G.; Danninger, T.M. Park Use among Youth and Adults: Examination of Individual, Social, and Urban Form Factors. *Environ. Behav.* **2014**, *46*, 768–800. [CrossRef]
42. Xiong, J. Research on the relationship between green growth and urbanization efficiency. *Fresenius Environ. Bull.* **2021**, *30*, 1764–1770.
43. Nordh, H.; Hartig, T.; Hagerhall, C.M.; Fry, G.L. Components of small urban parks that predict the possibility for restoration. *Urban For. Urban Green.* **2009**, *8*, 225–235. [CrossRef]
44. Cohen, D.A.; Marsh, T.; Williamson, S.; Han, B.; Derose, K.P.; Golinelli, D.; McKenzie, T.L. The Potential for Pocket Parks to Increase Physical Activity. *Am. J. Health Promot.* **2014**, *28*, S19–S26. [CrossRef]
45. Peschardt, K.K.; Schipperjin, J.; Stigsdotter, U.K. Use of small public urban green spaces (SPUGS). *Urban For. Urban Green.* **2012**, *11*, 235–244. [CrossRef]
46. Ly, T.P.; Zhang, C.Z. Why public-private cooperation is not prevalent in national parks within centralised countries. *Asia Pac. J. Tour. Res.* **2019**, *24*, 1109–1125. [CrossRef]
47. Ma, X.J.; Xie, J.; Jennings, G. Health, identification and pleasure: An ethnographic study on the self-management and construction of Taijiquan Park culture space. *Int. J. Environ. Res. Public Health* **2021**, *18*, 8860. [CrossRef] [PubMed]
48. Özgüner, H. Cultural differences in attitudes towards urban parks and green spaces. *Landsc. Res.* **2011**, *36*, 599–620. [CrossRef]
49. Gómez, E.; Baur, J.W.R.; Hill, E.; Georgiev, S. Urban parks and psychological sense of community. *J. Leis. Res.* **2015**, *47*, 388–398. [CrossRef]
50. Brown, G.; Rhodes, J.; Dade, M. An evaluation of participatory mapping methods to assess urban park benefits. *Landsc. Urban Plan.* **2018**, *178*, 18–31. [CrossRef]
51. Department of Statistics Jinan City Demographic Statistics: Jinan Seventh National Population Census Bulletin. Available online: http://jntj.jinan.gov.cn (accessed on 1 November 2021).

52. Jasmani, Z.; Ravn, H.P.; van den Bosch, C.C.K. The influence of small urban parks characteristics on bird diversity: A case study of Petaling Jaya, Malaysia. *Urban Ecosyst.* **2017**, *20*, 227–243. [CrossRef]
53. Strum, R.; Cohen, D. Proximity to urban parks and mental health. *J. Ment. Health Policy Econ.* **2014**, *17*, 19–24.
54. Menec, V.H.; Brown, C.L.; Newall, N.E.; Nowicki, S. How important is having amenities within walking distance to middle-aged and older adults, and does the perceived importance relate to walking? *J. Aging Health* **2016**, *28*, 546–567. [CrossRef]
55. Veitch, J.; Christian, H.; Carver, A.; Salmon, J. Physical activity benefits from taking your dog to the park. *Landsc. Urban Plan.* **2019**, *185*, 173–179. [CrossRef]
56. Chiang, Y.C.; Li, D.Y. Metric or topological proximity? The associations among proximity to parks, the frequency of residents' visits to parks, and perceived stress. *Urban For. Urban Green.* **2019**, *38*, 205–214. [CrossRef]
57. Tu, X.Y.; Huang, G.; Wu, J.; Guo, X. How do travel distance and park size influence urban park visits? *Urban For. Urban Green.* **2020**, *52*, 126689. [CrossRef]
58. Rigolon, A. Parks and young people: An environmental justice study of park proximity, acreage, and quality in Denver, Colorado. *Landsc. Urban Plan.* **2017**, *165*, 73–83. [CrossRef]
59. Luo, J.M.; Vu, H.Q.; Li, G.; Law, R. Topic modelling for theme park online reviews: Analysis of Disneyland. *J. Travel Tour. Mark.* **2020**, *37*, 272–285. [CrossRef]
60. Richardson, E.; Pearce, J.; Mitchell, R.; Day, P.; Kingham, S. The association between green space and cause-specific mortality in urban New Zealand: An ecological analysis of green space utility. *BMC Public Health* **2010**, *10*, 240. [CrossRef]
61. Madureira, H.; Nunes, F.; Oliveira, J.V.; Cormier, L.; Madureira, T. Urban residents' beliefs concerning green space benefits in four cities in France and Portugal. *Urban For. Urban Green.* **2015**, *14*, 56–64. [CrossRef]
62. Wang, J.; Choe, Y.; Song, H. Brand behavioral intentions of a theme park in China: An application of brand experience. *Sustainability* **2020**, *12*, 4500. [CrossRef]
63. Cabanas, E. Experiencing designs and designing experiences: Emotions and theme parks from a symbolic interactionist perspective. *J. Destin. Mark. Manag.* **2020**, *16*, 100330. [CrossRef]
64. Zhang, L.; Liu, S.; Liu, S. Mechanisms underlying the effects of landscape features of urban community parks on health-related feelings of users. *Int. J. Environ. Res. Public Health* **2021**, *18*, 7888. [CrossRef]
65. Rathore, M.M.; Ahmad, A.; Paul, A.; Rho, S. Urban planning and building smart cities based on the Internet of Things using Big Data analytics. *Comput. Netw.* **2016**, *101*, 63–80. [CrossRef]
66. World Health Organization. Global Recommendations on Physical Activity for Health. Available online: https://www.who.int/publications (accessed on 28 February 2022).
67. Veitch, J.; Rodwell, L.; Abbott, G.; Carver, A.; Flowers, E.; Crawford, D. Are park availability and satisfaction with neighbourhood parks associated with physical activity and time spent outdoors? *BMC Public Health* **2021**, *21*, 306. [CrossRef] [PubMed]
68. Peters, B.E.; Buijs, A. Social interactions in urban parks: Stimulating social cohesion? *Urban For. Urban Green.* **2010**, *9*, 93–100. [CrossRef]
69. Cengiz, C.; Cengiz, B.; Bekci, B. Environmental quality analysis for sustainable urban public green spaces management in Bartm, Turkey. *J. Food Agric. Environ.* **2013**, *10*, 938–946.
70. Sanesi, G.; Chiarello, F. Residents and urban green spaces: The case of Bari. *Urban For. Urban Green.* **2006**, *4*, 125–134. [CrossRef]
71. Loukaitou-Sideris, A. Urban form and context: Cultural differentiation in the uses of urban parks. *J. Plan. Educ. Res.* **1995**, *14*, 89–102. [CrossRef]
72. Gobster, P.H. Managing urban parks for a racially and ethnically diverse clientele. *Leis. Sci.* **2002**, *24*, 143–159. [CrossRef]
73. Burgess, J.; Harrison, C.M.; Limb, M. People, parks and the urban green: A study of popular meanings and values for open spaces in the city. *Urban Stud.* **1988**, *25*, 455–473. [CrossRef]
74. Staats, H.; Hartig, T. Alone or with a friend: A social context for psychological restoration and environmental preferences. *J. Environ. Psychol.* **2004**, *24*, 199–211. [CrossRef]
75. Mac Naghten, P.; Urry, J. Bodies in the woods. *Body Soc.* **2000**, *6*, 166–182. [CrossRef]
76. Sreetheran, M.; van den Bosch, C.C.K. A socio-ecological exploration of fear of crime in urban green spaces—A systematic review. *Urban For. Urban Green.* **2014**, *13*, 1–18. [CrossRef]
77. Wei, F. Greener urbanization? Changing accessibility to parks in China. *Landsc. Urban Plan.* **2017**, *157*, 542–552. [CrossRef]
78. Yasumoto, S.; Jones, A.; Shimizu, C. Longitudinal trends in equity of park accessibility in Yokohama, Japan: An investigation into the role of causal mechanisms. *Environ. Plan. A Econ. Space* **2014**, *46*, 682–699. [CrossRef]
79. Sonti, N.F.; Campbell, L.K.; Svendsen, E.S.; Johnson, M.L.; Novem Auyeung, D.S. Fear and fascination: Use and perceptions of New York City's forests, wetlands, and landscaped park areas. *Urban For. Urban Green.* **2020**, *49*, 126601. [CrossRef]
80. Chen, B.; Qi, X. Protest response and contingent valuation of an urban forest park in Fuzhou city, China. *Urban For. Urban Green.* **2018**, *29*, 68–76. [CrossRef]
81. Ministry of Public Security Urban People's Police Patrol Regulations. Available online: https://app.mps.gov.cn (accessed on 24 February 1994).
82. Van Hecke, L.; Van Cauwenberg, J.; Clarys, P.; Van Dyck, D.; Veitch, J.; Deforche, B. Active Use of Parks in Flanders (Belgium): An Exploratory Observational Study. *Int. J. Environ. Res. Public Health* **2017**, *14*, 35. [CrossRef]
83. Crenshaw, K. Mapping the margins: Intersectionality, identity politics, and violence against women of color. *Stanf. Law Rev.* **1991**, *43*, 1241. [CrossRef]

84. Koskela, H.; Pain, R. Revisiting fear and place: Women's fear of attack and the built environment. *Geoforum* **2000**, *31*, 269–280. [CrossRef]
85. Kruize, H.; van der Vliet, N.; Staatsen, B.; Bell, R.; Chiabai, A.; Muiños, G.; Higgins, S.; Quiroga, S.; Martinez-Juarez, P.; Aberg Yngwe, M.; et al. Urban Green Space: Creating a Triple Win for Environmental Sustainability, Health, and Health Equity through Behavior Change. *Int. J. Environ. Res. Public Health* **2019**, *16*, 4403. [CrossRef]
86. Ward-Thompson, C.; Roe, J.; Aspinall, P.; Mitchell, R.; Clow, A.; Miller, D. More green space is linked to less stress in deprived communities: Evidence from salivary cortisol patterns. *Landsc. Urban Plan.* **2012**, *105*, 221–229. [CrossRef]
87. Lin, W.; Chen, Q.; Jiang, M.; Zhang, X.; Liu, Z.; Tao, J.; Wu, L.; Xu, S.; Kang, Y.; Zeng, Q. The effect of green space behaviour and per capita area in small urban green spaces on psychophysiological responses. *Landsc. Urban Plan.* **2019**, *192*, 103637. [CrossRef]
88. Payne, L.L.; Mowen, A.J.; Orsega-Smith, E. An Examination of park preferences and behaviors among urban residents: The role of residential location, race, and age. *Leis. Sci.* **2002**, *24*, 181–198. [CrossRef]
89. Larsen, F.W.; Petersen, A.H.; Strange, N.; Lund, M.P.; Rahbek, C.A. Quantitative analysis of biodiversity and the recreational value of potential national parks in Denmark. *Environ. Manag.* **2008**, *41*, 685–695. [CrossRef]
90. Indrawan, M.; Lowe, C.; Sundjaya; Hutabarat, C.; Black, A. Co-management and the creation of national parks in Indonesia: Positive lessons learned from the Togean Islands National Park. *J. Environ. Plan. Manag.* **2014**, *57*, 1183–1199. [CrossRef]
91. Redford, T.; Banpot, S.; Padungpong, C.; Bidayabha, T. Records of Spotted Linsang Prionodon pardicolor from Thap Lan and Pang Sida National Parks, Thailand. *Small Carniv. Conserv.* **2011**, *44*, 22–24.
92. Vierikko, K.; Gonçalves, P.; Haase, D.; Elands, B.; Ioja, C.; Jaatsi, M.; Pieniniemi, M.; Lindgren, J.; Grilo, F.; Santos-Reis, M.; et al. Biocultural diversity (BCD) in European cities—Interactions between motivations, experiences and environment in public parks. *Urban For. Urban Green* **2020**, *48*, 126501. [CrossRef]
93. Williams, R.; Vogelsong, H.; Green, G.; Cordell, K. Outdoor Recreation Participation of People with Mobility Disabilities: Selected Results of the National Survey of Recreation and the Environment. *J. Park Recreat. Adm.* **2004**, *22*, 84–100.
94. van den Berg, P.; Weijs-Perrée, M.; Dane, G.; van Vliet, E.; Liu, H.; Sun, S.; Borgers, A. A Comparative Study of Urban Park Preferences in China and The Netherlands. *Int. J. Environ. Res. Public Health* **2022**, *19*, 4632. [CrossRef]
95. Smiley, K.T.; Sharma, T.; Steinberg, A.; Hodges-Copple, S.; Jacobson, E.; Matveeva, L. More inclusive parks planning: Park quality and preferences for park access and amenities. *Environ. Justice* **2016**, *9*, 1–7. [CrossRef]
96. Freudenberg, P.; Arlinghaus, R. Benefits and constraints of outdoor recreation for people with physical disabilities: Inferences from recreational fishing. *Leis. Sci.* **2009**, *32*, 55–71. [CrossRef]

International Journal of
Environmental Research and Public Health

MDPI

Article

How Do Landscape Heterogeneity, Community Structure, and Topographical Factors Contribute to the Plant Diversity of Urban Remnant Vegetation at Different Scales?

Xingzhao Liu [1,2], Guimei Yang [1], Qingmin Que [3], Qi Wang [4], Zengke Zhang [5] and Liujing Huang [1,*]

1 College of Landscape Architecture and Art, Fujian Agriculture and Forestry University, Fuzhou 350002, China
2 Innovation Center of Engineering Technology for Monitoring and Restoration of Ecological Fragile Areas in Southeast China, Ministry of Natural Resources, Fuzhou 350013, China
3 College of Forestry and Landscape Architecture, South China Agriculture University, Guangzhou 510642, China
4 School of Modern Agriculture and Environment, Weifang Institute of Technology, Weifang 262500, China
5 School of Ecological and Environmental Sciences, East China Normal University, Shanghai 200241, China
* Correspondence: huanglj@fafu.edu.cn

Abstract: In highly fragmented urban areas, plant diversity of remnant vegetation may depend not only on community structure and topographical factors, but also on landscape heterogeneity. Different buffer radius settings can affect the relative importance of these factors to plant diversity. The aim of this study was to examine the relative importance of landscape heterogeneity, community structure, and topographical factors on plant diversity under different buffer radii in biodiversity hotspots. We established 48 plots of remnant vegetation in Guangzhou city, one of the biodiversity hotspots. A buffer radius of 500 m, 1000 m, and 2000 m was established around the center of each sample plot, and 17 landscape heterogeneity indices in each buffer were calculated by FRAGSTATS 4.2 software. Combined with the community structure and topographical factors, the impact factors of plant diversity under different buffer radii were analyzed by multiple regression analysis. We found the following: (1) The combined explanatory power of the three factors accounted for 43% of the species diversity indices and 62% of the richness index at its peak. The three impact factors rarely act independently and usually create comprehensive cumulative effects. (2) Scale does matter in urban landscape studies. At a 500 m buffer radius, community structure combined with road disturbance indices was strongly related to diversity indices in herb and shrub layers. The stand age was negatively correlated with the tree-layer richness index. As the scale increased, the diversity indices and richness index in the three layers decreased or increased under the influence of comprehensive factors. (3) The richness index in the herb layer was more responsive to impact factors than other biodiversity indices. Except for the herb layer, the interpretation of landscape heterogeneity for each plant diversity index was more stable than that for the other two factors. Road disturbance indices, combined with the other six landscape pattern metrics, can well indicate species diversity and richness. We suggest that the vegetation area of remnant patches within a radius of 500–2000 m should be appropriately increased to protect plant diversity, and the negative effects of road disturbance should also be considered.

Keywords: landscape pattern metrics; community structure; topographic; diversity; richness; buffer radius

Citation: Liu, X.; Yang, G.; Que, Q.; Wang, Q.; Zhang, Z.; Huang, L. How Do Landscape Heterogeneity, Community Structure, and Topographical Factors Contribute to the Plant Diversity of Urban Remnant Vegetation at Different Scales? *Int. J. Environ. Res. Public Health* **2022**, *19*, 14302. https://doi.org/10.3390/ijerph192114302

Academic Editors: Yuan Li, Hongxiao Liu and Tong Wu

Received: 2 September 2022
Accepted: 31 October 2022
Published: 1 November 2022

1. Introduction

Urbanization is one of the main factors affecting plant diversity at the regional scale [1–4]. However, ecologists have avoided studying urban areas for most of the 20th Century [5]. Recently, there has been a growing research interest in studying plant diversity change caused by the acceleration of global urbanization [6,7]. Urbanization has rapidly spread across biodiversity hotspots [5,8], which often refer to urban vegetation that becomes a refuge for certain species [9]. Urban vegetation has been generally divided into three types: natural, seminatural, and cultivated vegetation [10]. Natural vegetation and seminatural vegetation, especially remnant secondary forests and urban weeds (dominated by spontaneously growing vascular species), are more likely to be a refuge for some species than cultivated vegetation, and the loss of plant diversity in these vegetation types is more likely to cause extinction debt [11]. Therefore, with the rapid development of global urbanization (68% of the global population will live in cities by 2050) [12], it is necessary to study the mechanism of the impact of urbanization on plant diversity in remnant vegetation, and to provide suggestions for protecting and restoring the high quality of urban vegetation.

There are two inconsistent views on the impact of urban landscape heterogeneity on plant diversity. On the one hand, landscape heterogeneity leads to the gradual loss of large areas of biological habitats, which directly reduces species diversity [13–16]. On the other hand, diverse habitats created by landscape heterogeneity are conducive to the settlement and survival of organisms from various sources, which support higher species diversity [17–20]. Landscape heterogeneity, with respect to area, shape, configuration, and quality of the effect, such as edge effect, cluster effect, and fragmentation effect, has been proven to influence the alpha (α), beta (β), and gamma (γ) of plant species diversity [21–25]. Hence, landscape heterogeneity around the habitat is one of the direct driving forces for the change in plant diversity caused by urbanization. However, how local biodiversity responds remains unclear [26,27]. Moreover, community structure, such as the height, coverage, and age of stands, has been demonstrated to affect plant diversity [28,29]. For vegetation with a community structure of three layers or more, the diversity distribution patterns of different levels are different. Topographical factors such as elevation, slope, and aspect have also been shown to affect plant diversity in specific habitats [30–32]. Therefore, more specifically, it is necessary to reveal the relative importance of landscape heterogeneity, community structure, and topographic factors in influencing plant diversity to protect plant diversity.

Evidently, the differences in landscape metrics and study scale can affect the changes and responses of biodiversity, but there is no clear conclusion. Most scholars consider that plant diversity change is caused by various urbanization factors at different scales [6,23,25]. On a large scale (urban, regional, and global), the urbanization process and socioeconomic factors are the dominant factors affecting urban plant diversity. On a small scale, landscape patterns and habitat conditions are the dominant factors affecting urban plant diversity [33–35]. Landscape heterogeneity has long been considered a key determinant of biodiversity. Previous studies have not considered the scale-dependent landscape heterogeneity of local habitats [23,27,36], but the research focus has been on the time lag in biodiversity response to landscape changes and the mechanisms causing extinction debt [37,38]. In the past, studies have addressed the biodiversity in natural forest landscapes [39], agricultural landscapes, and other types of landscapes [40]; however, studies on the biodiversity in urban landscapes are lacking in the literature. Therefore, the mechanism of the scale effect and the relative importance of landscape heterogeneity and habitat conditions on plant diversity needs to be investigated. Different landscape structure parameters, such as landscape heterogeneity, landscape connectivity, landscape complexity, and landscape matrix, may have different degrees of influence on regional biodiversity and have certain scale effects [25,41,42].

How are species richness and species diversity differentially driven by the spatial scale in landscape ecology? There is still a lack of unified conclusions about the response of different layers in stands to landscape heterogeneity. For the rich and sensitive hotspots of biodiversity on Earth, it is necessary to reveal the impact of specific landscape types on biodiversity and its mechanisms to provide a reference for the protection and utilization of sustainable biodiversity. Therefore, it is necessary to combine the method of multiscale analysis to explore the distribution patterns of biodiversity on local and even regional scales from a multidimensional perspective, and to explore the commonness and regularity of biodiversity distribution on the global scale. At the same time, research should elucidate key factors, explore the mechanisms, and provide a scientific basis for biodiversity protection and management against the background of rapid and high urbanization.

China's urbanization has attracted worldwide attention. Approximately 13% of China's urban land is located in these biodiversity hotspots, especially in Guangdong Province, which accounts for two-thirds of the urban land in this hotspot [43]. As the capital of Guangdong Province, Guangzhou is one of the most densely populated and highly urbanized cities, with a resident population of 14.5 million and an urbanization rate of 86.14% [44]. In addition, the city has abundant plant resources in its built-up area; there are 572 species belonging to 123 families and 386 genera of vascular plants, including 228 species belonging to 61 families and 171 genera of herbaceous plants [45]. Most of these plants are distributed in the remnant vegetation in the urban area. However, there is only 290 km^2 of natural vegetation in Guangzhou city, which is located in reserves, fengshui forests, and scenic forests, accounting for 9.4% of the total forest area (mainly plantation area) and 3.9% of Guangzhou city [46]; grassland and artificial forests occupy most of the urban vegetation. In this study, we considered remnant vegetation with less human disturbance scattered in different urban landscape patterns as research objects, and explored how landscape heterogeneity, community structure, and topographic factors affect plant diversity under different buffer radii in urbanization landscapes.

2. Data and Methods

2.1. Study Area

Guangzhou is at the heart of the most-populous built-up metropolitan area in mainland China, and is ranked as an alpha global city. It has 11 districts with an administrative area of 7434.4 km^2. Its permanent population is 14.5 million, and its permanent population density is 1950 person/km^2 [44], which is highly concentrated in central built-up areas. The study area is focused on the central built-up part of Guangzhou city. This core area is composed of two old districts (Yuexiu and Liwan), three medium-age developing districts (Baiyun, Tianhe, and Huangpu), and one young district (Luogang). The city's green rate of built-up area is 37.5%, and the forest stock volume is 17 million m^3 [47]. Guangzhou is located in the north subtropical-humid climatic zone, which is influenced by the regional dominant monsoon weather. The mean annual precipitation attained is 2035.2 mm, and the mean air temperature is 22.6 °C [44]. The mild climate enables vegetation to continuously grow all year round. The natural vegetation in this area before urbanization was subtropical evergreen broad-leaved forests, which have been destroyed by several millennia of agricultural activities and recent urbanization. The different original natural setup, development history, urban fabric, habitat conditions, and management systems have resulted in an uneven spatial pattern of urban vegetation among land uses and administrative districts [48]. In southern China, the climax vegetation is subtropical evergreen broadleaved forests.

2.2. Sampling and Vegetation Survey

To highlight the impact of urbanization (nonhuman subjective selection) on vegetation, remnant vegetation dominated by spontaneously growing vascular species was selected, including 3 types: remnant secondary coniferous and broad-leaved mixed forests, broad-leaved forests, and urban weeds. The sample sites were selected based on satellite images, land use status maps, and field surveys. A total of 16 sample sites under different landscape patterns in Guangzhou (Figure 1 and Table 1), including 4 urban weeds, 3 secondary coniferous and broad-leaved mixed forests, and 9 secondary broad-leaved forests, were selected.

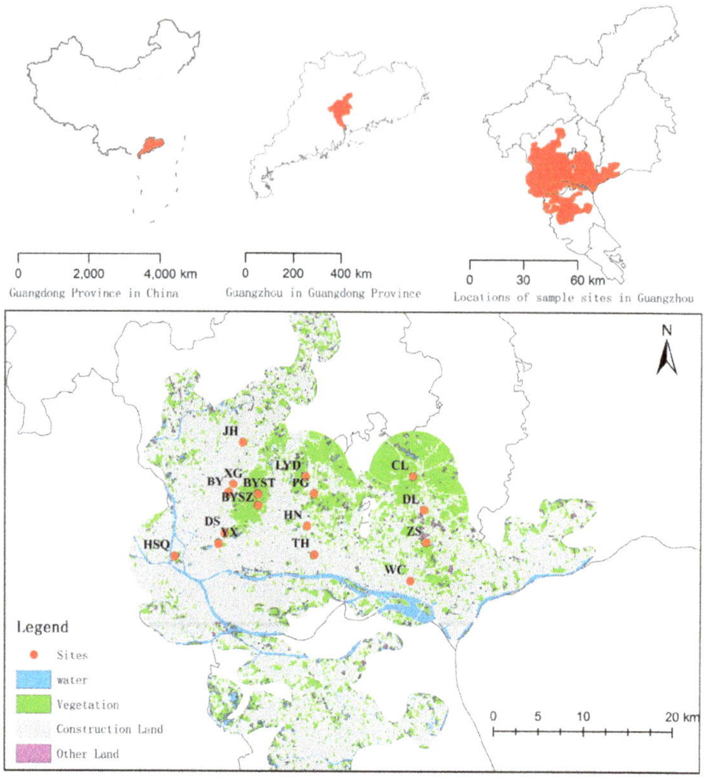

Figure 1. Map of 16 sample sites in different landscape patterns in Guangzhou, southern China.

Table 1. Description of 16 sample sites in Guangzhou, southern China.

Vegetation Type	Sample Sites	Longitude and Latitude	Altitude (m)	Slope (°)	Aspect (°)	Dominant Species	Age of Stand (Year)	Distance from the Road (m)
Grassland	JH	113°17′09″, 23°14′31″	13	-	-	*Neyraudia reynaudiana*	≈10	50
	XG	113°16′29″, 23°12′03″	11	-	-	*Neyraudia reynaudiana*	≈10	50
	BY	113°16′08″, 23°11′34″	20	-	-	*Neyraudia reynaudiana*	≈10	50
	HSQ	113°12′38″, 23°07′52″	9	-	-	*Neyraudia reynaudiana*	≈10	50
Secondary coniferous and broad leaved mixed forest	DS	113°15′53″, 23°09′11″	59	20	280	*Pinus massoniana* -*Celtis sinensis* -*Ottochloa nodosa*	40–60	250
	YX	113°15′28″, 23°08′32″	28	10	220	*Pinus massoniana*-*Cinnamomum burmanni* -*Piper sarmentosum*	≈60	50
	TH	113°21′42″, 23°07′50″	60	8	250	*Pinus massoniana* +*Psychotria rubra* +*Ottochloa nodosa*	≈60	40

Table 1. *Cont.*

Vegetation Type	Sample Sites	Longitude and Latitude	Altitude (m)	Slope (°)	Aspect (°)	Dominant Species	Age of Stand (Year)	Distance from the Road (m)
Secondary broad leaved forest	BYST	113°16′10″, 23°11′33″	180	45	13	*Schima superba* -*Psychotria rubra* -*Lophatherum gracile*	40–60	50
	BYSZ	113°18′05″, 23°10′52″	255	3	60	*Schima superba* -*Psychotria rubra* -*Lophatherum gracile*	40–60	50
	LYD	113°21′12″, 23°12′29″	60	25	230	*Schima superba* -*Psychotria rubra* -*Lophatherum gracile*	≈60	300
	PG	113°21′45″, 23°11′26″	34	5	200	*Schima superba* -*Psychotria rubra* -*Lophatherum gracile*	60–80	50
	HN	113°21′15″, 23°09′30″	60	5	210	*Schima superba* -*Cinnamomum burmanni* +*Lophatherum gracile*	60–80	250
	CL	113°28′17″, 23°12′24″	58	20	290	*Castanea henryi* -*Castanea henryi* -*Cibotium barometz*	60–100	200
	DL	113°28′58″, 23°10′21″	50	2	100	*Schima superba* -*Psychotria rubra* -*Lophatherum gracile*	>150	300
	ZS	113°29′06″, 23°08′27″	45	30	330	*Schima superba* -*Psychotria rubra* -*Adiantum capillus-veneris*	60–100	300
	WC	113°28′00″, 23°06′12″	60	30	210	*Schima superba* -*Rhaphiolepis indica* -*Dianella ensifolia*	60–100	150

Plant species in three sample plots (each 20 × 20 m^2) at each study site were surveyed from March 2014 to March 2015. According to the technical guidelines for biodiversity monitoring [49], the forest observation plot should be larger than 1 ha, so we set three plots of 400 m^2 (each 20 m × 20 m) in the sample sites for a total of 48 sample plots. The total survey plot was 1.92 ha, including 12 in urban weeds, 9 in secondary coniferous and broad-leaved mixed forests, and 27 in secondary broad-leaved mixed forests. Woody species and trees with a diameter breast height (DBH at 1.3 m) larger than 2 cm in each plot were counted and identified. Shrubs and herbs in the same plot were counted and identified in four randomly distributed 5 × 5 m^2 subplots and four 1 × 1 m^2 subplots, respectively. The species, numbers, height, density, canopy, and coverage of tree, shrub, and herb layers were individually counted and recorded. We used GPS for geographic coordinate positioning. The stand age, elevation, longitude and latitude, slope, and aspect were also recorded.

2.3. Landscape Pattern Metrics Combined with Community Structure and Topographic Factors

First, the landscape classification map of the research area was drawn based on the current land use status map (2011, 1:10,000), which was divided into six types: vegetation, residential area, road, construction land, river, and other lands (Figure 1). Taking each sampling point as the center of the circle, the landscape pattern metric within different buffer radii was calculated. The buffer radius was set to 500, 1000, and 2000 m. We used the vegetation coverage in the buffer zone, the road disturbance indices, the distance between the center of the plot and the edge of the forest, and some landscape pattern metrics to represent the landscape heterogeneity. The above indices of different radii in the buffer area were calculated by Fragstats 4.2. Through the collinearity analysis of the indices, a total of 17 indices were selected to characterize the landscape heterogeneity of different sampling plots, including matrix indices (1), road disturbance indices (2), distance from the edge of the forest (2), clustering indices (4), area-edge indices (1), edge indices (2), shape indices (2), and landscape diversity indices (3) (Table 2). Second, tree layer coverage, herb layer height, herb layer coverage, and stand age were selected to characterize the community structure. Third, elevation, slope, and aspect were selected to characterize the topographic factors.

Table 2. The selection indices of landscape heterogeneity, community structure, and topographical factors.

Types	Subtypes	Number of Indices	Indices Name
Landscape heterogeneity characteristics	Matrix indices	1	Vegetation coverage (Cv)
	Road disturbance indices	2	Road density (RD) Average distance from sample point to road (AD)
	Distance from the edge of the forest	2	Shortest distance from sample point to forest edge (SD) Farthest distance from sample point to forest edge (FD)
	Clustering indices	4	Number of patches (NP) Patch density (PD) Largest patch index (LPI) Contagion index (CONTA)
	Area-edge indices	1	Landscape shape index (LSI)
	Edge indices	2	Edge density (ED) Total edge length (TE)
	Shape indices	2	Shape Index (SHAPE) Fractal dimension index (FRAC)
	Landscape diversity indices	3	Shannon's diversity index (SHDI) Simpson's diversity index (SIDI) Shannon's Evenness index (SHE)
Community structure		4	Coverage of the herb layer (Ch) Height of the herb layer (Hh) Coverage of the tree layer (Ct) Age of stand (Age)
Topographic		3	Elevation, Slope, Aspect

2.4. Data Analysis

2.4.1. Data Analysis of Sampling and Vegetation Survey

Importance value: Importance values are important indicators for measuring the position and function of a population in a community and are also the basis for calculating species diversity. In this study, the calculation method of Qian et al. was used to measure the importance values of each plant population in the community. The formula is as follows:

$$IV_i = (rf_i + rd_i + rdo_i)/3$$

where IV_i is the importance value of species i; rf_i is the relative frequency of species i; rd_i is the relative density of species i; and rdo_i is the relative dominance of species i.

Species diversity: Four indicators, the species richness index (S), Shannon diversity index (H'), Simpson's index (D), and Pielou uniformity index (J), were used to measure the species diversity of arbor, shrub, and herbaceous in urban green space plant communities. In terms of plant diversity indices, four common alpha biodiversity indices were selected in this study (Table 3). The biodiversity indices were analyzed using R Language 2.11.0.

Table 3. The plant diversity indices selected in the study areas.

Types of Index	Subtypes of Index	Abbreviations	Formula	Description
Species diversity index	Shannon-Wiener index	H'	$H' = -\sum P_i ln P_i$	P_i is the relative abundance of the *i*th species at each plot, *ln* is the natural log, and *H* describes the species richness and the equitability of individual distribution within species.
	Simpson's index	D	$D = 1-\sum {P_i}^2$	P_i is the proportion of the individuals in species *i*, and *D* reflects the dominance in the community.
	Evenness	J	$J = H'/H'_{max}$	H' is Shannon-Wiener's biodiversity index, and H'_{max} is the maximum of H'.
Species richness index .	Patrick Richness	R	R = S	S is the number of species in the sample plot.

2.4.2. Data Analysis of Landscape Pattern Metrics Combined with Community Structure and Topographic Factors

Multiple regression model analysis (stepwise regression, $p < 0.05$) was used to explore the relationships between landscape heterogeneity, community structure features, topographic factors, and plant diversity indices under 500, 1000, and 2000 m buffer radii. The explanatory power (adjusted R^2) indicates the contribution of the impact of each factor to the plant diversity indices. Data were processed and analyzed in Microsoft 365 Excel and IBM SPSS Statistics 25.

3. Results

3.1. Species Composition in Different Remnant Vegetation in Guangzhou City

A total of 234 species (179 genera and 82 families) were recorded in the remnant vegetation at the 16 study sites. A total of 1277 trees (DBH $\geq$ 2 cm) representing 67 species (50 genera and 31 families), 141 shrub-layer species (103 genera and 53 families), and 128 herb-layer species (111 genera and 54 families) were recorded. A total of 38 species (37 genera and 19 families) were recorded in urban weeds, 110 species (98 genera and 53 families) were recorded in secondary coniferous and broad-leaved mixed forest, and 167 species (119 genera and 67 families) were recorded in secondary broad-leaved forest.

The dominant species (importance value >5%) in the three vegetation types are shown in Table 4. The dominant species in urban weeds belonged to *Gramineae* and *Asteraceae*, which were mainly composed of exotic species and generalists. *Bidens pilosa*, a common annual agricultural weed, was widely distributed in tropical and subtropical regions and accounted for more than 30% of the importance value among most urban weeds. It has a strong allelopathic effect and often forms a large area of monodominant population, resulting in a decrease in local biodiversity and a serious threat to the survival of local plants [50–52]. *Neyraudia reynaudian*, another dominant species, is a native grass species and is widely distributed in southern China, with developed roots, strong growth, and good resistance to stress [53]. The dominant herb-layer species in forests were native herb species, including shade or half-shade plants, and ferns that survive in forest or humid environments, such as *Eriachne pallescens, Pteris semipinnata*, and *Lophatherum gracile*. The dominant shrub-layer species were similar in the two different forest types, which were dominated by *Psychotria rubra*. The dominant tree-layer species in secondary coniferous and broad-leaved mixed forest was *Pinus massoniana*, while in secondary broad-leaved forest, it was *Schima superba*.

Table 4. The dominant species in three vegetation types in Guangzhou, southern China.

Vegetation Type	Code	Dominant Herb-Layer Species (Importance Value >5%)	Dominant Shrub-Layer Species (Importance Value >5%)	Dominant Tree-Layer Species (Importance Value >5%)
Urban weeds	I	*Bidens Pilosa* (24%) *Neyraudia reynaudiana* (22%) *Themeda villosa* (6%)	-	-
Secondary coniferous and broad-leaved mixed forest	II	*Eriachne pallescens* (51%) *Pteris semipinnata* (5%)	*Psychotria rubra* (25%) *Cinnamomum burmanni* (10%) *Celtis sinensis* (6%) *Ilex asprella* var. *asprella* (6%) *Desmos chinensis* (6%) *Trema cannabina* (5%)	*Pinus massoniana* (34%) *Cinnamomum burmanni* (11%) *Celtis sinensis* (10%) *Cinnamomum camphora* (6%)
Secondary broad-leaved forest	III	*Lophatherum gracile* (22%) *Dicranopteris dichotoma* (19%) *Eriachne pallescens* (8%) *Adiantum capillus-veneris* (7%)	*Psychotria rubra* (31%) *Desmos chinensis* (5%)	*Schima superba* (45%)

3.2. Plant Diversity in Different Layers of Remnant Vegetation in Guangzhou City

The differences in diversity indices and richness index were compared in remnant vegetation (Figure 2). In the herb layer, there were no significant differences in the Shannon index values among the three vegetation types ($p > 0.05$). The Simpson index was highest in II, while the Evenness index showed the opposite trend, which was highest in I and III. The order of the richness index values was II > III > I. In the shrub layer, there were no differences in the values of all indices ($p > 0.05$). In the tree layer, the order of the Shannon index, evenness index, and richness index values was II > III, while the Simpson index showed the opposite trend.

3.3. Relationships between Landscape Heterogeneity, Community Structure, Topographic Factors, and Plant Diversity Indices

Although diversity indices and richness indices in tree, shrub, and herb layers were affected by various factors and showed different trends under three buffer radii, they still had the following common characteristics: (1) At the 500 m scale, the diversity indices in the herb layer and shrub layer were primarily affected by community structure (herb layer coverage and tree layer coverage, respectively), followed by road disturbance indices, while the diversity indices in the tree layer had no relationship with any factors. (2) As the scale expanded, diversity indices in the herb layer declined under the influence of comprehensive factors. In contrast, the richness index in the herb layer increased with increasing buffer radius, which was affected by comprehensive factors. Shrub layer diversity was affected by different factors under different buffer radii. The richness of the shrub layer decreased with increasing scale and the influence of comprehensive factors. The diversity of the tree layer was most significantly affected by comprehensive factors on the 1000 m radius scale. The tree layer richness index was negatively correlated with stand age and was affected by comprehensive factors as the buffer radius increased.

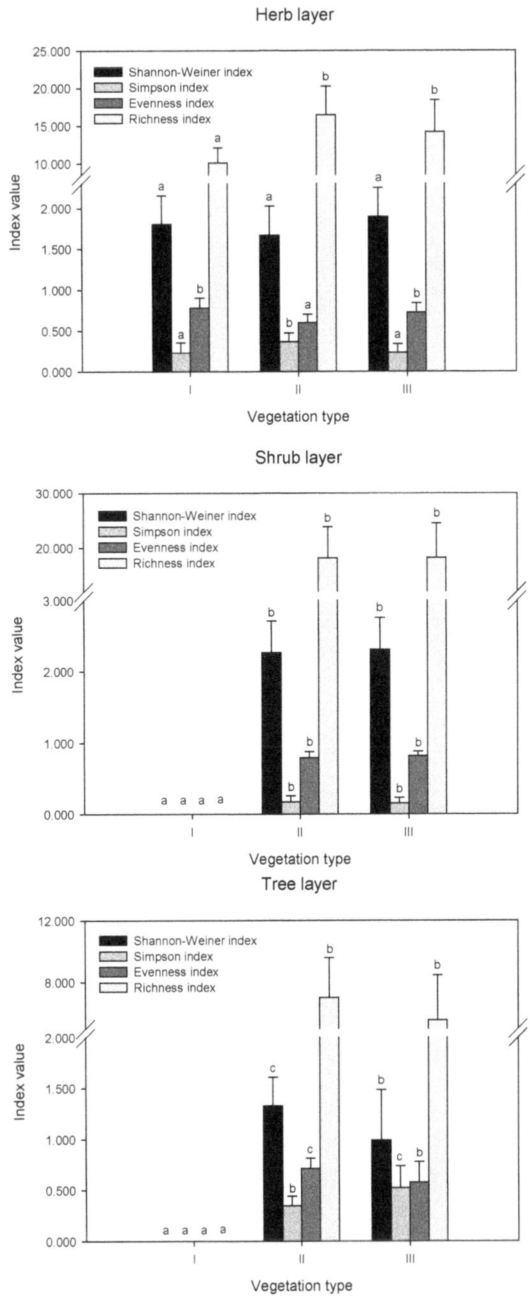

Figure 2. Biodiversity indices in different layers of remnant vegetation. I indicates urban weeds; II indicates secondary coniferous and broad-leaved mixed forest; III indicates secondary broad-leaved forest. Different letters indicate significant differences at the confidence level of $p < 0.05$ among the vegetation types (one-way ANOVA, $p < 0.05$).

Based on the multiple regressions (Table 5), we found that the relationships between plant diversity indices and richness index and impact factors in the tree, shrub, and herb layers were significantly different with the change in scale. The plant diversity indices in the herb layer were most closely related to impact factors at the 500 m scale, and their association with factors decreased as the scale increased. The herb-layer diversity indices were mainly impacted by Ch and were also affected by road disturbance (AD, PD, and RD) at the 500 m and 1000 m buffer radii. In contrast, the richness index in the herb layer was most closely related to impact factors at the 2000 m scale, and its association with factors decreased as the scale declined. The herb-layer richness index was significantly impacted by SHAPE and Ct at the 500 m scale, and as the scale increased, the association with SLOPE, ELEVATION, LPI, and SHDI also increased.

Table 5. The multiple regression of landscape heterogeneity, community structure, and topographic factors on diversity indices and richness index in tree, shrub, and herb layers.

Layer	Biodiversity Indices	Buffer Radius/m		
		500	1000	2000
Herb	Shannon index	-Ch *, -AD	-Ch *	-Ch *
	Simpson index	AD *, -PD, Ch	Ch, RD	Ch
	Evenness	-AD *, -SLOPE *, PD *, LPI	-RD	-Ct *, Ch
	Richness	-SHAPE *, Ct *	SLOPE *, ELEVATION *, -LPI *, -ED	SLOPE, -SHAPE *, Ct *, SHDI *, AD
Shrub	Shannon index	-RD *, Ct	-NP	-LPI *, -LSI *, PD
	Simpson index	RD *, -Ct *, FD	-SLOPE *, -Ct	-SLOPE *, -Ct
	Evenness	Ct *, -FD, SLOPE	Cv *, -SHDI *, -FD	Ct *, SLOPE
	Richness	-RD *, CONTAG	AD *, -FD	-FD *
Tree	Shannon index	Ct	Ct *, -PD *, -AD *	AD,Ct
	Simpson index	-Ct	PD *, AD, -Ct	-AD, -CONTAG
	Evenness	-	-FD, -NP	-SIDI, AD
	Richness	-AGE *, NP	-AGE*	-AGE *, AD *, Cv

* indicates significant at $p < 0.01$.

The shrub-layer diversity indices were mainly impacted by Ct. In addition, they were also affected by road disturbance (RD and FD) at the 500 m scale; by topographic factors (SLOPE), matrix (Cv), and landscape diversity indices (SHDI) at the 1000 m scale; and by topographic factors (SLOPE), clustering indices (LPI), and area-edge indices (LSI) at the 2000 m scale. The shrub-layer richness index was significantly impacted by road disturbance (RD, AD, and FD).

The diversity index in the tree layer was most closely related to Ct, PD, and AD at the 1000 m scale, while the tree-layer richness index was significantly negatively related to AGE under the three buffer radii, and was most closely related to -AGE*, AD*, and Cv at the 2000 m scale.

3.4. The Explanatory Power of the Regression Model

Based on the explanatory power (adjusted R^2) of various regression models (Figure 3), landscape heterogeneity, community structure, and topographic factors had the strongest explanatory power for the richness index, but not at the same scale. The explanatory power of the three layers was ranked as follows: herb layer (0.62) at the 2000 m scale > shrub layer (0.55) at the 500 m scale > tree layer (0.40) at the 2000 m scale. The explanatory power of factors on other diversity indices was relatively weak (<0.43). The combined explanatory power of the three factors accounted for 43% of the species diversity indices and 63% of the richness index at its peak.

In the herb layer, the explanatory power of factors on the richness and diversity indices showed an opposite trend with increasing scale. The impact factor had the strongest explanatory power on the richness index for the three buffer radii (0.47–0.62), and its explanatory power increased with increasing scale. The explanatory power of factors on diversity indices (the Evenness index, the Shannon diversity index, and the Simpson index) was relatively low (0.08–0.4), and the highest explanatory power appeared on the 500 m scale. Except for the Evenness index, the explanatory power of factors on the Shannon index and Simpson index decreased with increasing scale. In the shrub layer, the regression equations of each index showed that diversity indices and richness index can both be well fitted at the 500 m scale. Except for the Shannon index, the explanatory power of factors on other indices decreased with increasing scale. In the tree layer, the explanatory power of factors on all diversity indices was shown to be the weakest at the 500 m scale. The comprehensive factor had the strongest interpretation for the Shannon index (0.39) and Simpson index (0.36) at the 1000 m scale.

3.5. The Relative Importance of the Effects of Landscape Heterogeneity, Community Structure, and Topographic Factors on Diversity Indices and Richness Index

Based on the contribution of each index to the overall explanatory power of the regression model (Figure 4), and except for the herb layer, the interpretation of landscape heterogeneity for each plant diversity index was more stable than that for the other two factors. In general, the relationships between the three impact factors and the diversity indices and richness index under different buffer radii showed strong scale volatility, but lacked consistency.

In the herb layer, the explanatory power of community structure for each plant diversity index was more stable than that of the other two factors. Community structure combined with landscape heterogeneity had a relatively strong explanatory power for plant diversity indices at a 500 m buffer radius compared to the others. For the richness index, the superposition of the three factors led to a very high overall explanatory power of the model.

In the shrub layer, landscape heterogeneity had a relatively higher explanatory power for all diversity indices and richness indices across the three buffer radii than topographic factors. For the Shannon and Simpson indices, landscape heterogeneity had the highest explanatory power at the 2000 m scale. For the Evenness index, the landscape heterogeneity showed the same explanatory power at the 500 m and 1000 m buffer radii. For the richness index, landscape heterogeneity was the only explanatory power that weakened as the scale increased.

In the tree layer, except for the Evenness index, landscape heterogeneity and community structure had explanatory power for the other diversity indices and richness index across the three buffer radii. For the Shannon and Simpson indices, landscape heterogeneity combined with community structure had a high explanatory power at 1000 m. For the Evenness index, landscape heterogeneity combined with topographic factors had the highest explanatory power at the 2000 m scale. For the richness index, landscape heterogeneity combined with community structure had the highest explanatory power at the 500 m scale.

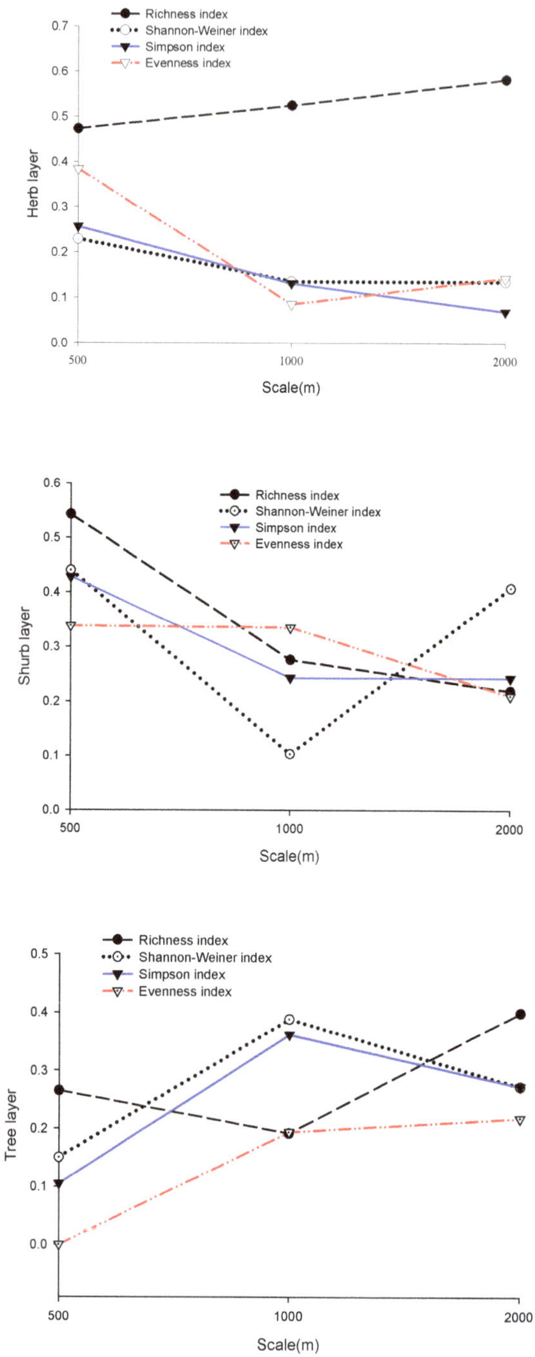

Figure 3. The explanatory power of landscape heterogeneity, community structure, and topographic factors on biodiversity and richness indices.

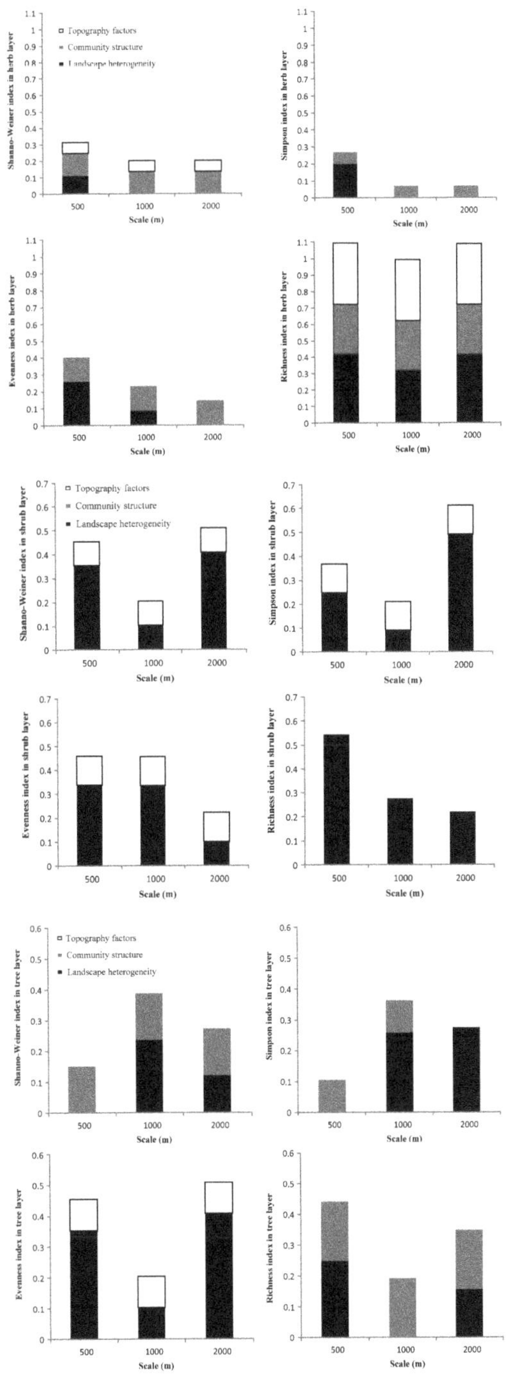

Figure 4. The variance decomposition of landscape heterogeneity, community structure, and topographic factors in the multiple linear regression models.

4. Discussion

4.1. The Response of the Diversity Index to the Impact Factors

From the different indices that characterize plant diversity, the diversity indices commonly used in the past, such as the Shannon diversity index (H′), Evenness index (E′), and dominance index (D′), were not well interpreted; however, the richness index in the herb layer was more responsive to impact factors than other biodiversity indices. This strong scale dependence is probably because the traditional diversity indices were also based on statistics, so the correlation is probably only statistical rather than ecological. Therefore, the richness index is more appropriate for analyzing the impact of landscape pattern changes on plant diversity. On the other hand, the richness index is additive and has the same measurement unit, so its ecological significance will be easier to understand and explain.

Therefore, the richness index can be used to effectively compare the differences in plant diversity in human activity intensity and landscape background in different spaces and temporal buffer radii [54].

4.2. The Relative Importance of Impact Factors on Species Biodiversity Varying with Buffer Radius

The analysis of biodiversity, without considering scale effects, may lead to erroneous conclusions [55]. Studies have shown that on a small spatial scale, changes in the environmental conditions of a community or ecosystem lead to changes in plant diversity and species composition [56,57]. However, on a large spatial scale, the effects of landscape pattern characteristics on species have not yet reached a consistent conclusion [58–62]. In general, our results found that the relationships between the three impact factors and the diversity indices and richness index under different buffer radii showed strong scale volatility, but lacked consistency.

According to the results of this study, the combined explanatory power of landscape heterogeneity, community structure, and topographic factors accounted for 43% of the species diversity indices and 62% of the richness index at its peak. Unexplained parts may be caused by factors that are not covered, such as spatial variation or the effects of random processes [63]. Although the factors that may affect the species diversity of urban remnant vegetation are complex and scale-dependent, and the tree, shrub, and herb layers may have different response mechanisms to them, relatively uniform research results may be obtained. Factors related to landscape heterogeneity, community structure, and topographic factors may not act alone, but they may create synergistic effects.

Our results found that as the scale increased, the richness index in the herb layer was affected by comprehensive factors that increased with increasing buffer radius. This is similar to many research results, indicating that the correlation between plant species richness and landscape pattern is enhanced with the expansion of spatial scale [42]. Species richness in the shrub layer and herb layer was under the influence of road disturbance and stand age, respectively. This may be because the species richness distribution of different functional groups is not affected by landscape structure factors, such as isolation degree, and only a few are affected by patch area, patch distance from road, and patch type. It may be that species at small buffer radii can spread to the target habitat in close proximity, and diffusion is limited to a small scale range, so isolation and other factors are not the maintenance mechanism of species richness [48,64].

4.3. The Impact of Landscape Heterogeneity on Plant Diversity

Some of the studies found that the relationships between landscape heterogeneity and plant species diversity are weak, and show strong scale volatility and lack consistency [65,66]. However, our results showed that the interpretation of landscape heterogeneity for each plant diversity index in shrub and tree layers was more stable than for the other two factors, which is consistent with the conclusion that buffer radii of 1000 m and 2000 m are the appropriate buffer radii to study the influence of various landscape patterns on forest, nonforest, and universal species diversity [67]. The reasons for this might be that (1) the 500 m scale is not suitable for analyzing the influence of landscape heterogeneity on the diversity in the tree layer of

20 m × 20 m patches; and (2) plants in the tree layer are mainly woody plants, whose response to landscape heterogeneity in the small-scale range is relatively lagging.

At a 500 m buffer radius, habitat conditions such as community structure and topographic factors may be the key factors determining species distribution [68,69]. However, our results showed that community structure combined with landscape heterogeneity had a relatively strong explanatory power for plant diversity indices in the herb layer at a 500 m buffer radius compared to the others. The reasons may be as follows: (1) α biodiversity is mostly affected by the matrix landscape characteristics around the habitat patches at the 500 m scale, which is mainly caused by adjacent human activities [67]. (2) The effect of the landscape complexity index on the diversity in the herb layer on a small scale (250–750 m) is higher than that on medium and large scales [70].

In general, the richness index and diversity index were influenced by different degrees of landscape heterogeneity (-SHAPE, -LPI, -ED, SHDI, -PD, Cv, and -LSI). Studies on tropical forests have shown that diversity indices of trees, shrubs, and lianas are mainly affected by patch shape (SHAPE) and patch diversity (SHDI) [71]. Our study found that SHAPE, ED, and LPI, which characterize the shape and edge of the landscape, had a negative impact on the richness index in the herb layer. Studies have reported that irregular landscape patches tend to maintain high plant diversity [72]. It is generally believed that narrow or irregular patches have large marginal lengths and marginal densities (ED), which lead to high landscape heterogeneity and relatively high plant diversity [42,73]. The reason may be that at different buffer radii, a higher marginal density (ED) of habitat patches helps maintain high species richness [21,74]. However, experimental studies have shown that the shape complexity index of landscape patches is inconsistent with plant diversity, or indicates either higher plant diversity, or lower plant diversity [71,73]. At the same time, studies in Western Europe have shown that plant diversity, especially woody plant species diversity, is inconsistent with different landscape shape indices [21], which may be related to landscape types and sampling scales.

With regard to landscape diversity, our study found that the SHDI was positively correlated with the herb-layer richness index. The consistent conclusion was that the species richness index is high in heterogeneous landscapes. Species richness has been reported to positively correlate with the diversity of landscape type patches [21], whose positive correlation is derived from abundant landscape landform types [75] or from landscape quality effects and adjacency effects that provide potential habitats for more species [39]. Studies in Western Europe have indicated that plant diversity, particularly the species diversity of woody plants, is significantly positively correlated with landscape diversity [21]. Moreover, studies in Austria [76], Belgium [77], and Spain [73] showed that the species richness of vascular plants, mosses, and birds positively correlated with landscape diversity. A study of vascular plant diversity in grasslands found that as the diversity of habitats around the plot increased, species diversity increased; conversely, as the habitat diversity decreased, the α diversity index decreased [78].

In addition to being related to landscape diversity, plant diversity indices are also associated with the aggregation degree between landscape patches. Regarding clustering indices (patches aggregation degree index), our study showed that patch density (PD) and largest patch index (LSI) had a negative effect on herb layer richness.

The results indicated that Cv was positively correlated with the diversity index in the tree layer. On the basis of the theory of island biogeography, in a certain range, the area of landscape patches is positively correlated with species richness and the diversity index, but this relationship weakens after exceeding a certain area [21,72]. The selective extinction hypothesis suggests that patch area is an important factor restricting species distribution; for example, the species with the smallest requirement area are more likely to become extinct in smaller patches [79]. Plant diversity has been shown to increase with the increase in habitat patch area and habitat similar to the surrounding landscape type, which would improve the habitat [80]. This is because the larger patch area contains more micro habitat types, and the amount and type can provide more species habitat. However, for different species, the relationship between landscape patch area and species richness is also different [81]. Studies on semiarid steppes have shown that plant richness and diversity indices were significantly positively correlated with patch area and habitat center area [67]. Many studies have found that the urban ecological system has a high diversity of vascular plants, especially in the green belt, of both native and exotic species, which means that the larger the patch area of greening is, the higher the diversity of vascular plants [82]. As global landscape fragmentation has become increasingly serious, small patches have become a common feature of the ecosystems, and at the same time, they are the main habitat in urban ecosystems. It is necessary to define the minimum patch area that allows native plants to survive. In many ecosystems, the minimum patch area controls the entire ecosystem. Therefore, the assessment of the size of patches in the maintenance ecosystem plays an important role in guiding researchers to determine the minimum patch area and number [83]. This is similar to many research results, indicating that species richness is significantly affected by patch area [70].

Our study showed that the road disturbance factors (- AD and - RD) and the farthest distance from the sample point to the forest edge (- FD) had different degrees of negative impact on the richness index in all three layers at different buffer radii. Roads, railways, and other artificial barrier edges often reduce biodiversity [84]. As patch area increases, the richness index of edge species and internal species increases, with the latter significantly increasing [72].

Finally, our results showed that road disturbance indices (AD and RD), farthest distance from the sample point to the forest edge (FD), area-edge indices (LPI), edge indices (ED), shape indices (SHAPE), and landscape diversity indices (SHDI and SIDI), i.e., eight indices in total, can well indicate species diversity and richness.

5. Conclusions

(1) The response mechanisms of the plant richness index and diversity indices in different layers under different buffer radii to impact factors were different. Compared to the biodiversity indices commonly used in the past, such as the Shannon diversity index (H′), evenness index (E), and dominance index (D′), the richness index in the herb layer was more direct and sensitive than the richness index in the tree and shrub layers and the diversity indices in the three layers to the impact factors.

(2) The combined explanatory power of landscape heterogeneity, community structure, and topographic factors accounted for 43% of the species diversity indices, and 62% of the richness index at its peak.

(3) The three impact factors that affect the species diversity indices and richness index of urban remnant vegetation rarely act alone, and often cause comprehensive cumulative effects and scale dependence.

(4) Scale does matter in urbanization landscape studies. At a 500 m buffer radius, community structure combined with road disturbance indices was strongly related to diversity indices in herb and shrub layers. The stand age was negatively correlated with the tree layer richness index. As the scale increased, the diversity indices and richness index in the three layers decreased or increased under the influence of comprehensive factors.

(5) Except for the herb layer, the interpretation of landscape heterogeneity for each plant diversity index was more stable than that for the other two factors. Road disturbance indices (AD and RD), farthest distance from the sample point to the forest edge (FD), area-edge indices (LPI), edge indices (ED), shape indices (SHAPE), and landscape diversity indices (SHDI and SIDI), a total of 8 indices, can well indicate species diversity and richness.

Under the background of rapid urbanization and increasingly fragmented urban vegetation, the buffer area of remnant patches, such as grassland and woodland, should be reasonably allocated. Furthermore, we suggest that the vegetation area of remnant patches within a radius of 500–2000 m should be appropriately increased to protect plant diversity in all layers, and the negative effect of landscape heterogeneity around remnant patches, such as road disturbance, should also be considered.

Author Contributions: Conceptualization, X.L. and L.H.; methodology, Q.Q., X.L. and L.H.; software, G.Y. and Q.W.; validation, X.L., Z.Z. and L.H.; formal analysis, X.L., G.Y. and L.H.; investigation, X.L. and Q.Q.; resources, X.L. and L.H.; data curation, Q.Q., Q.W. and Z.Z.; writing—original draft preparation, X.L. and G.Y.; writing—review and editing, X.L., G.Y. and L.H.; visualization, Q.W. and Z.Z.; supervision, L.H.; project administration, L.H.; funding acquisition, L.H. All authors have read and agreed to the published version of the manuscript.

Funding: This study was supported by the National Natural Science Foundation of China (32071578), the Science and Technology Innovation Project of Fujian Province (KY-090000-04-2021-012) and Subject Cross-Integration Project of College of Landscape Architecture, Fujian Agriculture and Forestry University (No. YSYL-xkjc-2).

Institutional Review Board Statement: Not applicable.

Informed Consent Statement: Not applicable.

Data Availability Statement: The data presented in this study are available on request from the author. The data are not publicly available due to privacy. Images employed for the study will be available online for readers.

Conflicts of Interest: The authors declare that they have no conflict of interest.

References

1. Mckinney, M.L. Urbanization as a major cause of biotic homogenization. *Biol. Conserv.* **2006**, *127*, 247–260. [CrossRef]
2. Rocha-Santos, L.; Pessoa, M.S.; Cassano, C.R.; Cassano, C.R.; Talora, D.C.; Orihuela, R.L.L.; Mariano-NetobJosé, E.; Morante-Filho, J.C.; Faria, D.; Cazetta, E. The shrinkage of a forest: Landscape-scale deforestation leading to overall changes in local forest structure. *Biol. Conserv.* **2016**, *6*, 1–9. [CrossRef]
3. Li, D.; Stucky, B.J.; Deck, J.; Baiser, B.; Guralnick, R.P. The effect of urbanization on plant phenology depends on regional temperature. *Nat. Ecol. Evol.* **2019**, *3*, 1661–1667. [CrossRef] [PubMed]
4. Meyer, S.; Rusterholz, H.P.; Baur, B. Urbanisation and forest size affect the infestation rates of plant-galling arthropods and damage by herbivorous insects. *Eur. J. Entomol.* **2020**, *117*, 34–48. [CrossRef]
5. Mao, Y.Y.; Chen, F.Q.; Guo, Y.; LiuM, J.X.; Jiang, F.; Li, X. Restoration dynamics of understory species diversity in Castanopsis hystrix plantations at different ages. *Chin. J. Appl. Environ. Biol.* **2021**, *27*, 930–937.
6. Ye, Y.H.; Lin, S.S.; Wu, J.; Li, J.; Zou, J.F.; Yu, D.Y. Effect of rapid urbanization on plant species diversity in municipal parks, in a new Chinese city: Shenzhen. *Acta Ecol. Sin.* **2012**, *32*, 221–226. [CrossRef]
7. Dos Santos, J.S.; Silva-Neto, C.M.; Silva, T.C.; Siqueira, K.N.; Ribeiro, M.C.; Collevatti, R.G. Landscape structure and local variables affect plant community diversity and structure in a Brazilian agricultural landscape. *Biotropica* **2021**, *54*, 239–250. [CrossRef]
8. Nakagawa, M.; Momose, K.; Kishimoto, K.; Kamoi, T.; Tanaka, H.; Kaga, M.; Yamashita, S.; Itioka, T.; Nagamasu, H.; Sakai, S.; et al. Tree community structure, dynamics, and diversity partitioning in a Bornean tropical forested landscape. *Biodivers. Conserv.* **2012**, *22*, 127–140. [CrossRef]
9. Goddard, M.A.; Dougill, A.J.; Benton, T.G. Scaling up from gardens: Biodiversity conservation in urban environments. *Trends Ecol. Evol.* **2010**, *25*, 90–98. [CrossRef]
10. Forman, R.T.T. Urban ecology principles: Are urban ecology and natural area ecology really different? *Landsc. Ecol.* **2016**, *31*, 1653–1662. [CrossRef]
11. Huang, L.; Zhu, W.; Ren, H.; Chen, H.; Wang, J. Impact of atmospheric nitrogen deposition on soil properties and herb-layer diversity in remnant forests along an urban–rural gradient in Guangzhou, southern China. *Plant Ecol.* **2012**, *213*, 1187–1202. [CrossRef]
12. UN: China's Urban Population Will Increase by Another 255 Million by 2050. Available online: https://www.un.org/development/desa/zh/news/population/2018-world-urbanization-prospects.html (accessed on 22 October 2021).
13. Stevens, M.H.H.; Carson, W.P. Resource quantity, not resource heterogeneity, maintains plant diversity. *Ecol. Lett.* **2002**, *5*, 420–426. [CrossRef]
14. Pausas, J.G.; Carreras, J.; Ferré, A.; Font, X. Coarse-scale plant species richness in relation to environmental heterogeneity. *J. Veg. Sci.* **2003**, *14*, 661–668. [CrossRef]
15. Dufour, A.; Gadallah, F.; Wagner, H.H.; Guisan, A.; Buttler, A. Plant species richness and environmental heterogeneity in a mountain landscape: Effects of variability and spatial configuration. *Ecography* **2006**, *29*, 573–584. [CrossRef]
16. Malkinson, D.; Kopel, D.; Wittenberg, L. From rural-urban gradients to patch-matrix frameworks: Plant diversity patterns in urban landscapes. *Landsc. Urban Plan.* **2018**, *169*, 260–268. [CrossRef]

17. Chávez, V.; Macdonald, S.E. The influence of canopy patch mosaics on understory plant community composition in boreal mixed wood forest. *For. Ecol. Manag.* **2010**, *259*, 1067–1075. [CrossRef]
18. Redon, M.; Bergès, L.; Cordonnier, T.; Luque, S. Effects of increasing landscape heterogeneity on local plant species richness: How much is enough? *Ecol. Evol.* **2014**, *29*, 773–787. [CrossRef]
19. Yang, Z.; Liu, X.; Zhou, M.; Zhou, M.; Ai, D.; Wang, G.; Wang, Y.; Chu, C.; Lundholm, J. The effect of environmental heterogeneity on species richness depends on community position along the environmental gradient. *Sci. Rep.* **2015**, *5*, 15723. [CrossRef]
20. Gastauer, M.; Mitre, S.K.; Carvalho, C.S.; Trevelin, L.C.; Sarmento, P.; Meira, N.J.; Caldeira, C.F.; Ramos, S.J.; Jaffé, R. Landscape heterogeneity and habitat amount drive plant diversity in Amazonian canga ecosystems. *Landsc. Ecol.* **2021**, *36*, 393–406. [CrossRef]
21. Schindler, S.; Wehrden, H.V.; Poriazidis, K.; Wrbka, T.; Kati, V. Multiscale performance of landscape metrics as indicators of species richness of plants, insects and vertebrates. *Ecol. Indic.* **2013**, *31*, 41–48. [CrossRef]
22. Priego-Santander, A.G.; Campos, M.; Bocco, G.; Ramirez-Sanchez, L.G. Relationship between landscape heterogeneity and plant species richness on the Mexican Pacific coast. *Appl. Geogr.* **2013**, *40*, 171–178. [CrossRef]
23. Fan, M.; Wang, Q.H.; Mi, K.; Peng, Y. Scale-dependent effects of landscape pattern on plant diversity in Hunshandak Sandland. *Biodivers. Conserv.* **2017**, *26*, 2169–2185. [CrossRef]
24. Kim, H.; Lee, C.B. On the relative importance of landscape variables to plant diversity and phylogenetic community structure on uninhabited islands, South Korea. *Landsc. Ecol.* **2020**, *36*, 209–211. [CrossRef]
25. Peng, Y.; Wang, W.T.; Lu, Y.T.; Dong, J.H.; Zhou, Y.Q.; Shang, J.X.; Li, X.; Mi, K. Multiscale influences of urbanized landscape metrics on the diversity of indigenous plant species: A case study in Shunyi District of Beijing, China. *Chin. J. Appl. Environ. Biol.* **2021**, *31*, 4058–4066.
26. Joern, F.; David, B. Landscape Modification and Habitat Fragmentation:A Synthesis. *Glob. Ecol. Biogeogr.* **2007**, *16*, 265–280.
27. Peng, Y.; Fan, M.; Qing, F.T.; Xue, D.Y. Study Progresses on Effects of Landscape Metrics on Plant Diversity. *Ecol. Environ. Conserv.* **2016**, *25*, 1061–1068.
28. Naidu, M.T.; Kumar, O.A. Tree diversity, stand structure, and community composition of tropical forests in Eastern Ghats of Andhra Pradesh, India. *J. Asia-Pac. Biodivers.* **2016**, *9*, 328–334. [CrossRef]
29. Pretzsch, H.; Schütze, G. Effect of tree species mixing on the size structure, density, and yield of forest stands. *Eur. J. For. Res.* **2016**, *135*, 1–22. [CrossRef]
30. Walker, S.J.; Grimm, N.B.; Briggs, J.M.; Gries, C.; Dugan, L. Effects of urbanization on plant species diversity in central Arizona. *Front. Ecol. Environ.* **2009**, *7*, 465–470. [CrossRef]
31. Zou, M.; Zhu, K.H.; Yin, J.Z.; Gu, B. Analysis on Slope Revegetation Diversity in Different Habitats. *Procedia Earth Planet. Sci. Lett.* **2012**, *5*, 180–187.
32. Qian, S.H.; Qin, D.; Wu, X.; Hu, S.W.; Hu, L.Y.; Lin, D.M.; Zhao, L.; Shang, K.K.; Song, K.; Yang, Y.C. Urban growth and topographical factors shape patterns of spontaneous plant community diversity in a mountainous city in southwest China. *Urban For. Urban Green.* **2020**, *55*, 126814. [CrossRef]
33. Arteaga, M.A.; Delgado, J.D.; Otto, R.; Fernández-Palacios, J.M.; Arévalo, J.R. How do alien plants distribute along roads on oceanic islands? A casestudy in Tenerife, Canary Islands. *Biol. Invasions* **2009**, *11*, 1071–1086. [CrossRef]
34. Schlesinger, M.D.; Manley, P.N.; Holyoak, M. Distinguishing stressors acting on land bird communities in an urbanizing environment. *Ecology* **2008**, *89*, 2302–2314. [CrossRef] [PubMed]
35. Mao, Q.; Ma, K.; Wu, J.; Tang, R.G.; Luo, S.H.; Zhang, Y.X.; Bao, L. Distribution pattern of allergenic plants in the Beijing metropolitan region. *Aerobiologia* **2013**, *29*, 217–231. [CrossRef]
36. Krishnamoorthy, M.A.; Webala, P.W.; Kingston, T. Baobab fruiting is driven by scale-dependent mediation of plant size and landscape features. *Landsc. Ecol.* **2022**, *37*, 226–250. [CrossRef]
37. Ryser, R.; Hirt, M.R.; Häussler, J.; Gravel, D.; Brose, U. Landscape heterogeneity buffers biodiversity of simulated meta-food-webs under global change through rescue and drainage effects. *Nat. Commun.* **2021**, *12*, 4716. [CrossRef]
38. Sultana, M.; Müller, M.; Meyer, M.; Storch, I. Neighboring Green Network and Landscape Metrics Explain Biodiversity within Small Urban Green Areas-A Case Study on Birds. *Sustainability* **2022**, *14*, 6394. [CrossRef]
39. Zeleny, D.; Li, C.F.; Chytry, M. Pattern of local plant species richness along a gradient of landscape topographical heterogeneity: Result of spatial mass effect or environmental shift? *Ecography* **2010**, *33*, 578–589. [CrossRef]
40. Fahrig, L.; Baudry, J.; Brotons, L.; Burel, F.G.; Crist, T.O.; Fuller, R.J.; Sirami, C.; Siriwardena, G.M.; Martin, J.L. Functional landscape heterogeneity and animal biodiversity in agricultural landscapes. *Ecol. Lett.* **2011**, *14*, 101–112. [CrossRef]
41. Syrbe, R.U.; Michel, E.; Walz, U. Structural indicators for the assessment of biodiversity and their connection to the richness of avifauna. *Ecol. Indic.* **2013**, *31*, 89–98. [CrossRef]
42. Amici, V.; Rocchini, D.; Filibeck, G.; Bacaro, G.; Santi, E.; Grei, F.; Landi, S.; Scoppola, A.; Chiarucci, A. Landscape structure effects on forest plant diversity at local scale: Exploring the role of spatial extent. *Ecol. Complex.* **2015**, *21*, 44–52. [CrossRef]
43. Güneralp, B.; Perlstein, A.S.; Seto, K.C. Balancing urban growth and ecological conservation: A challenge for planning and governance in China. *Ambio* **2015**, *44*, 532–543. [CrossRef] [PubMed]
44. China Statistics Press. Statistics Bureau of Guangzhou Municipality. In *Guangzhou Statistical Yearbook*; China Statistics Press: Beijing, China, 2018; pp. 6–214.

45. Luo, Z. The Study of Plant Diversity of Green Space in Urban Built-Up Area of Guangzhou City. Master's Thesis, Zhongkai University of Agriculture and Engineering, Guangzhou, China, 2013.
46. Zhang, X.C.; Pan, Q. Research on spatial structure changes of urban land use. GIS and Remote Sensing in Hydrology. *Water Environ. Res.* **2004**, *289*, 141–150.
47. Interpretation of Relevant Greening Statistical Data of Guangzhou Forestry and Garden Bureau in 2017. Available online: http://lyylj.gz.gov.cn/zwgk/sjfb/content/post_3038151.html (accessed on 10 April 2022).
48. Tian, Y.; Jim, C.Y.; Liu, Y. Using a Spatial Interaction Model to Assess the Accessibility of District Parks in Hong Kong. *Sustainability* **2017**, *9*, 1924. [CrossRef]
49. Li, G.; Li, J.S.; Guan, X.; Wu, X.P.; Zhao, Z.P. *Shengwu Duoyangxing Jiance Jishu Shouce*; China Environmental Science Press: Beijing, China, 2014; pp. 15–45.
50. Wu, W.C.; Ning, J.Z. The morphological characteristics and classification of the main Asteraceae weeds in Guangzhou. *J. South China Agric. Univ.* **1993**, *2*, 32–46.
51. Yu, S.H.; Mao, R.B.; Tang, W.; Ma, W. Study of the areal types of seed weeds in China. *Southwest China J. Agric. Sci.* **2008**, *4*, 1189–1192.
52. Wang, J.H.; Wang, J.W.; Wang, D.X.; Christensen, S.; Zhang, Y.W.; Lu, D.Y.; Chen, K.R. An Eco-Field Simulation Analysis on Normal Patch Distribution of Annual Weed. *Guizhou Sci.* **2012**, *30*, 1–2.
53. Zeng, X.L.; Liao, F.L.; Wen, G.R.; Li, W.D. Investigation of the distribution and growth of major energy grasses in Meizhou city. *Guangdong Agric. Sci.* **2008**, *7*, 25–28.
54. Crist, T.O.; Veech, J.A. Aditive partitioning of rarefaction curves and species-area relationships: Unifying α-, β- and γ-diversity with sample size and habitat area. *Ecol. Lett.* **2006**, *9*, 923–932. [CrossRef]
55. Turtureanu, P.D.; Palpurina, S.; Becker, T.; Dolnik, C.; Ruprecht, E.; Sutcliffe, L.M.E.; Szabó, A.; Dengler, J. Scale- and taxon-dependent biodiversity patterns of dry grassland vegetation in Transylvania. *Agric. Ecosyst. Environ.* **2014**, *182*, 15–24. [CrossRef]
56. Loreau, M.; Naeem, S.; Inchausti, P.; Bengtssonet, J.; Grime, J.P.; Hector, A.; Hooper, D.U.; Huston, M.A.; Raffaelli, D.; Schmid, B.; et al. Biodiversity and ecosystem functioning: Current knowledge and future challenges. *Science* **2001**, *294*, 804–808. [CrossRef] [PubMed]
57. Hautier, Y.; Tilman, D.; Isbell, F.; Seabloom, E.W.; Borer, E.T.; Reich, P.B. Anthropogenic environmental changes affect ecosystem stability via biodiversity. *Science* **2015**, *348*, 336–340. [CrossRef] [PubMed]
58. Bender, D.J.; Contreras, T.A.; Fahrig, L. Habitat loss and population decline: A meta-analysis of the patch size effect. *Ecology* **1998**, *79*, 517–533. [CrossRef]
59. Hanski, I.; Ovaskainen, O. The metapopulation capacity of a fragmented landscape. *Nature* **2000**, *404*, 755–758. [CrossRef]
60. Tscharntke, T.; Klein, A.M.; Kruess, A.; Steffan-Dewenter, I.; Thies, C. Landscape perspectives on agricultural intensification and biodiversity-ecosystem service management. *Ecol. Lett.* **2005**, *8*, 857–874. [CrossRef]
61. Lomba, A.; Bunce, R.G.H.; Jongman, R.H.G.; Moreira, F.; Honrado, J. Interactions between abiotic filters, landscape structure and species traits as determinants of dairy farmland plant diversity. *Landsc. Urban Plan.* **2011**, *99*, 248–258. [CrossRef]
62. Talaga, S.; Petitclerc, F.; Jean-François, C.; Dézerald, O.; Leroy, C.; Cereghino, R.; Dejean, A. Environmental drivers of community diversity in a neotropical urban landscape: A multi-scale analysis. *Landsc. Ecol.* **2017**, *32*, 1805–1818. [CrossRef]
63. Yongjiu, F.; Yang, L.; Xiaohua, T. Spatiotemporal variation of landscape patterns and their spatial determinants in Shanghai, China. *Ecol. Indic.* **2018**, *87*, 22–32.
64. Stiles, A.; Scheiner, S.M. A multi-scale analysis of fragmentation effects on remnant plant species richness in Phoenix, Arizona. *J. Biogeogr.* **2010**, *37*, 1721–1729. [CrossRef]
65. Dong, C.F.; Liang, G.F.; Ding, S.Y.; Lu, X.L.; Tang, Q.; Li, D.K. Multi-scale effects for landscape metrics and species diversity under the different disturbance: A case study of Gongyi City. *Acta Ecol. Sin.* **2014**, *34*, 3444–3451.
66. Wang, Y.T.; Ding, S.Y.; Liang, G.F. Multi-scale effects analysis for landscape structure and biodiversity of seminatural habitats and cropland in a typical agricultural landscape. *Prog. Phys. Geogr.* **2014**, *33*, 1704–1716.
67. Monteiro, A.T.; Fava, F.; Goncalves, J.; Huete, A.; Gusmeroli, F.; Parolo, G.; Spano, D.; Bocchi, S. Landscape context determinants to plant diversity in the permanent meadows of Southern European Alps. *Biodivers. Conserv.* **2013**, *22*, 937–958. [CrossRef]
68. Siefert, A. Spatial patterns of functional divergence in old-field plant communities. *Oikos* **2012**, *121*, 907–914. [CrossRef]
69. Reitalu, T.; Helm, A.; Pärtel, M.; Bengtsson, K.; Gerhold, P.; Rosén, E.; Takkis, K.; Znamenskiy, S.; Prentice, H.C. Determinants of fine-scale plant diversity in dry calcareous grasslands within the Baltic Sea region. *Agric. Ecosyst. Environ.* **2014**, *182*, 59–68. [CrossRef]
70. Rocchini, D.; Delucchi, L.; Bacaro, G.; Cavallini, P.; Feilhauer, H.; Foody, G.M.; He, K.S.; Nagendra, H.; Porta, C.; Ricotta, C.; et al. Calculating landscape diversity with information-theory based indices:A GRASS GIS solution. *Ecol. Inform.* **2013**, *17*, 82–93. [CrossRef]
71. Hernandez-Stefanoni, J.L. The Role of Landscape Patterns of Habitat Types on Plant Species Diversity of a Tropical Forest in Mexico. *Biodivers. Conserv.* **2006**, *15*, 1441–1457. [CrossRef]
72. Krauss, J.; Steffan-Dewenter, I.; Tscharntke, T. Landscape occupancy and local population size depends on host plant distribution in the butterfly Cupido minimus. *Biol. Conserv.* **2004**, *120*, 355–361. [CrossRef]

73. Saura, S.; Torras, O.; Gil-Tena, A.; Pascual-Hortal, L. Shape Irregularity as an Indicator of Forest Biodiversity and Guidelines for Metric Selection. In *Patterns and Processes in Forest Landscapes*; Lafortezza, R., Sanesi, G., Chen, J., Crow, T.R., Eds.; Springer: Dordrecht, The Netherlands, 2008; pp. 167–189.
74. Mitchell, M.S.; Rutzmoser, S.H.; Wigley, T.B.; Loehle, C.; Gerwin, J.A.; Keyser, P.D.; Lancia, R.A.; Perry, R.W.; Reynolds, C.J.; Thill, R.E.; et al. Relationships between avian richness and landscape structure at multiple scales using multiple landscapes. *For. Ecol. Manag.* **2006**, *221*, 155–169. [CrossRef]
75. Wellstein, C.; Otte, A.; Waldhardt, R. Seed bank diversity in mesic grasslands in relation to vegetation type, management and site conditions. *J. Veg. Sci.* **2007**, *18*, 153–162. [CrossRef]
76. Wrbka, T.; Schindler, S.; Pollheimer, M.; Schmitzberger, I.; Peterseil, J. Impact of the Austrian Agri-environmental scheme on diversity of landscapes, plants and birds. *Community Ecol.* **2008**, *9*, 217–227. [CrossRef]
77. Honnay, O.; Piessens, K.; Van Landuyt, W.; Hermy, M.; Gulinck, H. Satellite based land use and landscape complexity indices as predictors for regional plant species diversity. *Landsc. Urban Plan.* **2003**, *63*, 241–250. [CrossRef]
78. Janisova, M.; Michalcova, D.; Bacaro, G.; Ghisla, A. Landscape effects on diversity of semi-natural grasslands. *Agric. Ecosyst. Environ.* **2014**, *182*, 47–58. [CrossRef]
79. Wang, R.X.; Yang, X.J. Nestedness theory suggests wetland fragments with large areas and macrophyte diversity benefit waterbirds. *Ecol. Evol.* **2021**, *11*, 12651–12664. [CrossRef] [PubMed]
80. Horak, J.; Peltanova, A.; Podavkova, A.; Safarova, L.; Bogusch, P.; Romportl, D.; Zasadil, P. Biodiversity responses to land use in traditional fruit orchards of a rural agricultural landscape. *Agric. Ecosyst. Environ.* **2013**, *178*, 71–77. [CrossRef]
81. Blaauw, B.R.; Isaacs, R. Larger patches of diverse floral resources increase insect pollinator density, diversity, and their pollination of native wildflowers. *Basic Appl. Ecol.* **2014**, *15*, 701–711. [CrossRef]
82. Matthies, S.A.; Ruter, S.; Prasse, R.; Schaarschmidt, F. Factors driving the vascular plant species richness in urban green spaces: Using a multivariable approach. *Landsc. Urban Plan.* **2015**, *134*, 177–187. [CrossRef]
83. Tulloch, A.I.T.; Barnes, M.D.; Ringma, J.; Fuller, R.A.; Watson, J.E.M. Understanding the importance of small patches of habitat for conservation. *J. Appl. Ecol.* **2016**, *53*, 418–429. [CrossRef]
84. Stejskalova, D.; Karasek, P.; Tlapakova, L.; Podhrazska, J. Sinuosity and Edge Effect-Important Factors of Landscape Pattern and Diversity. *Pol. J. Environ. Stud.* **2013**, *22*, 1177–1184.

International Journal of
Environmental Research and Public Health

Article

Optimization of Work Environment and Community Labor Health Based on Digital Model—Empirical Evidence from Developing Countries

Shiya Gao [1,†], Zeyu Wang [2,*,†], Shaoxiang Jiang [3], Wen Ding [4], Yuchen Wang [5] and Xiufang Dong [2]

1 School of Management, Wuhan Polytechnic University, Wuhan 430023, China
2 School of Public Administration, Guangzhou University, Guangzhou 510006, China
3 National School of Development, Peking University, Beijing 100871, China
4 Faculty of Engineering, University of Waterloo, Waterloo, ON N2L 6J5, Canada
5 Carey School of Business, Johns Hopkins University, Baltimore, MD 21202, USA
* Correspondence: theowang@gzhu.edu.cn
† These authors contributed equally to this work.

Abstract: As far as we know, for large manufacturing enterprises, there is often a community of labor gathered around such enterprises, which is especially used as a place for the enterprise to place the labor force. This paper aimed to update the industry model of Chinese Manufacturing Enterprises (CMEs) to improve workers' health management. This work first discusses the value, mode, and process of Enterprise Digital Transformation (EDT) and Worker Health and Safety Management (WHSM). Then, it proposes the CMEs-oriented EDT model and WHSM system based on Big Data Technology (BDT) and the Internet of Things (IoT). The proposed model and system are verified through a case study on the Shanghai BYD manufacturing enterprise (short for BYD) using the Fuzzy Comprehensive Evaluation Method (CFEM). The EDT model verification considers the adaptation and performance of enterprises after EDT. The WHSM system considers workers' oxygen inhalation status to evaluate their heart and cardiovascular health. The results show that EDT improves the enterprise's revenue and reshuffles the revenue structure. The EDT model has absolute adaptability to BYD. It has greatly improved BYD's indexes, especially financial performance, market capability, and technical capability.

Keywords: FCEM manufacturing enterprises; enterprise digital transformation; health and safety management; internet of things; big data technology; FCEM

Citation: Gao, S.; Wang, Z.; Jiang, S.; Ding, W.; Wang, Y.; Dong, X. Optimization of Work Environment and Community Labor Health Based on Digital Model—Empirical Evidence from Developing Countries. *Int. J. Environ. Res. Public Health* **2022**, *19*, 13114. https://doi.org/10.3390/ijerph192013114

Academic Editors: Yuan Li, Hongxiao Liu and Tong Wu

Received: 25 August 2022
Accepted: 8 October 2022
Published: 12 October 2022

1. Introduction

Chinese Manufacturing Enterprises' (CMEs) profit margin is still meager compared with developed nations. Enterprise Digital Transformation (EDT) faces many difficulties, such as a lack of capital investment [1]. First, the CME executives are not aware of the necessity, urgency, and complexity of EDT, and the management concept stays on the deployment of common Information Technology (IT) systems. The IT department cannot achieve Enterprise Digital Transformation (EDT) alone. It must be led by the enterprise decision-makers and promoted from top to bottom [2,3]. Secondly, CMEs have applied many information systems. However, the basic data are inaccurate, and the coding system is not unified, failing to promote the EDT [4]. Many CMEs believe that promoting an automation system can accelerate the effect of EDT, and there is a general situation of emphasizing automation and neglecting digitalization. Additionally, many CMEs only emphasize production line automation and labor reduction without a solid foundation for equipment networking and data acquisition. The workshop has not genuinely realized visualization, failing to monitor equipment and manage workers' health conditions. Moreover, the CMEs-oriented EDT investment has not achieved remarkable results, restricting their motivation to further transformation. For example, some CMEs invest in self-built e-commerce platforms. Nevertheless, most customers still favor mainstream e-commerce platforms, so self-built platforms have not achieved the expected results. Meanwhile, sub-industries under manufacturing vary greatly. Enterprises from different industrial chains feature high personalization. Thus, no universal implementation path is applicable, and no ready-made EDT models are available [5,6].

In March 2022, Zhengzhou Haier Water Heater Interconnection Factory (Zhengzhou Haier) was selected by virtue of the wide application of advanced industrial 4.0 technology in the whole industrial chain, becoming the world's first end-to-end "lighthouse factory" of water heaters and the fourth landing application of the Haier Group. The World Economic Forum (WEF) commented: "facing the vigorous development of the water heater market and the increasing demand for high-end products and services, Zhengzhou Haier, a new factory, uses Big Data Technology (BDT), and 5th Generation (5G), Edge Computing (EC), and ultra-wideband solutions to establish close ties with suppliers, factories, and users." From 2020 to early 2021, Haier's water heater's order response speed increased by 25%, production efficiency increased by 31%, and product quality increased by 26%. According to the report of the WEF, each lighthouse factory has unique and diversified success cases. More broadly, it mainly includes the digital realization within a single factory (focusing on digital assembly, processing, maintenance, performance management, quality management, and sustainable development) and opening up the end-to-end value chain (focusing on the supply network, product development, planning, delivery, and customer connectivity). Obviously, EDT has brought a new business model to Zhengzhou Haier, improved its core competitiveness, reduced operating costs by digital means, increased revenue and efficiency, and upgraded management mode and operation efficiency [7].

In fact, EDT uses digital technology to promote a series of changes in production mode, organizational structure, and corporate culture [8]. After EDT, enterprise management should explore new sources of revenue and develop new products, services, and business models. Therefore, EDT is the deep integration of technology and business models, and it promotes digital technological development and technical support capabilities. EDT revolutionizes the traditional business model by creating a new and dynamic digital business model [9,10]. In terms of the research on EDT, Zimmerman et al. (2019) studied the transformation and upgrading of enterprises based on GVC. They proved that cross-industrial upgrades could improve enterprises in all aspects [11]. Kaletnik et al. (2020) explored the method and process of EDT and put forward a new path and implementation EDT scheme [12]. Correani et al. (2020) examined EDT and upgrading from internal and external aspects [13]. Shen et al. (2021) believed that innovative technologies could be used to realize the CMEs-oriented EDT [14]. Meanwhile, Kane et al. (2019) argued that the EDT could realize the data connection between all manufacturing enterprises and the digital structure up-gradation and improve operation efficiency [15]. Son et al. (2019) contended that changing enterprise operation thinking could promote EDT. For example, the new operation modes could be realized by expanding consumer groups [16]. The above literature carries out EDT aiming at a certain point of the enterprise business model. They do not comprehensively study the enterprise organizational structure, enterprise resource allocation method, production, and operation mode. Thus, the above research is not comprehensive and mono-directional.

At the same time, in recent years, workers' sub-health and even sudden death have attracted much attention. All social sectors generally call for strengthening Worker Health and Safety Management (WHSM). Workers have promoted the rapid development of enterprises. However, the problem of overwork also affects workers' physical and mental health. The claim settlement data of corporate group insurance also presents the younger trend of workers' illnesses. The 2021 Group Health Insurance Claim Settlement Report released by Taikang Endowment Insurance Co., Ltd. (Beijing, China) shows that 54% of workers with hypertension are young adults aged 30–49; 45% of workers with intervertebral disease are concentrated in middle-aged adults, 30–39 years old; and 43% of sudden death cases are concentrated among young people in the 40–49 age group. In fact, most sub-healthy people often do not realize what risks they are facing. They often need to go to the hospital to realize the seriousness of their health problems. This traditional concept of emphasizing "medicine" over "prevention" easily leads to workers' health status deterioration, increases the medical expenses of enterprises and society, and exacerbates the shortage of medical resources. Therefore, enterprise managers gradually have formulated strategies for WHSM.

The keywords of health management involve many aspects of research. At present, the actual development of health management, especially in community health services, WHSM, and the health insurance industry, still lacks systematic research and governance [17]. Therefore, this work applies the Internet of Things (IoT) technology to establish the CMEs-oriented EDT model and WHSM system. The aim is to provide solutions for the CMEs' development and promote workers' health conditions.

Firstly, this work discusses the value, mode, and process of EDT and WHSM in CMEs. Secondly, based on the BDT and IoT, it establishes the CMEs-oriented EDT model and the WHSM system. Finally, Shanghai BYD Manufacturing Enterprise (short for BYD), it evaluates the adaptation and performance of the proposed EDT model through Fuzzy Comprehensive Evaluation Method (FCEM). Under the proposed WHSM system, workers' health is analyzed by considering their heart and cardiovascular situation when inhaling hyperbaric oxygen.

2. Materials and Methods

2.1. Significance of EDT

Both "digitization" and "digitalization" are used in EDT. The first term means explicitly converting analog information into digital information (for example, manually filled documents are automatically recognized as digital information). The other term refers to integrating digital information into enterprises, deepening the application of various business software and emerging technologies, such as the IoT. In the research process, it is believed that the digital transformation of manufacturing enterprises is not a single-link improvement process, but more like the penetration of digital kinetic energy into all components of manufacturing enterprises to stimulate the aggregation effect of enterprises. In this process, it is not to apply the emerging digital technology surface to the production and circulation of products. It not only improves production efficiency, but also reconstructs business processes through new digital thinking concepts, generates new economic benefits, better opens up new markets, and improves the competitiveness of enterprises. The essence is to apply digitalization to the strategic layer of the enterprise, to promote the reform of the organizational structure and management methods of the enterprise, and to enhance the core competitiveness of the enterprise through endogenous and external growth. Feng et al. (2022) defined digitization as the transformation of organizations that integrate mathematical techniques and business processes into the digital economy [18]. Shpak et al. (2020) defined digital transformation as the use of technology to radically improve business performance or reach [19]. It helps realize data-driven decision analysis and completely changes the business process of enterprises. Essentially, EDT is the process of enterprises' real digitalization. Lin et al. (2020) studied the digital transformation of manufacturing enterprises, and believed that the digital transformation of manufacturing enterprises is not the further improvement of the evolution of various business processes, but a comprehensive reform of the value proposition, operation mode, enterprise organization mode, resource allocation mode, R&D mode, production mode, and marketing mode [20]. Figure 1 is the value system of EDT.

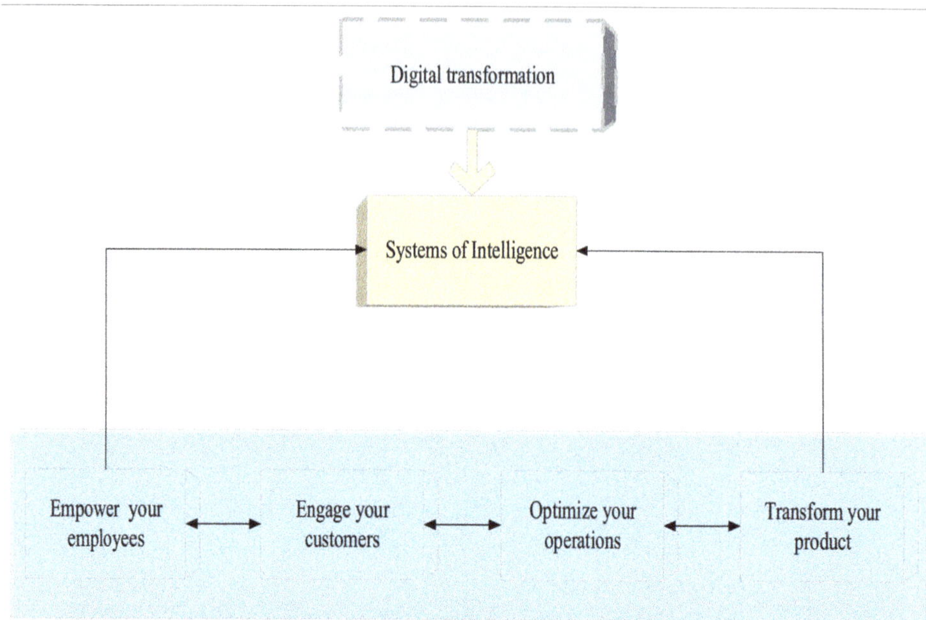

Figure 1. Value system of EDT.

In Figure 1, EDT can realize data acquisition, state perception, remote control, increased product added value, and service revenue. At the same time, EDT can help manufacturing enterprises cope with increasingly complex compliance requirements, especially in industries involving people's livelihood, such as medicine, food, and import–export. EDT helps them realize the traceability of the whole production process. The complexity of manufacturing enterprises' organization, business, products, and value chain has brought many obstacles to the EDT [21]. Maciag (2022) studied the basic model of digital transformation and believed that digital transformation of enterprises should use the Internet's way of thinking to reconstruct enterprises from four aspects: business model, capital model, management model, and mental model, so that enterprises can obtain stronger vitality through transformation [22]. Zhang et al. (2022) studied the basic model of digital transformation of Chinese manufacturing enterprises and believed that China's manufacturing industry had many problems such as excess capacity, low resource utilization, and unreasonable structure. The technology and thinking of the Internet should be used in the whole life cycle of the manufacturing industry. Traditional industries should be transformed into science and technology, intelligence and efficiency, redefine R&D design, manufacturing, operation management, sales and services, and promote the in-depth development of the industry and industrial transformation and upgrading [23]. Figure 2 gives common modes of EDT for manufacturing enterprises. If EDT (Enterprise Digital Transformation) is regarded as a theoretical framework model, we need to define its components. Then, considering the basic theory of digital transformation, it should include changes in the business operation mode, service mode, research and development mode, organizational structure management mode, and production mode. After completing the above changes, the decision-making model of the enterprise will be comprehensively changed.

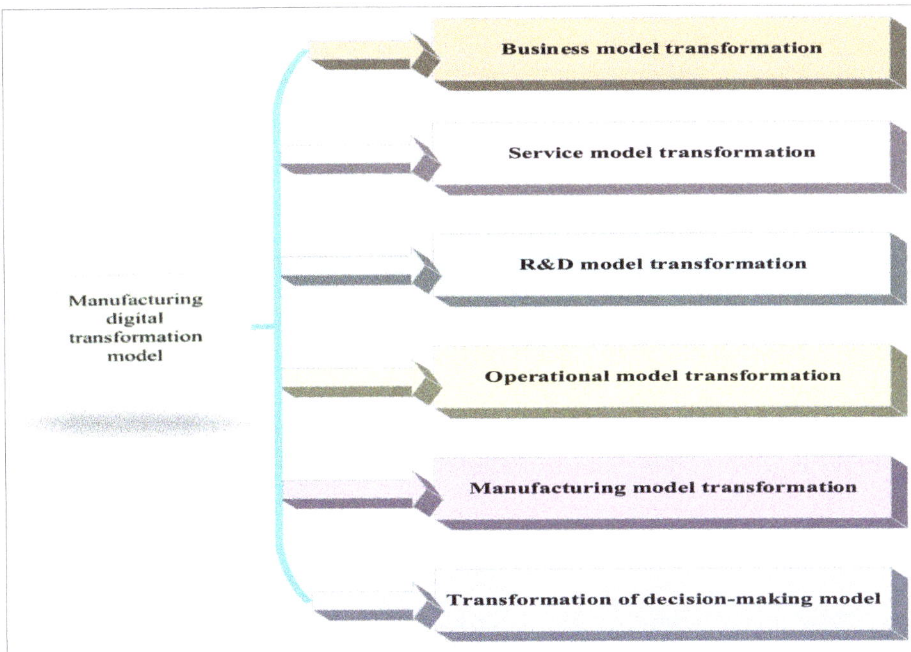

Figure 2. Common modes in EDT for manufacturing enterprises.

In Figure 2, the common modes of manufacturing enterprises' EDT are dissected in detail. First, the new business model is a Pay-For-Service mode based on digital technology. Enterprises no longer sell products but sell the services they use. To realize the Pay-For-Service mode, enterprises must realize product digitalization. The products themselves should be integrated into a Cost Per Sales (CPS) system with communication, calculation, and control capabilities. At the same time, a cloud platform must be established to monitor product operation to achieve predictive maintenance of the operating products. Additionally, manufacturing enterprises should also promote the customization of online products and the combination of online and offline experience marketing, which is also part of the transformation of the business model. Second, the service mode transformation enables customers to realize self-service by developing products and service-specific Applications (APPs) to improve service efficiency. Some of the world's leading equipment manufacturing enterprises can also develop service APPs to realize remote condition monitoring and predictive maintenance. Using Augmented Reality (AR) technology to maintain the equipment and display the equipment sensor data and the assembly process can greatly improve the efficiency of equipment maintenance. Third, the Research and Development (R&D) model transformation can reduce the testing of actual manufactured products through simulation-driven design. By managing the R&D data and process of the whole product life cycle, enterprises can improve the reuse rate of parts and the R&D efficiency while reducing costs. In addition, it can realize remote and collaborative R&D. Fourth, operation mode transformation can help manufacturing enterprises realize fine management. Fifth, the manufacturing mode transformation can be associated with the technology of other excellent enterprises to realize fully automatic processing of different mechanical parts. Lastly, enterprises can conduct multi-dimensional big data analysis by transforming the decision-making model. It improves the real-time visualization of data analysis. As such, enterprises realize data governance, decision-making based on the data drive, and analyzing the key information behind the data using AI and BDT [24].

In order to promote EDT, manufacturing enterprises must define the digital transformation strategy, formulate the digital transformation plan, and then implement it. In this process, they need to rely on professional consulting service institutions to complete processes such as EDT state diagnosis, demand analysis, process review, and overall framework [25]. Bican et al. (2020) studied the process of digital transformation and believed that digital transformation is a process of upgrading strategic thinking. Although this process may involve the upgrading of enterprise Information Technology (IT) infrastructure, it is not a technology in nature, but a strategy [26]. Figure 3 illustrates the flow of EDT of manufacturing enterprises. After setting the basic elements to be considered for the transformation of the Enterprise Digital Model. In fact, the figure defines the object of digital transformation from the perspective of goal setting, and analyzes the motivation and necessity of digital transformation from the perspective of Motivation Analysis Theory.

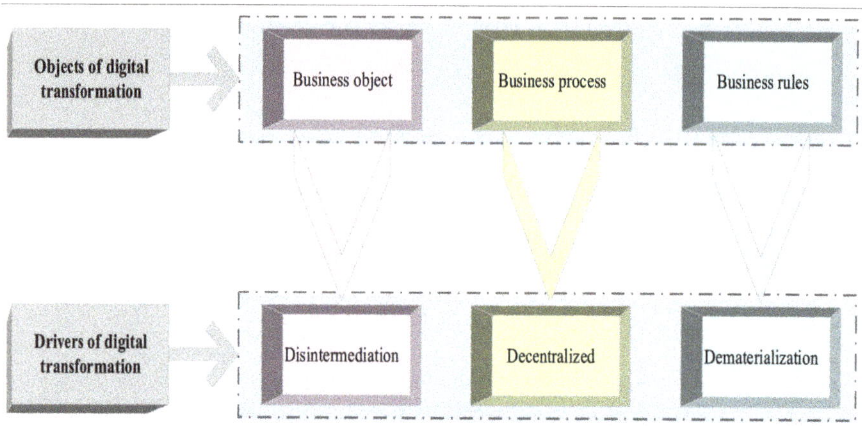

Figure 3. EDT flow of manufacturing enterprises.

2.2. CME and WHSM

According to the 2020 China Workplace Worker Health Risk Report survey results, 29% of workers are overweight and obese, 43% have substandard diets, and 48% lack sleep. Among these respondents, more than 60% believe that their sub-health status affects their normal life and activities, and more than half of the workplace workers lack vitality [27]. It shows that the enterprise's good employee health management is a necessary guarantee for the development of the enterprise. Whether it is to ensure the growth and development of enterprises and build harmonious labor relations from the macro level of national policies, social economy, and meso level, or from the perspective of the health needs of employees and their families at the micro level, it is the general trend for enterprises to do a good job in employee health management. The research believes that the concept of personal mobile health management is a way to monitor and record the whole process of human health by using mobile network technology. Mobile health includes smart terminals, health data management, related matching resources, and collaborative systems. Whether it is a pure digital medical concept or a health management IoT architecture based on the IoT and cloud computing, its core business is still in the field of mobile health management. There are four main points in mobile health: First, is the collection of human data. Some supplementary programs are related to human health and fitness, and others do medical consultations that do not require a lot of diagnosis and treatment methods and data. Follow-up can be performed after discharge, including disease investigation and remote testing. Priyono et al. (2020) integrated the resources and transformation strategies of enterprise digital transformation, and proposed a theoretical framework of resource adaptation for enterprise digital transformation. They believed that the prerequisite variables for the success of the digital transformation of enterprises are internal and external resources and internal and external capabilities, and external capabilities are an important factor for the success of digital transformation [28]. The attention to enterprise workers' physical and mental health has also been improved. Managing workers' health and welfare has become an important consideration for many people's career choices. Figure 4 describes the process of WHSM in manufacturing enterprises. Figure 4 is in Figure 3. On the basis of the Business Process mentioned above, specific examples were analyzed. Further, we analyzed how Business Process is designed and works in WHSM based on our research objectives.

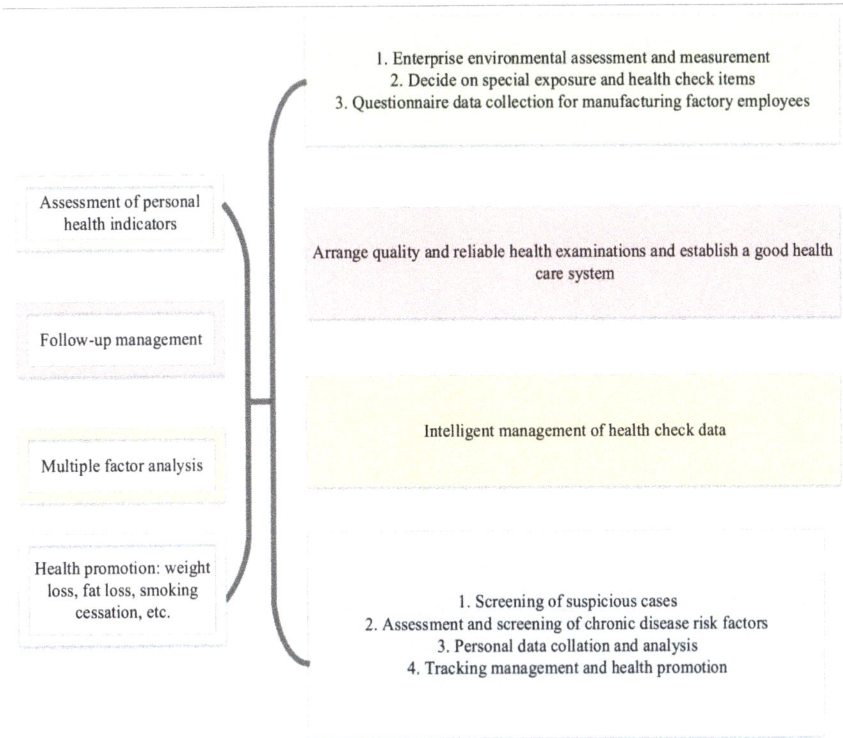

Figure 4. Process of WHSM in manufacturing enterprises.

2.3. IoT Technology

With the development of Cloud Computing (CC), BDT, and other IT, the IoT+cloud platform has come into being. It is the value cohesion of the IoT industry. Moreover, the IoT+cloud platform sees a vast market potential with more technological innovation. The IoT+cloud platform is currently in the exploration of post-precipitation mode. It is about to enter the inflection point before rapid development, from focusing on the underlying hardware to the multi-scenario business capability of the software platform [29]. Mehmood et al. (2019) studied the characteristics of the IoT platform and believed that the core and foundation of the IoT was still the Internet, which was an extended and expanded network based on the Internet. The IoT client extends to any item for information exchange and communication [30]. Figure 5 demonstrates the architecture of the IoT+cloud platform. Figure 5 is the link to the product design. For example, based on how IoT plays a specific role in digital transformation, especially in the digital transformation of manufacturing enterprises. This step is equivalent to the design of the transformation path based on the theoretical design in Figures 2–4 and the actual situation.

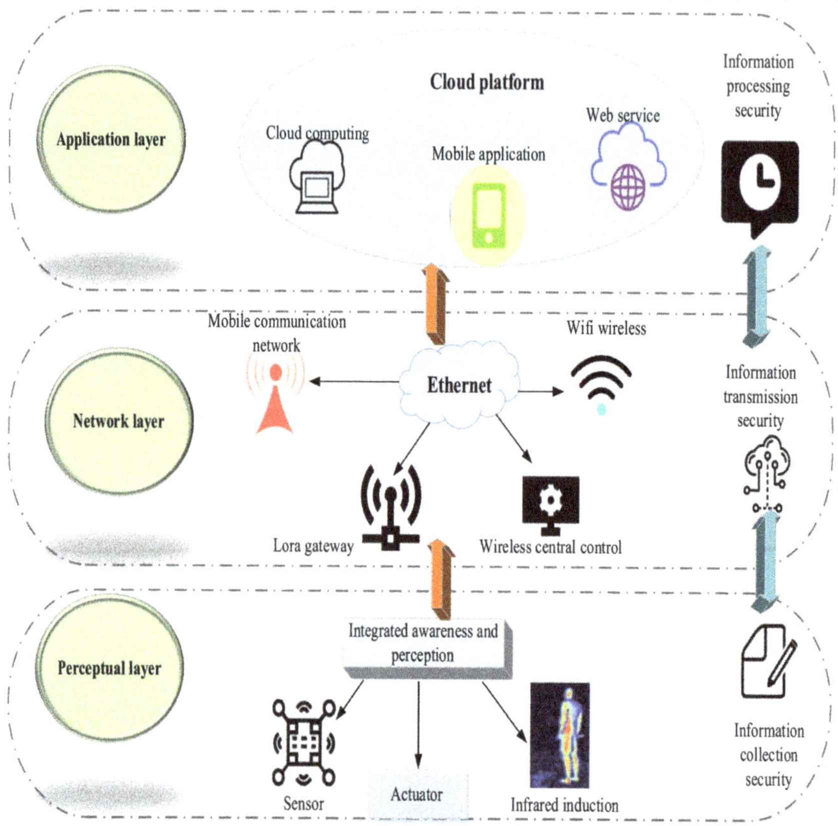

Figure 5. IoT+cloud platform architecture.

According to Figure 5, the IoT+cloud platform connects and manages IoT terminals through the network and perception layers. It collects and stores sensing data and provide standard interfaces and general tool modules for development and applications through the network and application layers. Finally, it indirectly reaches the end-user through Software as a Service (SaaS). The IoT+cloud platform is the technical integration of the IoT platform and CC. From the perspective of the concept of the IoT, the concept of the IoT must be the process control of the whole process, which must be supported by standards. Evolution is the use of complex IT technology to create a simple digital health management system. The standardization of the digital health management system is to improve the standardization level of the digital health management system and gradually promote the standardization of the digital health management process. Standardization of digital health management systems also solves IT challenges. The most difficult thing for manufacturing companies is the health management of employees. There is no perfect standard for employee health management. The emergence of the IoT essentially promotes the improvement of the entire medical informatization, and the digital health management system has broad application prospects in the development of enterprises.

Cloud computing is also a way to achieve infrastructure sharing, where a large number of resources are linked together in a public or private network to provide IT services. From an IT perspective, cloud computing is the provision of Internet-based software services. The software used by the user does not need to be on their own computer, but uses the Internet to access the software on an external machine through a browser. All work, user files, and data are also stored on these external machines [31]. Big data, on the other hand, enable the collection of raw, structured, or unstructured data from various sources. With the execution of machine learning techniques and big data predictive models, it is feasible to predict machine failures and customer behavior. Big data analytics helps businesses identify customers by understanding their thought processes and feedback in advance. Enterprises can adjust their strategies accordingly, providing deeper insights from disparate data sources in unique ways to help define the overall strategy. Big data analysis can promote customer acquisition and retention, effectively manage enterprises, and reduce enterprise management risks [32].

The large-scale application of edge computing can improve computing efficiency, reduce latency, and improve data security. In this way, various problems caused by the excessively centralized data processing mode of cloud computing in the current digital transformation can be solved. Massive amounts of data for digital transformation are aggregated from tiered networks to the cloud, where they are then analyzed and processed. In the era of massive data generation, the efficiency of this data processing method will become lower and lower. Sinking data to the "micro cloud" on the edge side for processing can not only improve efficiency, but also promote the transformation of traditional manufacturing from "hardware thinking" to "service thinking". However, edge computing drives digital transformation from product-oriented to service-oriented. Every terminal with edge computing capabilities can become a service provider, and all service providers are on the same plane. The application of edge computing in digital transformation drives the transformation of the life cycle from product to service operations, which in turn leads to product service and business model innovation. This has a profound impact on the development of value chains, supply chains, and ecosystems, and promotes their high-quality development [33].

2.3.1. Modeling CMEs-Oriented EDT Based on IoT

Applying IoT in manufacturing enterprises is called Manufacturing Internet of things (MIoT). It focuses on Machine to Machine (M2M) communication. It uses BDT, AI, CC, and other technologies to achieve more efficient and reliable manufacturing operations.

As an information technology term, big data can refer to the data set. These data sets are difficult to process and analyze by traditional computing methods. However, they can be systematically obtained by advanced technical means, including collection, storage, analysis, transmission, retrieval, and display. They can generate value. Therefore, big data is a collective term involving both the data sets and corresponding processing methods. The distributed file system is the first link in the big data storage system. Its basic idea is to establish a file system on multiple machines using the master–slave structure. The file system exposes the management functions and interface capabilities through the master node. The real data are stored on the data node. However, all read and write operations are scheduled through the master node. The file system has multiple backups to avoid data loss or single-node failure errors. Multiple backups can also be used for load balancing between nodes. Another type of big data storage is relational database clusters. The traditional relational database system includes two most important parts, the storage engine, and the Structured Query Language (SQL) service. The storage engine is responsible for data storage and reading. SQL services include SQL parsing, caching, and index calculation. The stand-alone version of the relational database is limited by the size of the stand-alone storage space. In contrast, the cluster version of the relational database can handle large-scale data sets. The storage engine is the key to breaking through the single-machine storage capacity. It can support multiple data backups and share data among multiple SQL service instances [34].

Big Data Analysis (BDA) plays a key role in the preventive maintenance of production equipment. Its core is the cyberspace entity system. The network entity system is designed through the 5C (Computer, Communication, Content, Customer, and Control) framework to realize the connection, transformation, networking, cognition, and configuration. It converts the collected data into useful information and optimizes the production process. Traditionally, service providers mainly obtain solutions through experiences in EDT, only referring partly to data and process analysis. However, the traditional EDT services are no longer applicable in the digital era, and many experiences and data connotations have changed quietly. Establishing a digital ecosystem is completely different from traditional informatization [35]. Willner et al. (2020) studied the core technologies of enterprise digital transformation, and believed that digital technology could improve the output or production technology of manufacturing enterprises, thereby realizing the improvement of the enterprise economy. The application of digital technology can link the data islands among all levels of the industry, improve the operational efficiency of enterprises, and build a new pattern of digital economy [36]. Figure 6 depicts the proposed CMEs-oriented EDT model based on the IoT.

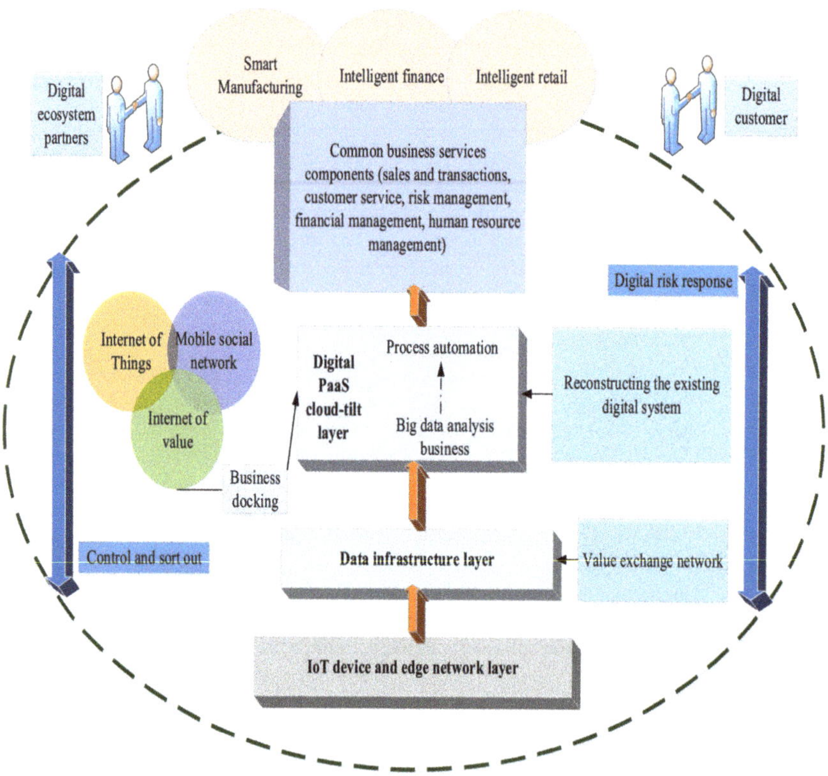

Figure 6. CMEs-oriented EDT model based on IoT.

Figure 6 provides a common CEMs-oriented EDT model built around vision, data, and organization. A comprehensive and systematic organizational capability change and transformation model will be completed by coordinating the strategy formulation, business functions, data, and implementation. Figure 6 is mainly based on Figure 5. The proposed IoT design is combined with the concept of CMES. This is mainly based on the analysis of the theory of information asymmetry between customers and enterprises. The reason why information asymmetry between them is greatly reduced by introducing CMES and IoT technology.

2.3.2. WHSM System Based on IoT

WHSM helps enterprises comprehensively monitor workers' health risk factors. Manufacturing enterprises can apply IoT-based visualization edge gateway, integration unit, micro base station, IoT Access Point (AP), and modules. In particular, visual edge gateways are used in low-frequency proximity sensing and positioning and long-distance signal transmission coverage in public areas. All units and departments use base stations and IoT modules (that support equipment access to IoT platforms with different frequency bands and protocols). These modules can expand the IoT to meet business needs with various application scenarios. Dayan et al. (2022) studied the development of health management systems and found that most pharmaceutical companies have established a standard system of equipment, frequency, operation, and coding for their own use. The standardization and compatibility are poor, affecting applications outside the system and within the supply chain. The IoT technology industry, which is closely related to health management, is lagging behind in China. Chip design and manufacturing, antenna design and manufacturing, label packaging and packaging equipment manufacturing, reader device development, data management software design, and other production links are a complete production process. In the IoT technology industry chain, there is a shortage of talent in RFID-related technology development, system design, experimental testing, and project implementation, and training is urgently needed [37]. By building a healthy MIoT environment, enterprises can invite more partner enterprises and avoid repeated construction. Moreover, it can simplify the network system architecture and reduce ownership and maintenance costs. The proposed CMEs-oriented WHSM system based on IoT is displayed in Figure 7.

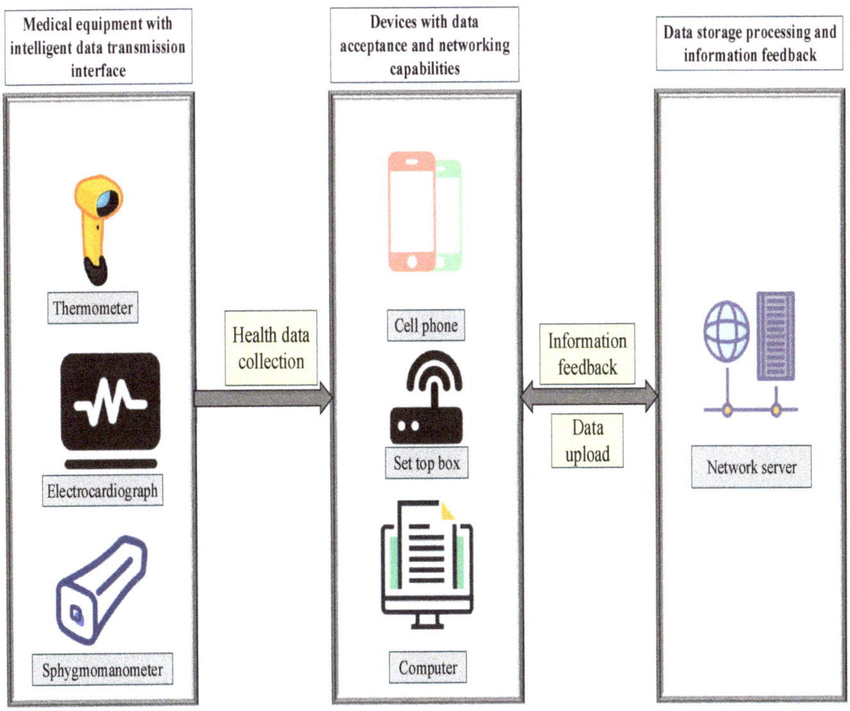

Figure 7. WHSM system based on IoT.

As in Figure 7, the WHSM is an information management system that tracks, records, analyzes, and processes the personal health parameters of enterprise workers over the long term. It feeds back workers' health suggestions on time. The system is based on the IoT technology and the Internet server and runs for a long time. Meanwhile, the health data of the employees of the enterprise is continuously updated on the Internet terminal, and the ability to analyze and process the data is improved to more comprehensively monitor the personal health of the enterprise labor force, provide more complete health guidance suggestions, and achieve the purpose of preventing working diseases and protecting the health of the labor force. Finally, as seen in Figure 7, the WHSM system based on IoT is actually the final result of the whole model design. Through analysis, we summarized s the types of IoT devices commonly used in Chinese manufacturing enterprises and how to incorporate these devices into a design framework of the WHSM system. This design module analysis diagram is actually a summary of the above logic derivation.

2.4. Data Source and Processing Method

Taking Shanghai BYD Manufacturing Company as an example, this paper analyzed the adaptability and effect of the enterprise's digital transformation, as well as the health management of employees after the enterprise's digital transformation. Due to the need to measure the effect of digital transformation on manufacturing enterprises from many aspects, it is difficult to determine the measurement standard among various indicators of the enterprise. Therefore, the Fuzzy Comprehensive Evaluation Method (FCEM) is adopted to evaluate the adaptability and effect of digital transformation of enterprises [38,39]. FCEM is a comprehensive evaluation method based on fuzzy mathematics, which comprehensively evaluates things or objects restricted by many factors. It has the characteristics of clear results and strong systematicness. It can better solve problems that are fuzzy and difficult to quantify, and it is suitable for solving various uncertain problems. It is assumed that there are n factors related to the evaluated indexes, and the index set is given below:

$$U = \{u_1, u_2, \cdots, u_i, \cdots, u_n\} \tag{1}$$

In (1), U is the total index factor. u_i indicates the single-index factor. The weight $A = \{a_1, a_2, \cdots, a_n\}$ is used to measure the importance of each factor. Set m shows the possible comments. (2) describes the comment set:

$$V = \{v_1, v_2, \cdots, v_i, \cdots, v_m\} \tag{2}$$

In (2), V is the total comment set. v_i represents a single comment on an index. The membership vector $r_i = (r_{i1}, r_{i2}, \cdots, r_{im})$ is obtained by single factor evaluation, finally forming a membership matrix, as shown in (3):

$$R = \begin{bmatrix} r_{11} & r_{12} & \cdots & r_{1m} \\ r_{21} & r_{22} & \cdots & r_{2m} \\ \vdots & \vdots & \vdots & \vdots \\ r_{n1} & r_{n2} & \cdots & r_{nm} \end{bmatrix} \tag{3}$$

Then, we determined the factor centralization vector, normalized the evaluation set, and finally calculated the comprehensive evaluation vector. The purpose is to make the comprehensive evaluation value according to the principle of maximum membership. The calculation of the comprehensive evaluation vector is shown in (4) (where $\circ$ is a fuzzy operator):

$$B = A \circ R \tag{4}$$

Based on this, a Comprehensive Fuzzy Evaluation (FCE) was conducted by factoring in the five indexes of the enterprise after the EDT. Specifically, the five indexes are enterprise operation (financial performance), market capacity, technical capacity, management capacity, and enterprise environment. Then, the comment sets v_1, v_2, v_3, and v_4 represent very adaptable, relatively adaptable, not much adaptable, and totally unadaptable. The membership vector is obtained by evaluating a single index. According to the capital distribution of BYD, the weight of the index vector is set as $\{0.3, 0.2, 0.3, 0.2, 0.1\}$.

In recent years, there have been more and more sudden death and stroke among manufacturing workers. After health management, the health status of workers is measured by the oxygen uptake of workers [40]. The calculation of blood oxygen saturation is shown in (5):

$$SpO_2 = HbO_2/(HbO_2 + Hb) \times 100\% \quad (5)$$

In (5), HbO_2 is oxygenated hemoglobin. Hb represents the hemoglobin in red blood cells. The capability of blood to carry and transport oxygen is measured by oxygen saturation. Clinically, the f oxygen absorption concentration is calculated by (6):

$$\text{Inhaled oxygen concentration}(\%) = 21 + 4 \times \text{Oxygen flow}(\text{L/min}) \quad (6)$$

3. Results

3.1. Adaptability and Performance Analysis of CMEs-Oriented EDT Model

According to the financial changes of BYD during EDT from 2012 to 2015, with the growth over time, BYD's Internet business revenue is far greater than the manufacturing revenue and net profit. Before EDT, BYD's financial revenue mainly came from manufacturing and net profit. After EDT, BYD's Internet business revenue accounts for 67.29% of the total revenue. Thus, EDT improves the revenue of manufacturing enterprises and reshuffles the revenue structure. Figure 8 manifests the enterprise financial data during EDT. Figure 9 portrays the changes in enterprise indexes during EDT.

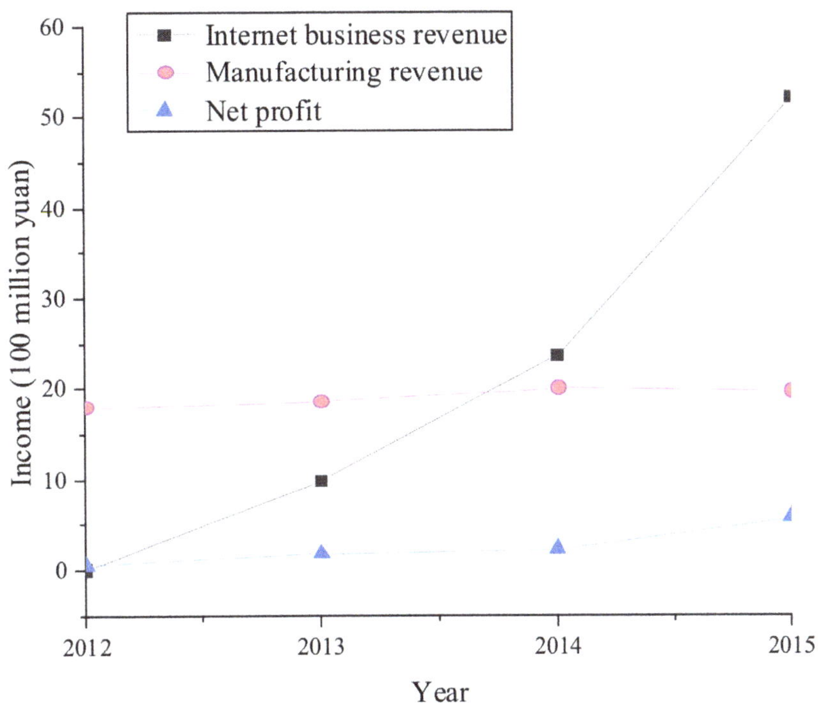

Figure 8. Changes in enterprise financial data during BYD's EDT.

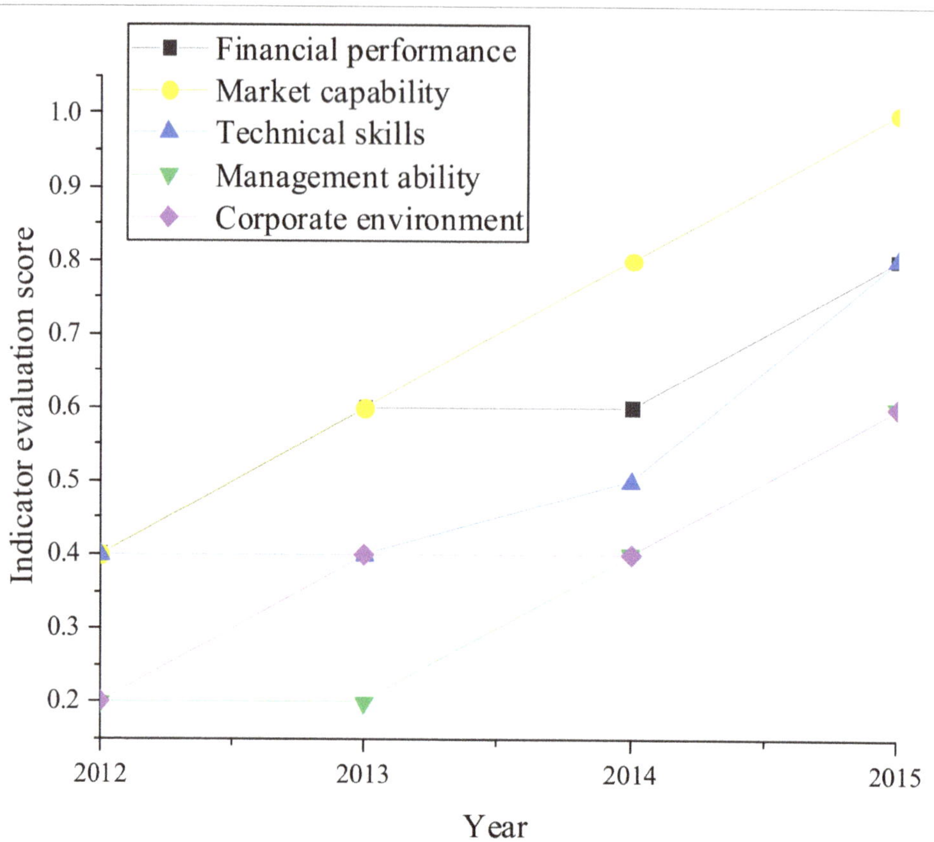

Figure 9. Changes in enterprise indexes during BYD's EDT.

According to Figure 9, all indexes have generally shown an upward trend during EDT, and the overall index increased by 42.11% from 2012 to 2015. From the change of the single index, the enterprise's financial performance, market capability, and technical capability have been greatly improved after EDT. Therefore, EDT has absolute adaptability to BYD and greatly improves BYD's index scores.

3.2. Analysis of Worker Health Status under the Proposed WHSM System

During EDT, BYD has carried out intelligent and comprehensive management of workers' health. Figure 10 plots the analysis results of the cardiopulmonary capability and cardiovascular health status of workers at different ages in BYD's EDT.

Apparently, after WHSM, workers' cardiopulmonary capability and cardiovascular capability have been improved. Therefore, the WHSM system can help improve workers' heart and cardiovascular health and effectively avoid sudden diseases during work.

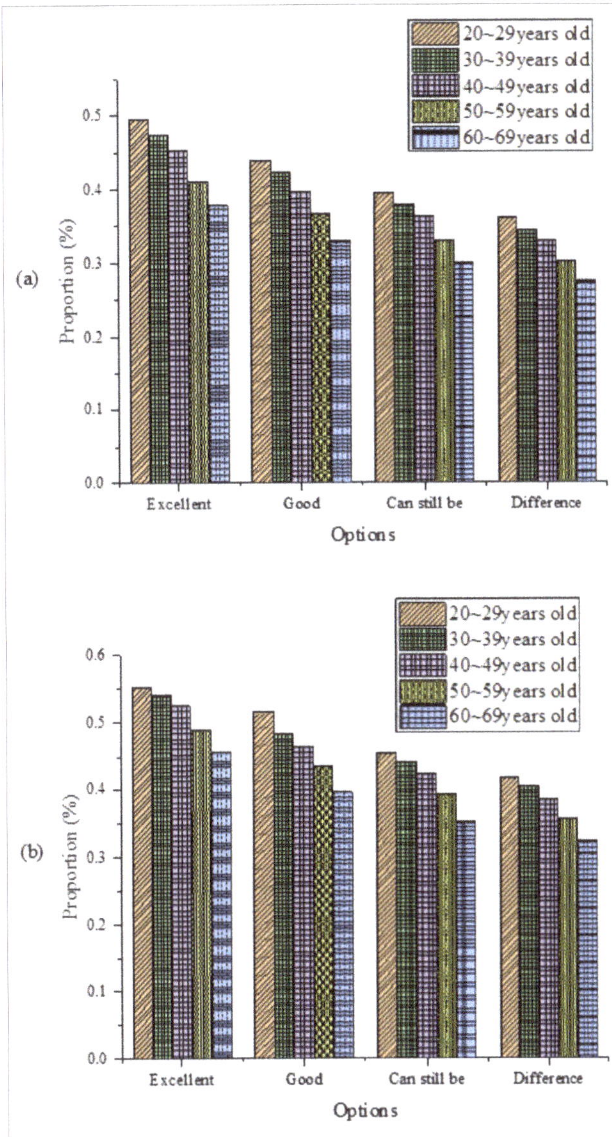

Figure 10. Analysis of cardiopulmonary capacity and cardiovascular health status of workers at different ages in BYD (**a**) before WHSM; (**b**) after WHSM.

4. Conclusions

This work conducted a case study on BYD using the proposed EDT model and WHSM system by considering the adaptation and performance of the enterprise and the health status of workers. The results show that from 2012 to 2015, the Internet business revenue of BYD in the EDT process is much greater than the manufacturing revenue and net profit. After EDT, the Internet business revenue reaches 29.67% of the total revenues. Each enterprise index generally shows an upward trend, and the overall index has increased by 42.11%. Therefore, digital transformation has absolute adaptability to BYD manufacturing enterprises. The scores of various indicators of the whole enterprise have been greatly improved, especially the financial performance, marketability, and technical ability of the enterprises after transformation. In addition, after the enterprise's digital transformation, employees' cardiorespiratory and cardiovascular capabilities have been improved through employee health management. The findings provide important theoretical support for applying BDT and IoT in EDT and model references for the manufacturing industry's EDT and WHSM. The proposed EDT model and WHSM system are based on the assumption that the attributes of manufacturing enterprises and other digital construction status quo share the same background. Thus, the proposed model and system cannot be applied to other industries. It is hoped that future research will improve the digital transformation mode of each enterprise.

Author Contributions: Conceptualization, Z.W. and S.G.; methodology, Y.W.; software, W.D.; validation, S.J. and W.D.; formal analysis, X.D.; investigation, Y.W.; resources, Z.W.; data curation, W.D.; writing—original draft preparation, Z.W.; writing—review and editing, Y.W.; visualization, S.G.; supervision, S.G.; project administration, Z.W.; funding acquisition, S.G., Z.W. contributed equally to this work as S.G., Z.W. should be considered as Co-first Author. All authors have read and agreed to the published version of the manuscript.

Funding: This research was funded by Wuhan Polytechnic University to introduce (train) talents scientific research start-up fund (No. 2022RZ076).

Institutional Review Board Statement: Not applicable.

Informed Consent Statement: Not applicable.

Data Availability Statement: The further data can be reached by contacting the correspondence author.

Conflicts of Interest: The authors declare no conflict of interest.

References

1. Giraldo, S.; la Rotta, D.; Nieto-Londoño, C.; Vásquez, R.E.; Escudero-Atehortúa, A. Digital transformation of energy companies: A colombian case study. *Energies* **2021**, *14*, 2523. [CrossRef]
2. Wrede, M.; Velamuri, V.K.; Dauth, T. Top managers in the digital age: Exploring the role and practices of top managers in firms' digital transformation. *Manag. Decis. Econ.* **2020**, *41*, 1549–1567. [CrossRef]
3. Kääriäinen, J.; Pussinen, P.; Saari, L.; Kuusisto, O.; Saarela, M.; Hänninen, K. Applying the positioning phase of the digital transformation model in practice for SMEs: Toward systematic development of digitalization. *Int. J. Inf. Syst. Proj. Manag.* **2021**, *8*, 24–43. [CrossRef]
4. García-Peñalvo, F.J. Avoiding the dark side of digital transformation in teaching. An institutional reference framework for eLearning in higher education. *Sustainability* **2021**, *13*, 2023. [CrossRef]
5. Radionova, N.; Skrypnyk, M.; Voronkova, T. Dual nature of industrial enterprise cost management system. *Balt. J. Econ. Stud.* **2019**, *5*, 184–190. [CrossRef]
6. Mugge, P.; Abbu, H.; Michaelis, T.L.; Kwiatkowski, A.; Gudergan, G. Patterns of digitization: A practical guide to digital transformation. *Res. Technol. Manag.* **2020**, *63*, 27–35. [CrossRef]
7. Schuh, G.; Kozielski, S.; Weber, W.; Gützlaff, A.; Sauermann, F.; Wlecke, S.; Maetschke, J. Digital Lean Lighthouse Factory—Nachhaltig effiziente Produktion. *Z. Für Wirtsch. Fabr.* **2020**, *115*, 175–177. [CrossRef]
8. Wessel, L.; Baiyere, A.; Ologeanu-Taddei, R.; Cha, J.; Blegind-Jensen, T. Unpacking the difference between digital transformation and IT-enabled organizational transformation. *J. Assoc. Inf. Syst.* **2021**, *22*, 102–129. [CrossRef]
9. Holmström, J. From AI to digital transformation: The AI readiness framework. *Bus. Horiz.* **2022**, *65*, 329–339. [CrossRef]
10. Szalavetz, A. Digitalisation, automation and upgrading in global value chains–factory economy actors versus lead companies. *Post-Communist Econ.* **2019**, *31*, 646–670. [CrossRef]
11. Zimmerman, P.; Gilbert, T.; Salvatore, F. Digital engineering transformation across the Department of Defense. *J. Def. Modeling Simul.* **2019**, *16*, 325–338. [CrossRef]
12. Kaletnik, G.; Lutkovska, S. Innovative Environmental Strategy for Sustainable Development. *Eur. J. Sustain. Dev.* **2020**, *9*, 89. [CrossRef]
13. Correani, A.; De Massis, A.; Frattini, F.; Petruzzelli, A.M.; Natalicchio, A. Implementing a digital strategy: Learning from the experience of three digital transformation projects. *Calif. Manag. Rev.* **2020**, *62*, 37–56. [CrossRef]
14. Shen, L.; Sun, C.; Ali, M. Role of servitization, digitalization, and innovation performance in manufacturing enterprises. *Sustainability* **2021**, *13*, 9878. [CrossRef]
15. Kane, G. The technology fallacy: People are the real key to digital transformation. *Res. Technol. Manag.* **2019**, *62*, 44–49. [CrossRef]
16. Son, K.J.; Son, H.R.; Park, B.; Kim, H.J.; Kim, C.B. A community-based intervention for improving medication adherence for elderly patients with hypertension in Korea. *Int. J. Environ. Res. Public Health* **2019**, *16*, 721. [CrossRef]

17. Rengasamy, D.; Jafari, M.; Rothwell, B.; Chen, X.; Figueredo, G.P. Deep learning with dynamically weighted loss function for sensor-based prognostics and health management. *Sensors* **2020**, *20*, 723. [CrossRef]
18. Feng, H.; Wang, F.; Song, G.; Liu, L. Digital Transformation on Enterprise Green Innovation: Effect and Transmission Mechanism. *Int. J. Environ. Res. Public Health* **2022**, *19*, 10614. [CrossRef]
19. Shpak, N.; Kuzmin, O.; Dvulit, Z.; Onysenko, T.; Sroka, W. Digitalization of the marketing activities of enterprises: Case study. *Information* **2020**, *11*, 109. [CrossRef]
20. Lin, T.C.; Sheng, M.L.; Jeng Wang, K. Dynamic capabilities for smart manufacturing transformation by manufacturing enterprises. *Asian J. Technol. Innov.* **2020**, *28*, 403–426. [CrossRef]
21. Hanelt, A.; Bohnsack, R.; Marz, D.; Antunes Marante, C. A systematic review of the literature on digital transformation: Insights and implications for strategy and organizational change. *J. Manag. Stud.* **2021**, *58*, 1159–1197. [CrossRef]
22. Maciag, R. Digital Transformation as an Epistemological Event: Predigital Transformation. *Dialogue Univers.* **2022**, *32*, 83–102. [CrossRef]
23. Zhang, T.; Shi, Z.Z.; Shi, Y.R.; Chen, N.J. Enterprise digital transformation and production efficiency: Mechanism analysis and empirical research. *Econ. Res. Ekon. Istraživanja* **2022**, *35*, 2781–2792. [CrossRef]
24. Solberg, E.; Traavik, L.E.; Wong, S.I. Digital mindsets: Recognizing and leveraging individual beliefs for digital transformation. *Calif. Manag. Rev.* **2020**, *62*, 105–124. [CrossRef]
25. Pronchakov, Y.; Prokhorov, O.; Fedorovich, O. Concept of High-Tech Enterprise Development Management in the Context of Digital Transformation. *Computation* **2022**, *10*, 118. [CrossRef]
26. Bican, P.M.; Brem, A. Digital business model, digital transformation, digital entrepreneurship: Is there a sustainable "digital"? *Sustainability* **2020**, *12*, 5239. [CrossRef]
27. Giorgi, G.; Lecca, L.I.; Alessio, F.; Finstad, G.L.; Bondanini, G.; Lulli, L.G.; Arcangeli, G.; Mucci, N. COVID-19-related mental health effects in the workplace: A narrative review. *Int. J. Environ. Res. Public Health* **2020**, *17*, 7857. [CrossRef]
28. Priyono, A.; Moin, A.; Putri, V.N.A.O. Identifying digital transformation paths in the business model of SMEs during the COVID-19 pandemic. *J. Open Innov. Technol. Mark. Complex.* **2020**, *6*, 104. [CrossRef]
29. Lv, Z.; Singh, A.K. Big data analysis of internet of things system. *ACM Trans. Internet Technol.* **2021**, *21*, 1–15. [CrossRef]
30. Mehmood, I.; Ullah, A.; Muhammad, K.; Deng, D.J.; Meng, W.; Al-Turjman, F.; Sajjad, M.; de Albuquerque, V.H.C. Efficient image recognition and retrieval on IoT-assisted energy-constrained platforms from big data repositories. *IEEE Internet Things J.* **2019**, *6*, 9246–9255. [CrossRef]
31. Sanchis, R.; García-Perales, Ó.; Fraile, F.; Poler, R. Low-code as enabler of digital transformation in manufacturing industry. *Appl. Sci.* **2019**, *10*, 12. [CrossRef]
32. Lutfi, A.; Alsyouf, A.; Almaiah, M.A.; Alrawad, M.; Abdo, A.A.K.; Al-Khasawneh, A.L.; Ibrahim, N.; Saad, M. Factors Influencing the Adoption of Big Data Analytics in the Digital Transformation Era: Case Study of Jordanian SMEs. *Sustainability* **2022**, *14*, 1802. [CrossRef]
33. Dall'Ora, N.; Alamin, K.; Fraccaroli, E.; Poncino, M.; Quaglia, D.; Vinco, S. Digital transformation of a production line: Network design, online data collection and energy monitoring. *IEEE Trans. Emerg. Top. Comput.* **2021**, *10*, 46–59. [CrossRef]
34. Zhang, C.; Liu, Z. Application of big data technology in agricultural Internet of Things. *Int. J. Distrib. Sens. Netw.* **2019**, *15*, 1550147719881610. [CrossRef]
35. Li, W.; Zhu, J.; Zhang, Y.; Zhang, S. Design and implementation of intelligent traffic and big data mining system based on internet of things. *J. Intell. Fuzzy Syst.* **2020**, *38*, 1967–1975. [CrossRef]
36. Willner, A.; Gowtham, V. Toward a reference architecture model for industrial edge computing. *IEEE Commun. Stand. Mag.* **2020**, *4*, 42–48. [CrossRef]
37. Dayan, V.; Chileshe, N.; Hassanli, R. A scoping review of information-modeling development in bridge management systems. *J. Constr. Eng. Manag.* **2022**, *148*, 03122006. [CrossRef]
38. Gu, H.; Fu, X.; Zhu, Y.; Chen, Y.; Huang, L. Analysis of social and environmental impact of earth-rock dam breaks based on a fuzzy comprehensive evaluation method. *Sustainability* **2020**, *12*, 6239. [CrossRef]
39. Peizhi, R.; Wei, L.; Ran, B.; Ping, M. A Simulation Credibility Assessment Method Based on Improved Fuzzy Comprehensive Evaluation. *J. Syst. Simul.* **2020**, *32*, 2469. [CrossRef]
40. Wang, Z.; Li, M.; Lu, J.; Cheng, X. Business Innovation based on artificial intelligence and Blockchain technology. *Inf. Process. Manag.* **2022**, *59*, 102759. [CrossRef]

International Journal of
Environmental Research and Public Health

MDPI

Article

Sustainable Development Efficiency of Cultural Landscape Heritage in Urban Fringe Based on GIS-DEA-MI, a Case Study of Wuhan, China

Han Zou [1,2,*], Yang Liu [1], Baihao Li [3] and Wenjing Luo [4]

1 School of Civil Engineering, Architecture and Environment, Hubei University of Technology, Wuhan 430068, China
2 Innovation Demonstration Base of Ecological Environment Geotechnical and Ecological Restoration of Rivers and Lakes, Wuhan 430068, China
3 School of Architecture, Southeast University, Nanjing 210096, China
4 Wuhan Planning and Design Institute, Wuhan 430014, China
* Correspondence: zouhan@hbut.edu.cn

Abstract: Cultural landscape heritage refers to the rare and irreplaceable cultural landscapes recognized by UNESCO and the World Heritage Committee. It is recognized as a "common works of nature and human beings" of outstanding significance and universal value, and is a type of world heritage. Dueto construction, land isincreasingly limited in urban and rural areasin the process of urbanization, and cultural landscape heritage faces a huge threat, especially larger culturallandscapeheritagelocated at the edgesof cities. However, most of the existing studies have mainly focused on the material protection of heritage but have not paid enough attention to the non-material aspects of heritage sites, failing to reveal the inseparable nature of heritage and land. Therefore, this study takes sustainable development efficiency as its analysis tool, examines two pieces of cultural landscape heritage (the Panlongcheng site and the Tomb of the King of the Ming Dynasty) in the urban edge area of Wuhan, China as examples, innovates and establishes a multidimensional evaluation method based on the GIS-DEA-MI model, and compares the dynamic changes of the spatial development efficiency and non-spatial development efficiency of the above two cultural landscape heritage cases. The results show that: both the spatial development efficiency and non-spatial development efficiency of Panlongcheng from 2010 to 2019 are significantly higher than that of the Tomb. This method makes up for the deficiency of traditional subjective qualitative analysis. It can be used to study the development efficiency of cultural landscape heritage more objectively and comprehensively, and promote the overall sustainable development of material and intangible cultural heritage. It can provide the basis for early decision-making and post-implementation evaluation for the preservation and utilization of cultural landscape heritage under the background of urban renewal.

Keywords: GIS-DEA-MI model; urban fringe; cultural landscape heritage; sustainable development efficiency; heritage preservation and utilization

Citation: Zou, H.; Liu, Y.; Li, B.; Luo, W. Sustainable Development Efficiency of Cultural Landscape Heritage in Urban Fringe Based on GIS-DEA-MI, a Case Study of Wuhan, China. *Int. J. Environ. Res. Public Health* **2022**, *19*, 13061. https://doi.org/10.3390/ijerph192013061

Academic Editors: Hongxiao Liu, Tong Wu and Yuan Li

Received: 30 August 2022
Accepted: 4 October 2022
Published: 11 October 2022

1. Introduction

J. Jackson believed that landscapes are never a simple natural space or a natural environment, and that each landscape is the space-time place of human beings [1]. In the first half of the 20th century, German geographer Otto Schluetter put forward the theory of cultural landscapes. He believed that landscapes can be divided into two categories; one is the original landscape, and the other is the cultural landscape (the landscape changed by human activities). He thought that cultural landscape is the form of human phenomena on the ground, and should be studied as a landscape created by human beings and their labor that can reflect the culture and economy of human groups. In 1906, Otto Schluetter formally put forward the concept of "cultural landscape form", emphasizing that landscape has not only its appearance, but also social, economic and spiritual power. It also points out the difference between cultural landscapes and natural landscapes, and requires the study of cultural landscapes as a phenomenon evolved from natural landscapes, and the form of landscapes as cultural products [2].

American geographer C. O. Sauer inherited and developed Otto Schluetter's theory of cultural landscape. He believed that cultural landscape is "a region composed of significant forms of connection including nature and culture". C. O. Sauer believed that a special human group, under the influence of culture, can create suitable surface features in areas that are long-lived and active. This feature is caused by the different ways in which human beings intervene with, apply and transform the environment, and is also a result of the transformation from natural landscape to cultural landscape [3]. In Sauer's paper on landscape morphology in 1925, the natural landscape is the result of the effect of human culture on the natural landscape [4]. In the Recent Development of Cultural Geography, published by Sauer in 1927, cultural landscape was first clearly defined for the first time as "the form of human activities appended to the natural landscape" [5].

In 1992, the World Heritage Committee included cultural landscape heritage in the World Heritage List. The study of cultural landscape heritage has attracted extensive attention from scholars all over the world. David Jacques called the 1990s "the rise of cultural landscape" [6]. The distribution of word culture landscape heritage in various regions is extremely uneven. The world cultural landscape heritage in Europe accounts for more than half of the world total, which is three times that of Asia [7]. Walliams Logan presented a report at the 10th UNESCO University and Heritage Symposium with the theme of "The Cultural Landscapes in the 21st Century", noting that "the conservation and utilization of cultural landscapes is much more smooth in Europe than in Asia" [8]. The 2004 ICOMOS report also pointed out the shortage of the word cultural landscape heritage sites in Asia [9]. Many of Asia's heritage sites have been designated as natural heritage sites, but in fact many of the cultural associations have been obliterated, although the number of world cultural landscape heritage sites in Asia is insufficient [2]. However, Professor P. Fllower analyzed the number and value of cultural landscapes in Asia and confirmed the value of cultural landscapes in Asia [10]. In Asian world heritage sites or candidate lists, the most important are human sites, architectural monuments and religious heritage. However, the World Heritage Committee did not consider cultural landscapes, vernacular architecture, technological or agricultural sites as a whole. These belong to the category of cultural landscape sequences, leading to a missed opportunity of demonstrating Asian local spirit [2]. For the sustainable and healthy development of Asian cultural landscape heritage, Ken Taylor recommends making efforts in the following areas [11]: give a clear Asian definition of cultural landscapes, define the content of Asian cultural landscape types according to Asian values and a regional cultural landscape list, grasp the construction and practice of cultural quantitative concepts, connect nature and culture, take man as an integral part of nature, and establish a unique protection mode, so as to avoid the simple imitation of the Western national park protection system of pure nature protection. Professor Ken Taylor believes that Asia's cultural landscapes can make a significant contribution to world heritage, and China should play a leading role [12]. Mitchell believes that China has a long history of civilization and unique man-land relationships, and covers a large part of its natural heritage. Therefore, it has special value to observe and study heritage protection from the perspective of cultural landscapes. The Asian cultural landscape perfectly reflects the interactive relationship between man and nature. It is not only a tangible cultural work, but also the result of the cultural effect of associative intangible values. In this respect, different from Western cultural landscapes, the Asian cultural landscape is a part of a "dynamic process of identity formation" [2]. Han Feng pointed out that in Eastern culture, nature and culture are an inseparable integral whole, and cultural landscape is the link between nature and culture heritage, revealing the inseparable nature of heritage, land and landscape. He once pointed out that China has difficulty understanding the concept of "cultural landscape" [13]. He believes that the landscape in the Eastern tradition itself has strong cultural attributes. Different from the West, landscape in Eastern culture is a landscape with special significance.

Some natural heritage scholars in the West regard nature and culture as two completely opposite things. In their view, human beings are not a part of nature and the landscape is not a product of culture. However, China pays attention to the overall characteristics of the landscape and believes that the protection of cultural landscape heritage not only involves simple material protection, but also involves the cultural and social dimensions. Therefore, non-material factors such as society, economy and land development and utilization need to be further considered in the protection and utilization of cultural landscape heritage and the study of sustainable development in China. Cultural landscape is the result of the interaction between human beings and nature. Its integrity and importance require us to achieve sustainable development in the management of cultural landscape heritage. The current management of China's cultural landscape heritage takes sustainability as the basic principle and overall requirement, and coordinates multiple fields such as society, economy and environment. The effectiveness of management is directly related to the status quo of cultural landscape heritage and the possibility of sustainable development, and the effectiveness evaluation of management refers to the objective evaluation of the process, results, efficiency and rationality of project implementation under the background of management. By setting reasonable evaluation indexes, the planning, management system and implementation in the process of regional management are evaluated scientifically.

Scholars from different countries have different perspectives on cultural landscape heritage. For example, in research on the protection and utilization of cultural landscape heritage, Shizhenjia et al. explored the protection and management methods of Chinese classical royal garden cultural heritage based on three-dimensional digital technology, which improved the management efficiency of cultural landscape heritage and promoted the sustainable development of cultural landscape heritage [14]. Xuhui Wang et al. constructed the ecological security model of ancient capital landscape based on cellular automata theory [15]. Alessandra Capuano explored more advanced Roman archaeological sites and urban sustainability based on sustainable concepts [16]. Janjira Sukwai used GIS and computer-generated 3D modeling to support a visual integrity assessment of change management in the Chiang Mai Historical City buffer zone [17]. In the study of the sustainable coordinated development of cultural landscape heritage and tourism, UNESCO launched the World Heritage Sustainable Tourism Program in 2001 in response to the opportunities and threats posed by tourism activities in World Heritage sites [18]. Joanne Connell et al. studied the monitoring of local government planning in sustainable tourism planning in New Zealand [19]. Muhamment Yasarata et al. studied the coexistence of sustainable tourism development and politics [20]. The contradiction between tourism development and sustainable heritage has always been a topic of academic debate. Di Zuo et al. explored the relationship between tourism, residents' practice and traditional cultural landscapes in cultural heritage sites by taking Hongcun Village in China as an example, and proposed the possibility of capital-driven and sustainable development and local cultural balance, which is of great significance to the sustainable development of cultural heritage [21].

The above research emphasizes the sustainable development of cultural landscape heritage, but mostly only focuses on the protection of cultural landscape heritage, cultural landscape heritage and the sustainable development of tourism. The sustainable development of cultural landscape heritage and the surrounding environment research perspective is insufficient, given the lack of quantitative research on the efficiency of the sustainable development of cultural landscape heritage. In addition, cultural landscape heritage is mostly located in urban edge areas, and the protection, utilization and sustainable development of the cultural landscape heritage are threatened by urban expansion. Therefore, it is particularly important to study the coordination and sustainable development of cultural landscape heritage sites and the surrounding environment, and cultural landscapes should be included in current territorial space planning and used as a resource element for scientific control. Under the background of China's territorial space planning, the protection and control of cultural landscape heritage in urban fringe areas and its overall environment are the focus of current research.

Urban fringe was first proposed in 1936 by German geographer, Herbert Louis, who found that originally rural areas had been gradually occupied by urban construction regions and had blended into urban areas during his study of Berlin urban structures. He then defined the zone as different from the type of urban land used as Stadtrandzonen.

In 1968, R. J. Pryor explained urban fringe as the fringe built-up areas and suburban areas with significant differences in urban-rural land use, society, and demography. Urban and rural areas are not well-delineated, bounded, or self-sufficient spaces [22]. During the period of rapid urbanization, the urban fringe area is the area where urban expansion occurs first, and land use change is the most active [23]. The urban fringe is quite unstable, influenced by urbanization activities, which leads to the strong dynamic fluctuations of multiple aspects such as economics, social structures, nature, and human landscape. With the expanding of urban areas and rapid development of the economy, urban fringes will be the first to be eaten away and deteriorated by urban areas, which therefore must be given importance in urban sustainable development. Major urban challenges facing China and Europe are related to changes in climate, environment, and to decision-making that makes urban and rural landscapes more susceptible to environmental pressures. Urbanisation is one of the most challenging processes we face in the 21st century. This calls for new thinking to create robust collaboration across urban–rural areasthat fosters sustainable development in both [24]. In Italy, a decisive steering role is entrusted to regional landscape planning, as introduced through the current Code of Cultural Heritage and Landscape. Regional landscape plans define the rules to which all municipal plans must conform [25].

K. E. Boulding et al. believe that more and more people have realized the importance of protecting cultural landscapes and thenatural environment under the increasingly serious situations of population explosion, rapid industry expansion, intensified natural resource consumption, ecological destruction, and the environmental pollution caused by the great industry and modern agriculture [26].

As cultural landscape heritage becomes part of the world's heritage, its effective management has been causing concerns worldwide.In 1969, L. McHarg published Design with Nature, in which he discusses environmental issues from the perspectives of nature, history and humanity, and explains how natural procedure guides land development [6]. After decades of work, various countries have combined the regional protection of cultural heritage with the national and local development of culture and ecology, society, and other aspects, and they have also rationally set up relevant cultural heritage reserves. These measures, in advance, offer the basis for scientific decision-making and new ideas in the protection and management of closely relevant departments, such as urban planning, government management, and land use. Additionally, such measures have provided strong support for the strategy of regional overall coordinated development, and have gained much successful experience.

Sustainable urban planning is essential in mediating the natural and built environments globally, yet there is little progress as regards its attainment in developing countries [4]. With different national conditions, each country has its own strategy of the protection and use of cultural landscape heritage. For example, the U.S. was the first to establish the National Park System so as to maintain the non-profit, complete, scientific, and sustainable protection of its cultural landscape heritage. Unlike most Western developed countries, China has an urban-rural dualism where there are enormous differences between urban and rural areas. Besides, the fact that many cities remain in the process of urbanization has made urban fringe even more unstable. China's large cultural landscape heritages are mostly located in urban fringe areas, which makes it easier for them to be influenced by the process of urbanization, and such special locations have given them special qualities different from other types of cultural landscape heritage. As a result, cultural landscape heritagesites in urban fringe areas are often developed into archaeological site parks with conservative features and historical memories, which can offer public cultural service and help the protection of relics as well as the strategy of rural vitalization, becoming a destination of cultural tourism and key supporter of beautiful rural construction.

However, since China has a behindhand archaeological site park construction, a mature model has not been formed in practice, and there are situations of "mass input with tiny output" in some cities' archaeological site park construction. Meanwhile, the planning and construction of an archaeological site park requires the comprehensive consideration of land use, population distribution, industrial distribution, and matching public facilities with base facilities so as to promoting the rural-urban development by protecting historical sites. Hence, it is necessary to conduct research on the sustainable development efficiency of cultural landscape heritages.

In 2000, the UNESCO (United Nations Educational, Scientific, and Cultural Organization) proposed the Hoi An Protocols for Best Conservation Practice in Asia at the Bangkok conference, which stressed the inner links between tangible heritage, intangible heritage, and cultural landscape protection, and can serve as professional guidance to guarantee and maintain Asian cultural heritage sites in the context of Asian cultures. The European Landscape Convention has set out a much more holistic understanding of landscape than was previously the case within Europe. The Convention embodies thinking that is beginning to be reflected in the work of governments, environmental agencies and a wide range of interested parties within the landscape field in Europe. The European Landscape Convention helpfully does not value any one landscape above another, indeed it recognises that local and degraded landscapes are as likely to be of importance to the communities or cultures who visit them as those which are commonly found to be labelled as globally important.

In 2005, the ICOMOS (International Council on Monuments and Sites) held an international conference and passed the Xi'an Declaration, where they emphasized the importance of the environment to cultural heritage preservation in ever-changing cities and landscapes, and that the implementation of specific protection needs to consider the overall environment's dual attributes of tangibility and intangibility [2].

The Chinese government attaches great importance to heritage preservation and has therefore published several national policy documents, including the Notice on Strengthening and Improving the Work of Cultural Relics, the Cultural Relics Protection Law of the People's Republic of China, and The 14th Five-Year Plan for the Protection and Utilization of Great Sites. Besides, the Chinese government has also laid much focus on its intangible cultural heritage preservation, and published Interim Measures for the Protection and Management of National Intangible Cultural Heritage in 2006. This proves again that heritage is both tangible and intangible in Chinese traditional culture.

From the above, we suggest that sustainable development efficiency is suitable for assessing the current situation of cultural landscapes in urban fringes and predicting their potential for sustainable development. Sustainable development efficiency, referring to the ratio of resource elements' effective gross output to the input, is also seen as the comprehensive reflection of effective allocation of factor input, running status, and operating management. The proper coordination of social economic development and ecological environment development is an inevitable requirement of sustainable development [27].

Since the concept of efficiency was proposed in 1957 by a British economist, who also put forward its measurement standard and model, efficiency study have stepped into a brand new stage. DEA (Data Envelopment Analysis), created by A. Charnes, W. W. Cooper and E. Rhodes in 1978, is the most widely used method in efficiency study [28–35]. DEA is a new field of crossover study involving operations research, management, and mathematical economics, which is a quantitative analysis method for evaluating the relative effectiveness of comparable units of the same kind using linear programming according to multiple input and output indicators.

At present, there have been many studies about applying DEA to efficiency studies, which therefore formed relatively normative research methods and ideas. For instance, Yang Li et al. studied the total factor energy efficiency of sustainable development in China based on DEA [36]. Zhu, S., Zhou, Z., Li, R. and Li, W. used the city-level panel data of the three urban agglomerations from 2006 to 2019 to construct the slacks-based measure integrating data envelopment (SBM-DEA) model for calculating each city's carbon dioxide emission efficiency [37].

This study uses sustainable development efficiency as a tool of analysis, takes two cultural landscape sites in the Wuhan urban fringe as the objects of study, creates and sets a multi-dimensional evaluation based on the GIS-DEA-MI model, and conducts a comparative analysis on the differences and dynamic changes of the developmental efficiency of the above two cultural heritage sites. This method makes up for the deficiency of traditional subjective qualitative analysisand fits in the integrality of cultural heritage sites, so as to study the development efficiency of cultural landscape sites from a more objective and comprehensive perspective, to improve the overall sustainable development of tangible and intangible cultural heritage, and to provide a basis for early decision-making and post-implementation evaluation for the preservation and utilization of cultural landscape heritage under the background of urban renewal.

2. Materials and Methods

Many previous studies about development efficiency tended to use the DEA model only to study its non-spatial development efficiency, or to analyze spatial development efficiency solely by spatial aspects, such as land use efficiency, planning operational efficiency and urban updating efficiency, while urban issuesare compound, which requires comprehensive perspectives to summarize the overall development efficiency objectively and roundly. Additionally, such a study would better be designed as a comparative one, which can lead to precise knowledge of the differences of the overall resource utilization among different archaeological parks through comparing the development efficiency of the research objects so as to redefine their future directions and positions, adapt to new situations, and make reasonable decisions according to our new duties.

The evaluation of spatial efficiency is proceeded by assessing urban operational efficiency according to the planning and variation of urban space in other words, various types of urban construction land, while non-spatial efficiency is evaluated by attributive standards including urban natural environment, economy and society [38]. Therefore, the developmental efficiency of archaeological site parks consists of both spatial and non-spatial development efficiency, as shown in Figure 1.

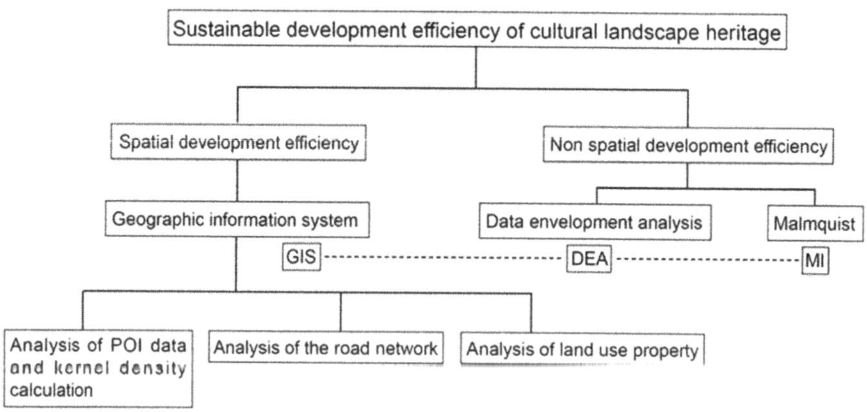

Figure 1. Knowledge mapping of the study on the developmental efficiency of cultural landscape heritage.

A GIS (geographic information system), has strong capacities of analyzing and modeling geographic space as well as supporting making spatial decisions. POI (point of interest) data was imported to ArcGIS to generate core density diagrams, which can be used to calculate the density of the key factors in their surroundings so as to reflect the degree of concentration in certain ranges of POI distribution.

The DEA-MI model was chosen to study the non-spatial development efficiency of the above archaeological parks. The DEA (data envelopment analysis) model, created by A. Charnes, W. W. Cooper and E. Rhodes in 1978, has been widely used for evaluating projects and policies.

DEA, a non-parametric examination based on the concept of relative efficiency evaluation, in which the evaluated units and organizations are called the DMU (decision-making units), constructs a data envelope curve through selecting various input and output data of DMU, using linear programming, and taking optimal input and output as production frontiers.

In this method, the efficient point is on the leading surface and the efficiency value is set as 1, while the invalid point is on the outside, and is given a relative efficiency value standard within the range from 0 to 1.

The traditional DEA model can only offer longitudinal comparison of the production efficiency of decision making units at the same timing, while the DEA-Malmquist exponential model can investigate the dynamic variation of the efficiency of DMU in different stages, thus being able to analyze panel data with a relatively wide application.

To sum up, the evaluation based on the GIS-DEA-MI model can both be applied to the development efficiency of archaeological parks and other fields, such as evaluating urban and rural planning operational efficiency and evaluating urban updating efficiency. This method combines multiple research methods of management, geographic information systems, urban and rural planning, and other subjects to provide strategies and advice to improve the development efficiency of archaeological parks, optimize the planning of urban and rural space, and achieve the integrated development of urban and rural areas, and it also offers reference to the development efficiency of other archaeological parks.

3. Empirical Research

3.1. Research on Spatial Efficiency

3.1.1. Analysis of POI Data and Kernel Density Calculation

POI (point of interest) data, based on the key data of location services, plays an essential role in urban planning studies and the preliminary planning of commercial development projects.

The distribution density of POI data is the expression of spatial phenomena, since human economic activities often manifest as the gathering of several joints and form gathering centers of different grades in different spatial statistics units. As a result, its kernel density can reflect the efficiency and potential of space development to some extent.

At present, research on the spatial efficiency of archaeological site parks still lacks a widely accepted standard. Professor Jonh Radke from the University of California, Berkeley, first put forward the theory of a two-step floating catchment, which was then improved by two scholars, Luo and Wang, who combined the gravity model method and the floating catchment area method. The former calculates the service potential of housing estates, while the latter calculates the availability of public service facilities, and the combination is named 2SFCA (two-step floating catchment area method). This method can be used to study the distribution features of the population and public service facilities in different searching areas through constructing different motion states of people in cities and the reachable range finitely formed by the service capacity of urban public service facilities as the searching areas, and to further measure the opportunities people seize to enjoy public service [39]. It can be expressed by mathematical expressions such as Equation (1).

$$A_i^F = \sum_{j=\{d_{ij}\leq d_0\}} R_j = \sum_{J\in\{d_{ij}\leq d_0\}} \left(\frac{S_j}{\sum_{k\in\{d_{ij}\leq d_0\}} D_k} \right) \quad (1)$$

One of the weaknesses of this method is that it uses the same searching radius for all the facilities and citizens. Therefore, McGrail put forward the idea of a dynamic radius according to differences in population density in different areas in 2014, and it was discovered through many experiments that increasing the searching radius gradually is feasible with the decrease of population density. The Huangpi District, where Panlongcheng is situated, and the Jiangxia District, where the Tomb of the King of the Ming Dynasty is situated, have significantly less population density than central areas, so there should be a larger searching radius. By praxiology, the walking speed of an adult is about 5 km/h, whose normal maximum psychological endurance is 30 min, calculated as 2.5 km.

Meanwhile, with an archaeological site park as our research object, which has a larger coverage and more service objects, including local and non-local visitors apart from its community residents, we also need to consider the searching radius under non-walking conditions.

According to big data, Liusong conducted research on the service radius of community parks in Shanghai based on the signaling data of cell phones, which indicated that the average was 5879 m long [40]. Additionally, an archaeological site park has a significantly larger scale than normal parks, so a small range of a study may fail to objectively reflect the surrounding infrastructure and space characteristics. In conclusion, the range of this research was considered appropriate within a 5 km radius centered in the park.

By adding the POI data of Wuhan in 2020 to ArcGIS, and conducting research within the range of 5 km centered, respectively, on Panlongcheng and the Tomb of the King of the Ming Dynasty, the POI data of the above two objects, including the surrounding traffic facilities, public facilities and service facilities, were calculated to investigate kernel density and were analyzed according to their current situation.

As shown in Figure 2, the traffic facilities surrounding Panlongcheng are uniformly distributed and are densely distributed in various spots, while that of the Tomb of the King of the Ming Dynasty are mainly concentrated on the north side of the highway, connecting Shanghai and Chongqing, which gathers cross-border e-commerce industries, and along Fenglian Avenue, and no dense distribution has formed.

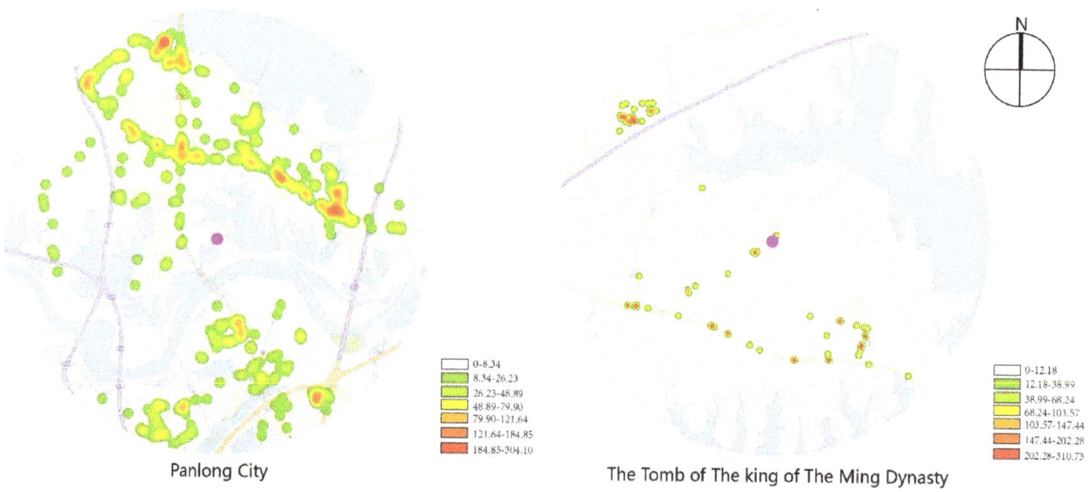

Figure 2. Kernel density analysis of traffic facilities.

As shown in Figure 3, the traffic facilities surrounding Panlongcheng are relatively uniformly distributed, but no large-scale distribution has formed in most districts except in Outlets Plaza near Houhu Bridge. With few public facilities nearby, the public services of the Tomb of the King of the Ming Dynasty are mainly its archaeological station and village committee, which serve a quite limited range of areas where the objects are mostly local citizens.

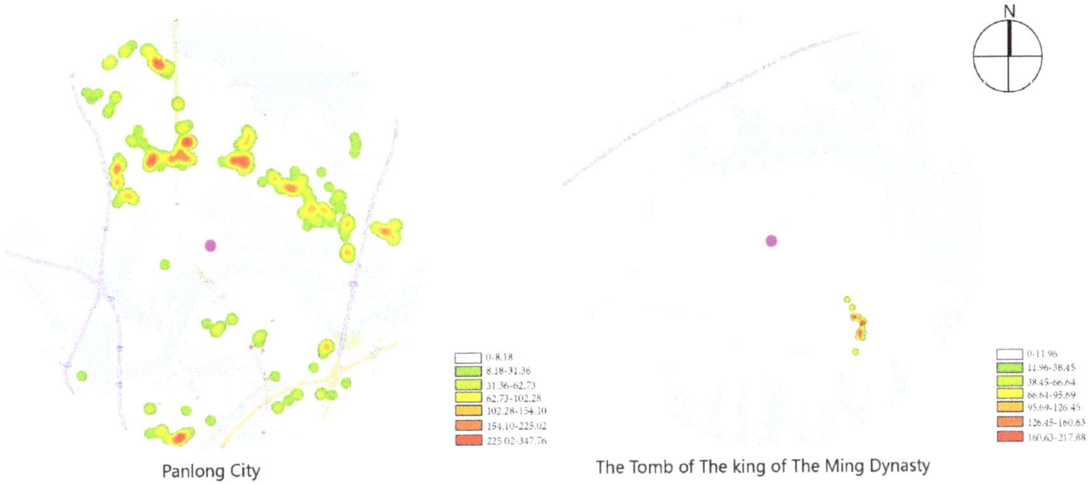

Figure 3. Kernel density analysis of communal facilities.

Figure 4 indicates that service facilities near Panlongcheng are mainly north-south distributed with relatively few service facilities in the west part, but they are uniformly distributed on the whole, and have formed dense distributions in multiple areas, which reveals relatively well-equipped service facilities. As for the Tomb of the King of the Ming Dynasty, its surrounding service facilities are near Wuhan Qingchuan University in the southeast part, which serve mostly teachers and students in the school with a limited range of service.

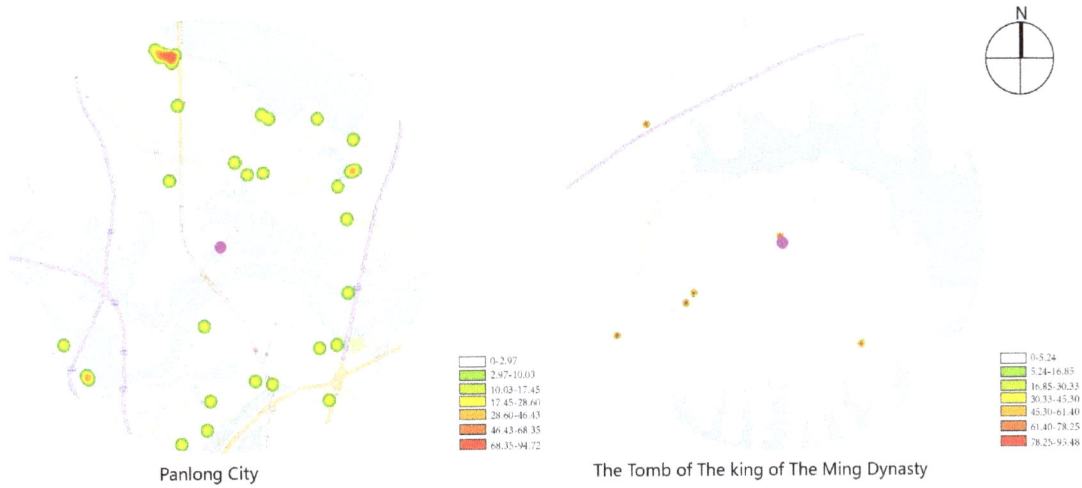

Figure 4. Kernel density analysis of service facilities.

3.1.2. Analysis of the Road Network

Traffic is the precondition of the development of an archaeological site park, and its convenience reflects the potential of space development. The study modified the road network analysis according to A Map of Wuhan Planning, respectively, centered in the two research objects within the range of the 5 km radius.

Figure 5 proves that road facilities near Panlongcheng have been relatively well-equipped and are distributed uniformly. There are three rail traffic lines in the range of research, and

37 intersections in main roads. When it comes to the Tomb of the King of the Ming Dynasty, whose road facilities have notbeen fully constructed and uniformly distributed, there is no rail traffic, and its main roads have 19 intersections. According to such an analysis of the road network, obviously, Panlongcheng has a better-built road network than the Tomb of the King of the Ming Dynasty with higher space accessibility. In 2019, the construction of Line 19 of the rail traffic in Wuhan was officially started, which connects Wuhan Railway Station and the Optical Valley Bonded Zone, significantly increasing the accessibility of the Tomb of the King of the Ming Dynasty Archaeological Park.

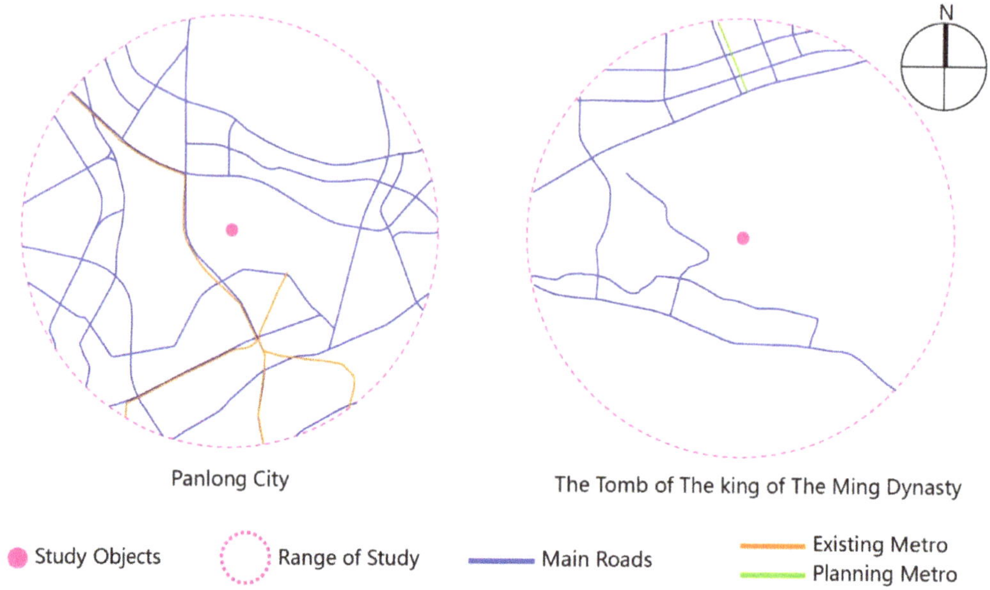

Figure 5. Analysis of road network.

3.1.3. Analysis of Land Use Property

A Map of Wuhan Planning, published by Wuhan Natural Resources and Planning Bureau, specified the nature of the land use in Wuhan, and it is a reflection of how the space in a district develops and its potential. It is based on this map that the the figure of the analysis of land use property is modified.

As seen from Figure 6, Panglongcheng has abundant land use properties, which are mainly used for residents and businesses, while the Tomb of the King of the Ming Dynasty has planning for land use property in the north part only, which is mainly for industrial use.

The efficiency of space development and utilization near Panlongcheng is significantly higher than that of the Tomb of the King of the Ming Dynasty. With Longquan Mountain, Liangzi Lake and other natural resources nearby, the development and utilization of the Tomb of the King of the Ming Dynasty should fulfil the principle of ecology first, lay a focus on ecological protection, and develop some businesses, industries and logistics only in the northern comprehensive bonded areas. Its unique strength in ecology helps to carve out a different and distinctive path of development than that of Panlongcheng.

The land use property near the Tomb of the King of the Ming Dynasty can be adjusted according to the space planning of national land, and it will benefit ecological protection and the longterm development of archaeological parks.

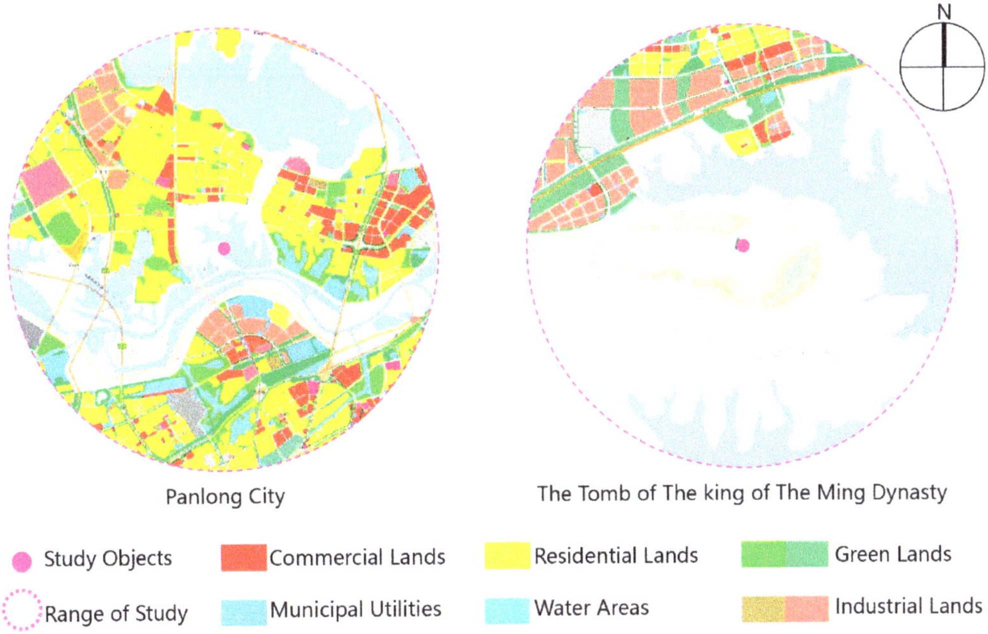

Figure 6. Analysis of land use property.

The efficiency of space development largely depends on the spatial pattern of urban and rural areas, such as the density of the road network, the properties of construction land in urban and rural areas, and the distribution of public service facilities. The application of GIS can directly reflect these features, which benefit the reasonable arrangement and adjustment of construction land in urban and rural areas, further improvement of equipment, and the cultivation of physical space for urban industries, policy measures and other immaterial planning objects.

3.2. Research on Non-Spatial Efficiency

3.2.1. The Selection of the Index

The evaluation index of the development efficiency of archaeological site parks still lacks a common standard, but it can be roughly classified into the input index and the output index. The selection of the index should be scientific, which means the input index should be logically related to the output index ona quantitative basis [41–44].

As is shown in Table 1, five indexes were selected in this study, including the general income and general reception of tourism as the output indexes, and industrial structures (ratio of tertiary industries), public utility input and the investment in human capital as the input indexes. Besides, the DEA method demands that DMU should be twice as large as the sum of output indexes and input indexes, thus the study is compliant since the DMU counted as 12. Table 1 presents how the indexes were selected and quantified.

Table 1. Evaluation index and quantitative standard.

Evaluation Index	Quantitative Criteria	Units
Total income from tourism	Annual tourism revenue of the administrative region	CNY 100 million
Total tourist arrivals	Annual tourism reception of the administrative region	Person-time
Industrial structure	Proportion of tertiary industry	percentage
Public service input	Annual total input of public utilities in the administrative region	CNY 100 million
Human capital input	Total number of tertiary industry employees	Person

The data comes from the Statistical Yearbook of Hubei Province, the Statistical Yearbook of Wuhan, and online open-source data over the ten years from 2010 to 2019. Affected by the COVID-19 pandemic, the data of tourism since 2020 has dramatic fluctuations and was therefore not involved in this research.

3.2.2. Data Analysis

Statistical data from 2010 to 2019 was added to Deap2.1 and the results are presented in Table 2, which can be drawn into a line chart as in Figure 7.

Table 2. Results of non-spatial development efficiency.

	2010	2011	2012	2013	2014	2015	2016	2017	2018	2019	Mean Value
Pan	0.95	1.21	1.15	0.95	1.05	1.07	1.17	1.09	1.15	1.26	1.21
Ming	0.88	0.79	0.85	0.96	0.87	0.90	0.92	0.93	0.95	0.91	0.89

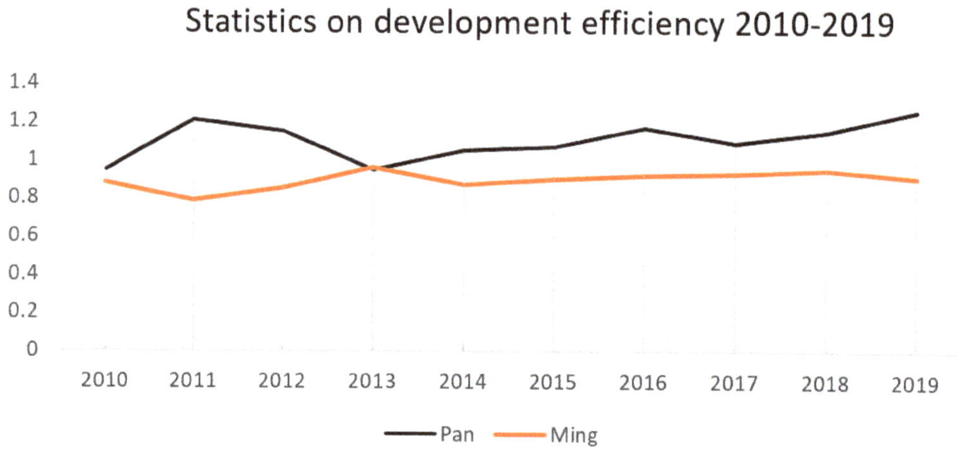

Figure 7. Statistical graph of non-spatial development efficiency.

Generally, Panlongcheng and the Tomb of the King of the Ming Dynasty share a relatively high development efficiency, with an average value above 0.8, and they have small standard deviations of tourism efficiency over the years. The standard deviation of Panlongcheng Archaeological Park is about 0.09, while that of the Tomb of the King of the Ming Dynasty is about 0.04, but both of their development efficiencies narrows the distinction, and the standard deviation decreases year by year.

The results accord with the socio-economic development of Wuhan, where the archaeological park of the Tomb of the King of the Ming Dynasty in the Jiangxia District has relatively deficient and late-constructed foundations, and its development efficiency over the years has been left behind by Panlongcheng Archaeological Park; on the other hand, however, because Panlongcheng Archaeological Park has been primarily constructed and its relevant equipment has been implemented, its development efficiency has small fluctuations in general. In recent years, with the smooth implementation of the archaeological work of the Tomb of the King of the Ming Dynasty and the increasing investment in various aspects, its development efficiency is increasing, narrowing the gap between it and Panlongcheng Archaeological Park.

Based on the results of the above tourism efficiency from 2010 to 2019, MI method is applied to calculate the variation of two parks' non-spatial development efficiency to more objectively and precisely picture the dynamic variation of the non-spatial development efficiency. The results are shown in Table 3, and its data is drawn into another line chart, as shown in Figure 8.

Table 3. Degree of change in non-spatial development efficiency.

	2010	2011	2012	2013	2014	2015	2016	2017	2018	2019	Mean Value
Pan	1.04	1.12	1.07	0.98	1.15	1.06	1.19	1.05	1.09	1.11	1.09
Ming	0.84	0.75	0.86	1.12	1.01	1.07	0.98	1.02	1.05	1.01	0.97

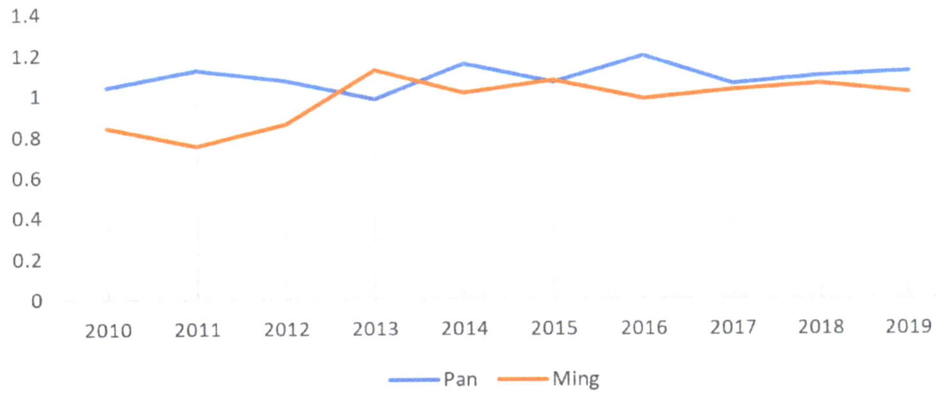

Figure 8. Statistical graph of the degree of change in non-spatial development efficiency.

The above data show that there has been a distinctive difference between the non-spatial development efficiency of Panlongcheng and the Tomb of the King of the Ming Dynasty. Over the past ten years, the MI of Panlongcheng Archaeological Park had an average standard deviation of 0.05, while that of the archaeological park of the Tomb of the King of the Ming Dynasty was 0.11, but both showeda main feature of increase. The non-spatial development efficiency of the two parks varied in accordance with their socio-economic development.

4. Conclusions

This study constructs a multidimensional evaluation based on GIS-DEA-MI, which was used to conduct research on the spatial and non-spatial sustainable development efficiency of the two cultural landscape heritagesites Panlongcheng Archaeological Park and the Tomb of the King of the Ming Dynasty, in the urban fringe of Wuhan, China. According to the study, we are able to conclude that:

1. Seen from spatial development efficiency, Panlongcheng has a much higher core density of POI data than the Tomb of the King of the Ming Dynasty, and the quantity and density of its road network are also superior than the latter's. Hence, more effort should be put into the construction of the Tomb's surrounding basic facilities. As for the type of land use, Panlongcheng has more varied types of its surrounding land use, which has been involved in Wuhan's urban-rural land use planning, while the surrounding land types of the Tomb are mostly ecological agriculture and forestry, with the obvious advantage of natural resources. Consequently, future work of urban planning and cultural landscape heritage preservation and utilization should stress the improvement of the Tomb's surrounding land use planning and the protection of its environmental integrality, so as to facilitate its sustainable and green development with its advantage in natural resources.
2. The non-spatial development efficiency of the two parks from 2010 to 2019 had small fluctuations, but that of the archaeological park of the Tomb of the King of the Ming Dynasty was low in general. As a national archaeological park with much better policy support, investment and human resources, Panlongcheng Archaeological Park enjoyed higher non-spatial development efficiency than the Tomb. As a result, the improvement

of the Tomb's non-spatial development efficiency should be achieved by digging deep into its cultural value, actively developing its protection and declaration of the world's cultural heritage, and seeking stronger support in policy, fund, and manpower.

3. The application of the GIS-DEA-MI model in research on the development efficiency of archaeological site parks in urban fringes is appropriate both in the early planning of construction and in the post-assessment stage. The study of the early planning can decide reasonable input and optimize the investment efficiency, while the post-assessment stage helps to improve resource allocation. Local government can make macroscopic adjustments according to the dynamic variation data of tourism efficiency to promote the development efficiency of archaeological parks in urban fringe and the harmonious development of urban and rural areas.

What remains to be further studied and discussed may include:

4. Research on spatial development efficiency still has room for a systematic study combining the theory of national land planning;
5. Research on non-spatial development still lacks common and widely accepted standards for the selection of output indexes and input indexes.

Author Contributions: Conceptualization, H.Z. and B.L.; methodology, H.Z. and Y.L.; investigation, H.Z., Y.L. and W.L.; data curation, Y.L.; formal analysis, H.Z. and Y.L.; writing—original draft preparation, Y.L. and H.Z.; writing—review and editing, H.Z., Y.L. and W.L.; visualization, Y.L.; supervision, H.Z. and B.L.; project administration, H.Z. All authors have read and agreed to the published version of the manuscript.

Funding: This research was funded by the Ministry of Education, Humanities and Social Sciences Youth Fund (Grant number: 20YJC760145), the Humanities and Social Sciences Fund of Hubei Education Department (Grant number: 20Y044), the Hubei University of Technology Green Industry Science and Technology Leading Program (Grant number: XJ2021005501) and the Innovation Demonstration Base of Ecological Environment Geotechnical and Ecological Restoration of Rivers and Lakes (2020EJB004).

Institutional Review Board Statement: Not applicable.

Informed Consent Statement: Written informed consent was obtained from all participants.

Data Availability Statement: The data presented in this study are available on request from the corresponding author.

Conflicts of Interest: The authors declare no conflict of interest.

References

1. Jackson, J.B. *Discovering the Vernacular Landscape*; Yale University Press: New Haven, CT, USA, 1984.
2. Taylor, K.; Feng, H.; Feng, T. Cultural Landscapes and Asian Values: Negotiating a Transition from an International Experience to an Asian Regional Framework. *Chin. Landsc. Archit.* **2007**, *23*, 4–9. (In Chinese)
3. Li, L. *Tradition and Change*; Jiangxi People's Publishing House: Nanchang, China, 2008; p. 8. (In Chinese)
4. Jixiang, S. From the Cultural Landscape to the Cultural Landscape Heritage (Part One). *Southeast Cult.* **2010**, *2*, 7–18.
5. Jixiang, S. From the Cultural Landscape to the Cultural Landscape Heritage (Part Two). *Southeast Cult.* **2010**, *16*, 7–12.
6. Banzhaf, E.; Anderson, S.; Grandin, G.; Hardiman, R.; Jensen, A.; Jones, L.; Knopp, J.; Levin, G.; Russel, D.; Wu, W.; et al. Urban-Rural Dependencies and Opportunities to Design Nature-Based Solutions for Resilience in Europe and China. *Land* **2022**, *11*, 480. [CrossRef]
7. Jacques, D. The Rise of Cultural Landscapes. *Int. J. Herit. Stud.* **1995**, *1-2*, 91–101. [CrossRef]
8. Hong, Y. Research on Cultural Landscape Conservation in China. Master's Thesis, Northwest A&F University, Yangling, China, 2009. (In Chinese).
9. Logan, W.S. Dien Bien Phu: Development and Conservation in a Vietnamese Cultural Landscape. In Proceedings of the Forum UNESCO University and Heritage 10th International Seminar, Newcastle upon Tyne, UK, 11–16 April 2005.
10. ICOMOS. *The World Heritage List: Filling the Gaps—An Action Plan for the Future*; ICOMOS: Charenton-le-Pont, France, 2004.
11. Peter, F. *World Heritage Cultural Landscapes, 1992–2002: A Review and Prospect*; UNESCO World Heritage Center: Paris, France, 2003.
12. Taylor, K. Cultural Landscapes and Asia: Reconciling International and Southeast Asian Regional Values. *Landsc. Res.* **2009**, *34*, 7–31. [CrossRef]

13. Taylor, K. From Physical Determinant to Cultural Construct: Shifting discourses in reading landscape as history and ideology. In Proceedings of the Fifteenth Annual Conference of the Society of Architectural Historians Australia and New Zealand, Melbourne, VIC, Australia, 28 September–1 October 1998; University of Melbourne: Melbourne, VIC, Australia, 1998; pp. 371–378.
14. Han, F. The Chinese View of Nature: Tourism in China's Scenic and Historic Interest Areas. Ph.D. Thesis, Queensland University of Technology, Brisbane, QLD, Australia, 2006.
15. Jia, S.; Liao, Y.; Xiao, Y.; Zhang, B.; Meng, X.; Qin, K. Methods of Conserving and Managing Cultural Heritage in Classical Chinese Royal Gardens Based on 3D Digitalization. *Sustainability* **2022**, *14*, 4108. [CrossRef]
16. Wang, X.; Li, J.; Zheng, T.; Diao, S.; Zhang, X.; Tian, Y. Constructing Landscape Ecological Security Patterns of an Ancient Capital Based on Cellular Automata Theory. *Urban Sci.* **2022**, *6*, 29. [CrossRef]
17. Capuano, A. Living Amidst the Ruins in Rome: Archaeological Sites as Hubs for Sustainable Development. *Sustainability* **2022**, *14*, 3180. [CrossRef]
18. Sukwai, J.; Mishima, N.; Srinurak, N. Balancing Cultural Heritage Conservation: Visual Integrity Assessment to Support Change Management in the Buffer Zone of Chiang Mai Historic City Using GIS and Computer-Generated 3D Modeling. *Land* **2022**, *11*, 666. [CrossRef]
19. Kaldun, B. Stakeholder Cooperation in Managing World Heritage Sites: The UNESCO Lijiang Models. In *World Tourism Organization Seminar Proceedings, Proceedings of the Sustainable Tourism Managementat World Heritage Sites-Enhancing Inter-Agency and Stakeholder Coordination for Joint Action, Huangshan, China, 24–27 March 2008*; World Tourism Organization: Madrid, Spain, 2009; pp. 113–119.
20. Connell, J.; Page, S.J.; Bentley, T. Towards sustainable tourism planning in New Zealand: Monitoringlocal government planning under the Resource Management Act. *Tour. Manag.* **2009**, *30*, 867–877. [CrossRef]
21. Yasarata, M.; Altinay, L.; Burns, P.; Okumus, F. Politics and sustainable tourism development—Can they co-exist? Voices from North Cyprus. *Tour. Manag.* **2010**, *31*, 345–356. [CrossRef]
22. Zuo, D.; Li, C.; Lin, M.; Chen, P.; Kong, X. Tourism, Residents Agent Practice and Traditional Residential Landscapes at a Cultural Heritage Site: The Case Study of Hongcun Village, China. *Sustainability* **2022**, *14*, 4423. [CrossRef]
23. Carta, M.; Gisotti, M.R.; Lucchesi, F. Settlements and Urban Morphological Quality in Landscape Planning–Analytical Models and Regulating Tools in the Landscape Plan of Regione Toscana. *Sustainability* **2022**, *14*, 1851. [CrossRef]
24. Dong, Q.; Qu, S.; Qin, J.; Yi, D.; Liu, Y.; Zhang, J. A Method to Identify Urban Fringe Area Based on the Industry Density of POI. *ISPRS Int. J. Geo-Inf.* **2022**, *11*, 128. [CrossRef]
25. UN-Habitat. *The Value of Sustainable Urbanization*; United Nations Human Settlements Programme: Nairobi, Kenya, 2020.
26. Sumari, N.S.; Xu, G.; Ujoh, F.; Korah, P.I.; Ebohon, O.J.; Lyimo, N.N. A Geospatial Approach to Sustainable Urban Planning: Lessons for Morogoro Municipal Council, Tanzania. *Sustainability* **2019**, *11*, 6508. [CrossRef]
27. Fang, X.C.; Chen, H.; Wang, X.L. Coordination of economic development and ecological environment in resource exhausted cities. *Ecol. Econ.* **2015**, *11*, 36–42.
28. Zheng, Y.; Rashid, H.U.; Siddik, A.B.; Wei, W.; Hossain, S.Z. Corporate Social Responsibility Disclosure and Firm's Productivity: Evidence from the Banking Industry in Bangladesh. *Sustainability* **2022**, *14*, 6237. [CrossRef]
29. Liu, W.; He, Z.; Chen, H.; Lin, C. Comparative Analysis Chinese Green Buildings' of Input-Output Effect Based on Data Envelope Analysis. *Buildings* **2022**, *12*, 659. [CrossRef]
30. Zou, W.; Zhang, L.; Xu, J.; Xie, Y.; Chen, H. Spatial-Temporal Evolution Characteristics and Influencing Factors of Industrial Pollution Control Efficiency in China. *Sustainability* **2022**, *14*, 5152. [CrossRef]
31. Tong, L.; Chen, Y.; Jin, L.; Zheng, X. Regional Sustainable Performance of Construction Industry in China from the Perspective of Input and Output: Considering Occupational Safety. *Buildings* **2022**, *12*, 618. [CrossRef]
32. Li, H.; Lin, Y.; Wang, Y.; Liu, J.; Liang, S.; Guo, S.; Qiang, T. Multi-Criteria Analysis of a People-Oriented Urban Pedestrian Road System Using an Integrated Fuzzy AHP and DEA Approach: A Case Study in Harbin, China. *Symmetry* **2021**, *13*, 2214. [CrossRef]
33. Li, D.; Wang, X. Land-Use Change and Efficiency in Laos' Special Economic Zones. *Land* **2021**, *10*, 1012. [CrossRef]
34. Yin, Q.; Wang, Y.; Wan, K.; Wang, D. Evaluation of Green Transformation Efficiency in Chinese Mineral Resource-Based Cities Based on a Three-Stage DEA Method. *Sustainability* **2020**, *12*, 9455. [CrossRef]
35. Xia, B.; Dong, S.; Ba, D.; Li, Y.; Li, F.; Liu, H.; Li, Z.; Zhao, M. Research on the Spatial Differentiation and Driving Factors of Tourism Enterprises' Efficiency: Chinese Scenic Spots, Travel Agencies, and Hotels. *Sustainability* **2018**, *10*, 901. [CrossRef]
36. Li, Y.; Liu, A.-C.; Wang, S.-M.; Zhan, Y.; Chen, J.; Hsiao, H.-F. A Study of Total-Factor Energy Efficiency for Regional Sustainable Development in China: An Application of Bootstrapped DEA and Clustering Approach. *Energies* **2022**, *15*, 3093. [CrossRef]
37. Zhu, S.; Zhou, Z.; Li, R.; Li, W. Impact of High-Speed Rail Construction on the Environmental Sustainability of China's Three Major Urban Agglomerations. *Sustainability* **2022**, *14*, 2567. [CrossRef]
38. Geng, H.; Shi, E.P.; Wang, L.Z.; Xie, R.; Yu, B.J. Evaluation of the Development Efficiency of Small Towns around Metropolis Based on GIS-DEA:Taking Wuhan as an Example. *Econ. Geogr.* **2018**, *38*, 72–79. (In Chinese)
39. Tao, Z.L.; Chen, G.Y. Research progress of the two-step floating catchment area method and extensions. *Prog. Geogr.* **2016**, *35*, 589–599. (In Chinese)
40. Liu, S.; Yang, Y.; Jia, H. Research on Service Radius and Influencing Factors of Community Parks in Shanghai Based on Mobile Phone Signaling Data. *Landsc. Archit.* **2021**, *28*, 88–93. (In Chinese) [CrossRef]

41. Yang, Y.Y.; Zhang, Y.; Zeng, D.Y.; Cheng, S. Spatial and temporal characteristics of tourism efficiency in Hainan Province based on DEA-MI model. *Tour. Overv.* **2019**, *14*, 157–158+160. (In Chinese)
42. Guo, T.Y.; Dong, G.P. Study on the Relationships between Metropolitan Spatial Compact Rations and Their Efficiencies in China in 1990 and 2000. *J. Geo-Inf. Sci.* **2009**, *11*, 482–490. (In Chinese)
43. Fang, X.C.; Chen, H.; Wang, X.L. Evaluation of tourism efficiency in resource-exhausted cities based on DEA-MI Model. *Stat. Decis.* **2015**, *7*, 55–57. [CrossRef]
44. Sheng, Y.C.; Liu, Q. Spatial Difference of Human Capital Promoting Regional Tourism Economic Efficiency: Empirical Re-search Based on "Hu Line". *Sci. Geogr. Sin.* **2020**, *40*, 1710–1719. [CrossRef]

International Journal of
Environmental Research and Public Health

MDPI

Article

The Impact of Rail Transit on Accessibility and Spatial Equity of Public Transit: A Case Study of Guangzhou, China

Huiling Chen [1], Wenyue Yang [2] and Tao Li [3,*]

1 School of Public Administration and Human Geography, Hunan University of Technology and Business, Changsha 410205, China
2 College of Forestry and Landscape Architecture, South China Agricultural University, Guangzhou 510642, China
3 Northwest Land and Resources Research Center, Shaanxi Normal University, Xi'an 710062, China
* Correspondence: ltao@snnu.edu.cn

Abstract: The urban rail transit network provides the possibility for people to shift from car to public transit for travel. This paper clarified the relationships among public transit, accessibility, and equity and studied the impact of rail transit on public transit accessibility that incorporates the measure of travel time and transit fare and the impacts' spatial equity. The results show that rail transit contributes to the similar distribution between high rate of changes of time-based accessibility communities and fare-based accessibility communities, which are located nearby the rail transit lines. The degree of inequity in travel time is higher than the degree in transit fare in two scenarios. Due to the well-connected bus transit in the city center, absolute changes in travel time are slight, while relative changes are high. The rail transit has promoted the improvement of public transit equity in some areas. The difference between the time-based accessibility of Conghua District, northern and southern Baiyun District, Huadu District, Nansha District and southern Panyu District, and other communities is getting smaller, which is conducive to the improvement of spatial equity. The results provide theoretical support for the development of an integrated multimodal public transit system.

Keywords: rail transit; public transit; travel time; transit fare; accessibility; equity

Citation: Chen, H.; Yang, W.; Li, T. The Impact of Rail Transit on Accessibility and Spatial Equity of Public Transit: A Case Study of Guangzhou, China. *Int. J. Environ. Res. Public Health* **2022**, *19*, 11428. https://doi.org/10.3390/ijerph191811428

Academic Editor: Paul B. Tchounwou

Received: 10 July 2022
Accepted: 4 September 2022
Published: 11 September 2022

1. Introduction

Conventional bus transit with limited traffic capacity cannot meet the public transit needs of big cities due to rapid expansion of the city, widening activity spaces, and lengthening travel distance [1,2]. The building of rail transit infrastructure has already provided an effective supplement to the public transit system. Urban rail transit is gradually dominating the public transit system due to characteristics of large capacity, speediness, reduced air pollution, and land use and is regarded as an important way to solve urban traffic congestion and achieve energy-saving and emission reduction targets [3]. However, the expensive construction cost of rail transit leads to its limited coverage rate in the transit network, which fails to meet the social needs of those dependent on public transit. From the economic view of travel cost, there is a gap between rail transit fare and bus fare. Individual consumption ability impacts the transport mode choice of riders, which affects the accessibility level. A household travel survey in Canada in 2001 showed that approximately 78% of households that earn less than $40,000 commute by public transit, among which only 13% of such income groups are rail commuters; the majority of people use a cheaper method of bus transit [4].

Urban rail transit construction in China is in progress. By 2017, 31 cities opened an underground railway, among which Shanghai has the greatest mileage. Studying the impacts of rail transit networks on the improvement and disparities of transit accessibility could help plan new rail transit lines and stations and help them coordinate better with the existing transit network, which attempts to increase the transformation possibility of the travel mode choice of the population from car to public transit [5]. This kind of research has an important implication for the improvement of the urban transport service level, the satisfaction of the diverse and growing public transit demands, and the realization of sustainable urban transport development. Rail transit significantly reduces the travel time between communities, generating both efficiency and spatial equity effects. It also leads to the obvious "time–space convergence" effect, enlarges the influence of the city center, creates transport advantage for more areas, reconstructs the urban structure, and influences the spatial distribution of economic activities [6,7]. However, it aggravates disparities of accessibility among regions and brings about the concern of transport equity [8–11].

Accessibility indicators are increasingly being used as supporting tools for transport infrastructure planning. Our study contributes to the existing literature by evaluating the influence of rail transit on transit accessibility and by answering the two following questions: (1) What are the spatial characteristics of the travel expense and travel time of communities by bus transit or integrated bus-rail transit and are there significant spatial differences between the two travel modes? (2) How are disparities for transit accessibility based on the rail transit network distributed across geographical areas and combined under an equity objective? The remainder of the paper is organized as follows: Section 2 presents the literature review on transit accessibility equity. The study area, data, and methodology are illustrated in Section 3. Section 4 evaluates the spatial influence of the rail transit in Guangzhou. The final section demonstrates conclusions and discussion about future research directions.

2. Literature Review

Equity, regarded as a policy target in public transit supply, initially appeared in the 1990s. One way to achieve goals about equity in the city is to provide transit services to the people who need them most frequently. Equity is a concept that involves multiple dimensions and multi-faceted aspects, and geographers mainly study equity in the distribution of public service facilities. Litman (2007) argued that public transport equity analysis is supposed to concentrate on two dimensions: horizontal equity and vertical equity [12]. Horizontal equity emphasizes the distribution of transport resources among various individuals. The studies that are related to the horizontal equity of public transit explore the minimum equity standard of "uniform" people without considering the socio–spatial differentiation. Meanwhile, vertical equity refers to the distribution of public transit resources among individuals differing in capabilities and needs. Thus, it emphasizes the needs of various social groups to alleviate social inequity. The allocation of public transit resources is a process of meeting the daily travel needs of people. The study on the equity of public transit has gone through three stages: geographical equality, spatial equity, and social equity. The studies on the spatial equity of public transit focus on the issue of equivalence, which is the minimum standard of "homogeneous persons". The studies on social equity began to pay attention to the in justice between different social groups.

The equity assessment of public transit supply is closely related to the measurement of accessibility [13,14]. Accessibility conventionally is defined as potential opportunities for interaction [15], which is an indicator commonly adopted to measure the geographical effects of the transport network and the relevance between transport and economic growth and to evaluate the impact of urban transport on travel and the land-use pattern [16–18]. Gleason (1975) analyzed a dataset covering bus stop locations by virtue of integer programming models, marking the beginning of studies on transit accessibility [19]. Existing studies on transit accessibility mainly focus on the measurement of accessibility and its application in different cities, among which the differences in the function adopted in accessibility application make the measurement differ significantly. There are three types of indicators for transit accessibility. The first type of indicator refers to physical accessibility based on a European geometrical distance, in which the common indicator is the walking distance from the starting site to the transit stop [20]. Such indicators are simple and easily calculated. However, they disregard the service provided by each transit station, desired destinations, and travel time to these destinations, so the accessibility reflected by such indicators is incomplete. Compared with the aforementioned indicator, the second type of indicator evaluates the service frequency of transit stations, which means that the importance of the station in the entire transit network has been considered [13]. The third type of indicator additionally accounts for the travel cost associated with destinations [21], as well as the temporal variability of public transit [22,23]. General Transit Feed Specification files, which construct a multimodal public transit network, are an important technical improvement. A Google Transit Feed Specification (GTFS) dataset has been recently applied to build a completely routable multimodal transit network, which enables the estimation of public transit travel times at different times of the day [24,25]. The fourth type of accessibility indicator refers to the log sum of various opportunities within the specific cost calculated from a regional travel demand model under public transit travel situations [26]. Most of these indicators are used for studies on the spatial equity impact of public transit. The social equity of public transit studies mainly focus on the limited accessibility issues that are affected by different individuals' space–time constraints [27]. Existing studies often firstly used specific criteria to distinguish transport-disadvantaged groups and then discussed the social equity of their travel behavior [28]. On the one hand, some studies explore the various impacts of public transit on the travel behaviour of residents in different regions, and it is unfair that some have fewer travel opportunities and transport resources than others [29]. On the other hand, some researches analyzed the relationship between the travel behavior of transport and the disadvantaged groups' socio–economic attributes under the influence of public transit accessibility [30].

In the studies reviewed above, several authors incorporated public transit travel expenses into the measurement of transit accessibility due to complex fare structures. Ford (2015) and Currie (2004) measured transit accessibility based on formal costs, which include travel time and travel expenses [31,32]. El-Geneidy et al. (2016) assessed the differences in accessibility resulting from transit fares by considering monthly or single fares [33]. The approaches they used to calculate the fares were not applicable to other rail transitpolitan areas because of the data. Different from the diverse ticket pricing strategies in many countries, the prices of public transit tickets in China are typically determined by the government. Although some special riders enjoy privileges (such as elderly individuals and students), the rest are charged the same when taking bus transit, and prices are based on travel distances for rail transit. To our knowledge, most authors have analyzed the positive effects of rail transit on increasing land values and accessibility in travel time [34,35], and literature on the multimodal public transit system mostly focuses on the joint optimization of rail transit lines and bus transit lines as well as the distribution of transfer stations [36–38], but the assessment of the spatial equity impacts caused by rail transit is limited. Therefore, Guangzhou is a typical empirical case for the study of the effects of the rail transit on public transit accessibility.

In this paper, travel time and transit fares are incorporated into the measurement of transit accessibility. We study disparities of the accessibility changes among communities under the influence of the rail transit and evaluate its spatial equity effects in an attempt to offer theoretical support to the development of an integrated multimodal urban public transit system.

3. Methodology

3.1. Study Area

This paper covers the study of Guangzhou city's 11 municipal districts, including an area greater than 7434.4 km^2 as well as more than 2000 communities (Figure 1). As one of the largest cities in China, Guangzhou city is located in the Pearl River Delta economic zone, known as a member of three significant and domestic economic circles. It is famous across the world for its leading role in the reform and opening-up to the rest of the world. Undoubtedly, Guangzhou city reflects the development of important cities in China to some extent. Therefore, we employ community-level statistics about the population gained from the sixth national population census results of Guangzhou city in 2011. The destinations of all trips are 11 major commercial complexes in Guangzhou. The type of destination is not that important at this point, as the basic aim of this paper is to analyze the impacts of rail transit on the accessibility by public transit—therefore, spatial distribution is the major criterion for their selection rather than the category of facility. Major commercial complexes are located evenly in the study region and thus provide a relatively typical sample for this study.

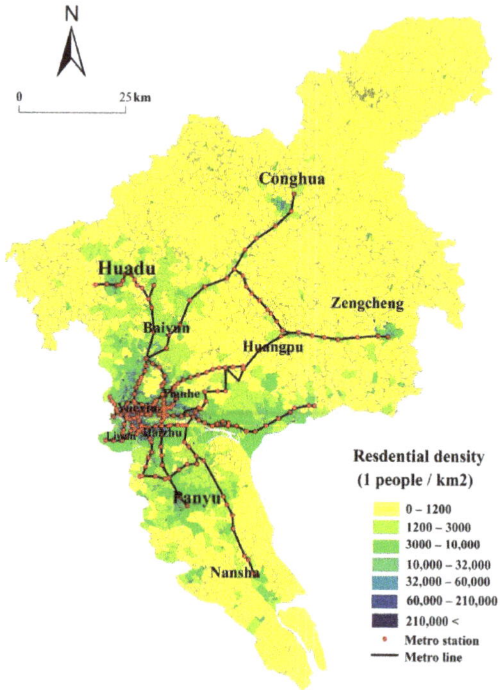

Figure 1. The spatial distribution of residential density and major commercial complexes in Guangzhou.

3.2. Data and Accessibility Calculation

The navigation function that the Baidu map provides automatically can implement the travel schemes between any two geographic points in the city based on the travel mode choice. We made iterate inquiries with the WebAPI based on the Baidu map LBS open platform in Pythonscript. Real-time transport information for the Baidu map is estimated every two minutes based on data from multiple sources, considering not only the speed of different levels of roads but also real-time traffic congestion, which can render an estimation with a high accuracy of travel distances and travel time [39–41]. The road network of the Baidu map is complete and updated constantly, in which travel distances and travel times are calculated with the total for all segments along the route [36]. To reflect the reliability of travel data based on the Baidu map, we invited 10 residents who live in different communities to record their daily travel time and travel distances on public transit (including the use of rail transit or bus and transfers involving rail transit and bus), while the same period of time and same distances were calculated with the Baidu map. Fifty tests were performed. The results show that the difference between the actual travel time/distances and those calculated based on the Baidu map ranged between −10% and 10%, so the travel time and travel distances based on the Baidu map can be considered reliable.

We took the centers of communities as the origins and 11 major commercial complexes as the destinations. We found the best paths between the starting points and the ending points using two different travel modes—the mode of the bus transit that is based on the bus transit system and the mode of the shortest travel time that is based on the integrated public transit system, here named the bus-only scenario or bus and rail transit scenario, respectively. The multiday and multiperiod methods were used for data collection to obtain the daily travel time and travel distance by public transit. The acquisition time included five working days and one weekend, and the average was calculated to represent the daily travel time and travel distance. The data of travel distances, travel times, and travel routes information between all communities and major complexes were obtained between 12 November and 19 November 2020. The dataset included 409,794 routes, 409,794 records of travel time, and 409,794 records of travel distance for each scenario.

As for the bus and rail transit scenario, the travel scheme with the shortest time would be selected if the rider could reach the destination by bus or rail transit directly. The scheme with the shortest total time would be selected if the rider could reach the destinations by rail transit involving transfers or they could reach the destination by bus transit directly. The scheme with the shortest total time would be chosen if the rider cannot reach the destination by either direct bus or rail transit. For the bus-only scenario, the travel scheme with the shortest time would be selected. Most regular bus transit fare is 2 yuan, and a few bus express lines connecting the main urban area with the marginal urban area (Zengcheng District, Conghua District, Nansha District) have higher prices, ranging from 10 yuan to 20 yuan. The ticket for rail transit is priced based on travel distances by rail transit. We first compared the travel time, transit fare, and travel speed among the CBD, major commercial complexes, and all communities under different scenarios. The impacts of rail transit on the equity of the spatial distribution of travel times improvement and increased transit fare cost were then discussed. The calculation of travel costs in different scenarios is shown in Figure 2. Residents who travel by public transit generally included three stages: walking from the starting point O to the public transit station stage, taking public transit to the station near the destination stage, and walking to the terminal D stage. The travel time and travel fare of these three stages were summed in this paper, and the time spent during transfer in the second travel stage was also considered.

This paper analyzed the accessibility changes based on the average travel time and transit fare from communities to major commercial complexes between the bus-only scenario and the bus and rail transit scenario. The expression is as follows:

$$T_i = \sum_{j=1}^{n} t_{ij} / n P_i = \sum_{j=1}^{n} p_{ij} / n$$

where T_i and P_i are the accessibility of community i, t_{ij} is the travel time to the destination of the commercial complex j, and p_{ij} is the transit fare to destination j. t_{ij} and p_{ij} adopt the minimal travel time or transit fare. The reduced value of T_i in the bus and rail transit scenario denotes the travel time saved in community i, and the increased value of P_i denotes the transit fare raised in the community. The community with the lowest average travel time or transit fare is considered to have the highest time-based accessibility or fare-based accessibility level among all communities.

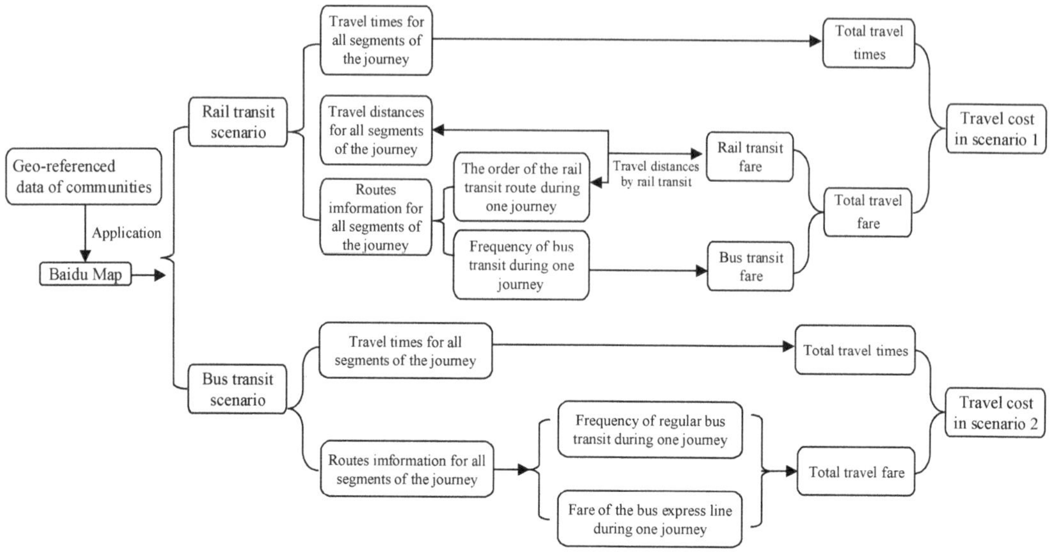

Figure 2. The flowchart of obtaining the OD matrix of travel costs in two scenarios based on the Baidu map.

3.3. Equity Analysis

The equity analysis in this paper is based on the differences in the spatial distribution of accessibility improvement between these two scenarios. Equity effects are usually measured by using a range of indicators of the spatial distribution of accessibility indicators. The selection for these indicators is based on the changes in accessibility between scenarios. An ideal equity indicator does not exist, and many scholars advise calculating a set of indicators to analyze their results as a complement. We utilized three steps for evaluation of the equity effects based on the findings. In the first analysis, the coefficient of variation is the ratio of the standard deviation to the average, which represents the relative change of the geographical data. The formula for CV is:

$$CV = \frac{1}{x} \times \sqrt{\frac{\sum\limits_{i=1}^{n} (x_i - x)^2}{n-1}}$$

In similar research, this indicator has been used frequently for the purpose of evaluating the equity effects. The increased CV value states the reduction in equity and negative equity effect, whereas a reduction in the CV value indicates the positive equity effect and more balanced spatial distribution of accessibility.

Then, the normalized value of relative and absolute accessibility improvement were calculated in each community. These two values are complementary because a community is able to obtain an absolute improvement but a relative low improvement if its initial accessibility value has a low level. The value of the absolute improvement and relative change ratios of travel time of communities after the operation were normalized (z-score) in this paper so that results could be comparable. A z-score of 0 is equivalent to the average accessibility benefit of the entire city. Accessibility benefits could be completely equivalent throughout the city if all communities have scores of 0. Positive values indicate a community has a larger than the city average in accessibility benefit, and negative values show it as lower than the average accessibility benefit. We illustrated how the values are distributed across each district to better assess the equity of accessibility changes.

Finally, we identified the role of different communities in the spatial equity of the rail transit by comparing two indicators referring to time-based accessibility and the rate of change between the two scenarios, which is beneficial for the subsequently targeted improvement of the regional transportation infrastructure and the promotion of spatial equity of transport accessibility. All communities were divided into two categories: ① The differences between the two scenarios tended to be smaller, which were conducive to spatial equity; this included two situations: low time-based accessibility (travel time higher than the average)—high rate of change (rate of change higher than the average), high time-based accessibility (travel time lower than the average)—low rate of change (rate of change lower than the average); ② The differences between the two scenarios tended to be larger, which was not good for spatial equity; this included two situations: low time-based accessibility (travel time higher than the average)—low rate of change (rate of change lower than the average), high time-based accessibility (travel time lower than the average) and high rate of change (rate of change higher than the average).

4. Results and Analysis

4.1. The Impacts of Rail Transit on Accessibility

4.1.1. CBD Accessibility Differences

From the selected origin (the commercial complex in the CBD) to all communities, the differences in the reachable area and population within periods, and the average travel speed and transit fare between the bus-only scenario and the bus and rail transit scenario are illustrated in Tables 1 and 2. In the bus-only scenario, 107 communities can be reached within a 30-min drive, and 637 communities can be reached within a 1-h drive. Within the same time, 476 and 1245 communities are reached in the rail transit scenario. The spatial coverage area and inhabited population within a 10- to 30-min isochronous circle from the CBD in the bus and rail transit scenario are more than double compared to that in the bus-only scenario (Figure 3a,b).

In the bus-only scenario, the isochronous circle of the CBD is distributed in a concentric mode. In the rail transit scenario, the isochronous circle of the CBD is extended and spread along the rail transit lines in a fingerlike style. Figure 3c shows the spatial distribution of speed ratios between the two scenarios in the communities. The maximum travel speed ratio is 3.75 when evaluating trips from all communities to the CBD. Communities with small differences in travel speed are found in the southern Nansha District, eastern Huadu District, most of the Zengcheng District, and part of the central city. Communities with a high variation of travel speed are near the rail transit line. There is a mixture of high ratio communities and communities with the ratio close to 1 in the city center. The fastest travel speed from communities to the CBD by rail transit is 3.75 times the speed of bus transit. The speed of travel in different scenarios significantly varies for communities along rail transit networks.

Table 1. A comparison of the average travel speed and average transit fare in the bus-only and rail transit and bus scenarios.

Travel Mode	Average Travel Speed (km/h)	Average Transit Fare (RMB)
Bus only	18.25	5.51
Integrated bus transit and rail transit	27.95	12.12

Table 2. The difference in the isochronous areal analysis of the CBD between the bus-only and bus and rail transit scenarios.

	Time Periods in the Bus-Only Scenario (Min)			Time Periods in the Bus and Rail Transit Scenario (Min)		
	0–30	30–60	60–120	0–30	30–60	60–120
Reachable population	78.94	307.27	510.94	258.93	691.49	2641.57
Spatial coverage area (km^2)	41.32	224.32	1292.26	99.55	481.93	430.58

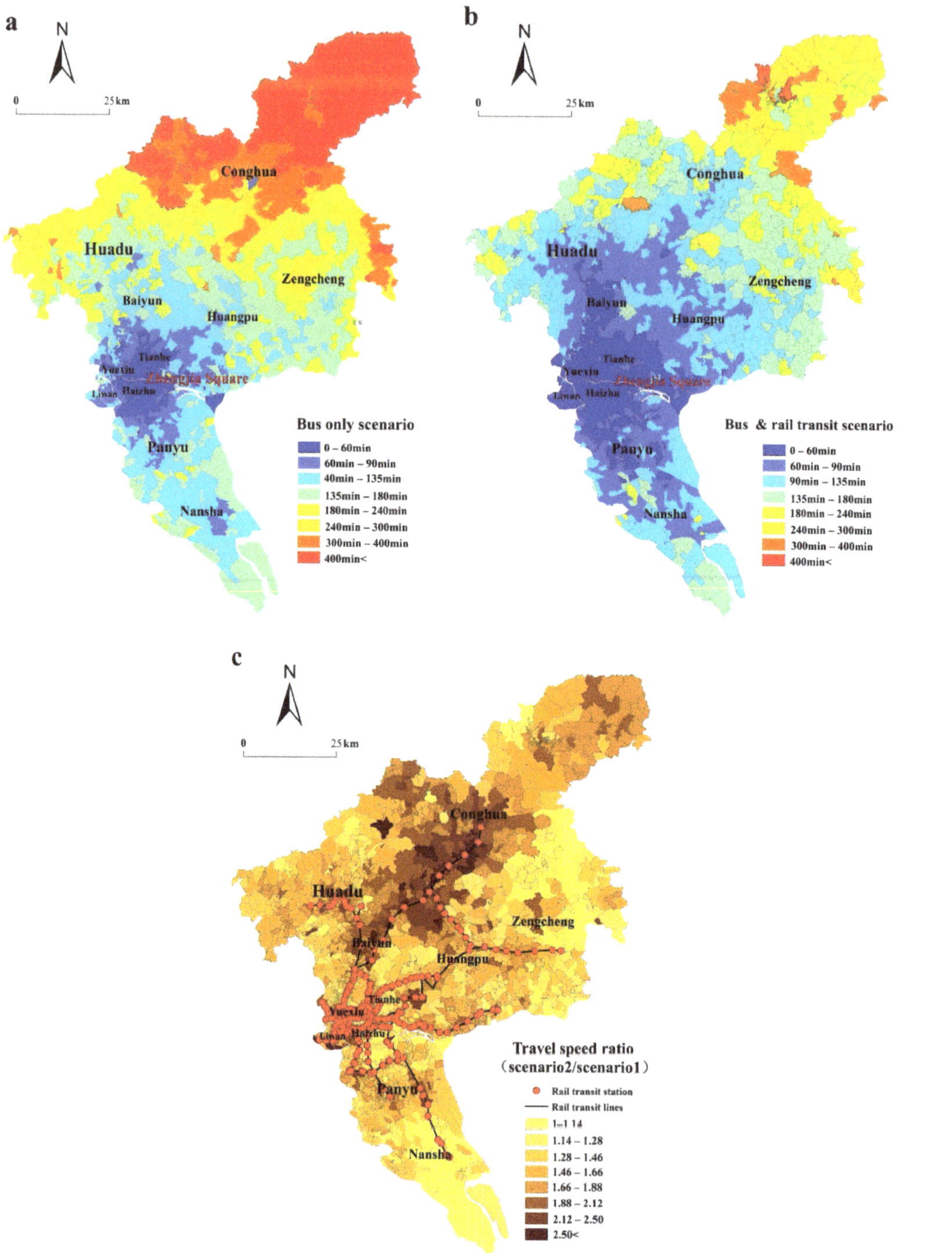

Figure 3. The spatial distribution of time-based accessibility of the CBD in the bus-only scenario (**a**), in bus and rail transit scenario (**b**), and the average travel speed ratio between the two scenarios (**c**).

4.1.2. Spatial Distribution of Time-Based Accessibility

We calculated the travel times from all communities to 11 major commercial complexes. The operation of rail transit lines leads to a decrease in mean time from 31.58 h to 19.68 h, and the total travel time decreases from 83,945.71 h to 52,314.46 h. The improvement of the total travel time is 37.68%, showing that the rail transit greatly reduces the time between communities and commercial complexes and promotes the communication of social economy and culture. As Figure 4a reveals, the spatial pattern of time-based accessibility shows the inner city is the core center with a low-value spatial distribution, and the value increases from the core to the fringe area. The spatial pattern of time-based accessibility shows the north–south extension in a "concentric" shape. The community with the longest average travel time is 8.65 times that with the shortest average travel time.

The spatial pattern of time-based accessibility is slightly reversed with the rail transit network (Figure 4b). The overall level of time-based accessibility in the city center is always the highest, indicating that the average time of reaching all major commercial complexes is the lowest, while the fringe areas need more time. The community with the longest travel time (519.42 min) is 11.3 times the community with the shrotest travel time (45.95 min) due to the improved link brought about by the rail transit lines. Figure 4c shows the percentage of changes in the time-based accessibility level between the two scenarios. This clearly highlights the transformation that would occur in communities of the Conghua District, as their time-based accessibility values in the bus-only scenario are low. The communities near the rail transit lines see high improvements, which offer virtually direct access to the commercial complexes. The lowest percentages of improvement occur in the communities of the northern Zengcheng District, which obtained an improvement value of less than 12.67%, demonstrating that they are less affected by the rail transit lines.

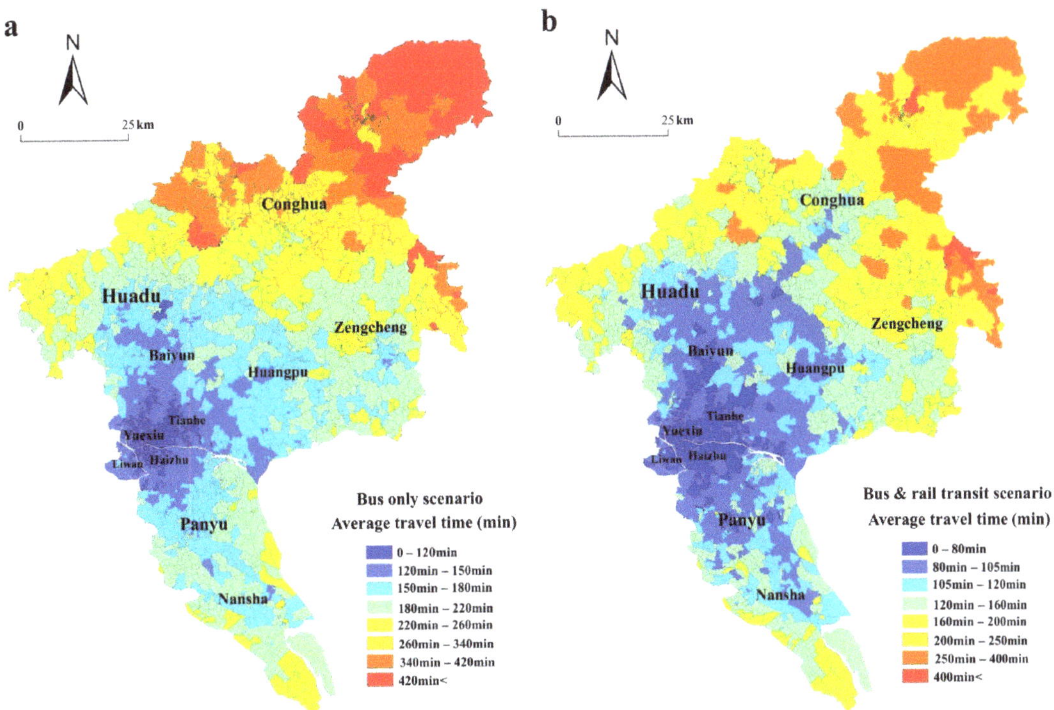

Figure 4. *Cont.*

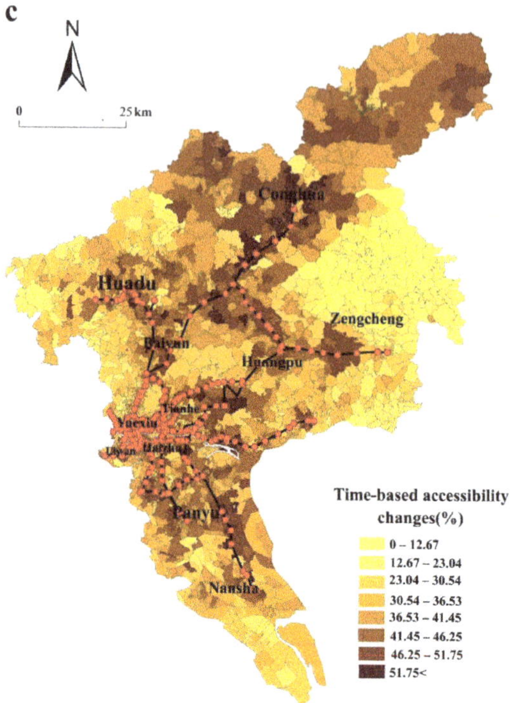

Figure 4. The spatial distribution of the time-based accessibility of communities in bus-only scenario (**a**), in bus and rail transit scenario (**b**), and accessibility changes between the two scenarios (**c**).

4.1.3. Spatial Distribution of Fare-Based Accessibility

Fare cost-based accessibility was calculated and compared between the two scenarios (Figure 5a,b). In the bus-only scenario, the average transit fare for Tianhebei residential block in Tianhe District is the lowest (3.27 RMB), while communities with high transit fares are in the western Huadu District and the southern Nansha District. The average fare between the Zhoudong village of the Conghua district and major commercial complexes is the highest (12 RMB). The highest average fare is approximately three times the lowest fare. In the bus and rail transit scenario, the average transit fare between the Lvhe residential block in the Tianhe District and major commercial complexes is the lowest (6.09 RMB), while the transit fares of communities near the rail transit lines are high. Figure 5c shows the rate of change of the transit fare bbetween the two scenarios. The rail transit significantly affects the spatial distribution of transit fare-based accessibility. The spatial pattern of the changes in the time-based accessibility and transit fare is similar for the two scenarios. Communities with a high rate of change are located near the rail transit lines.

Figure 6 shows the proportion of the population in the areas within certain travel times and fare costs. Approximately 80% of the population can reach the nearest commercial complex within 43 min by bus only with a travel cost lower than 2 RMB and can reach the nearest commercial complex within 30 min with travel costs lower than 4 RMB through combined bus and rail transit.

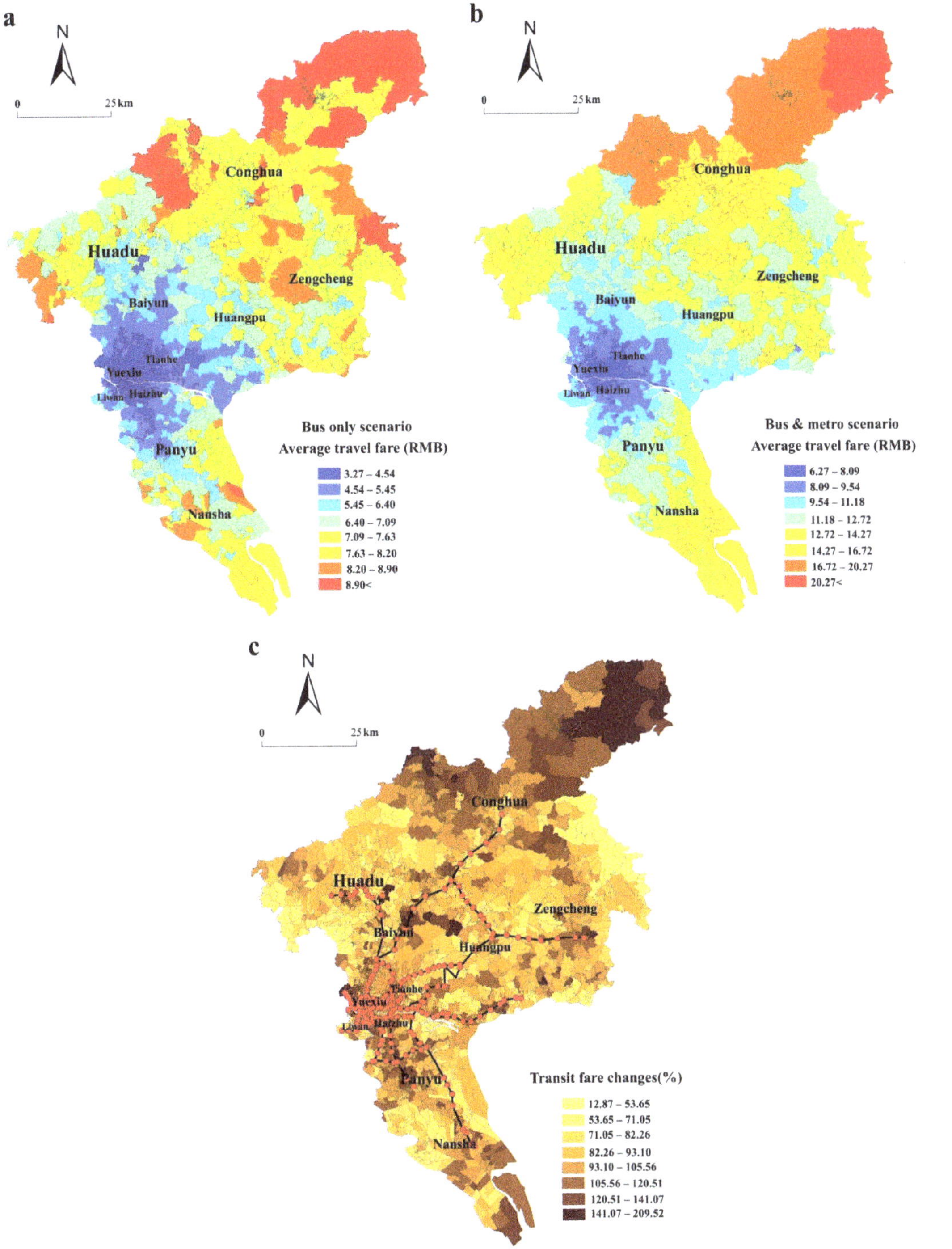

Figure 5. The spatial distribution of the transit fare of communities in the bus-only scenario (**a**), in the bus and rail transit scenario (**b**), and the fare changes between the two scenarios (**c**).

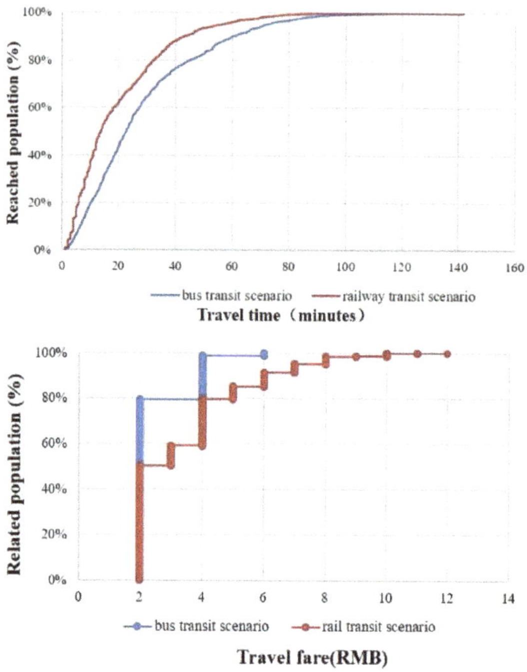

Figure 6. Cumulative share of the population accessing the closest urban commercial complex within a certain travel time and fare cost.

4.2. The Impacts of Rail Transit on Spatial Equity

Equity is one of the social impacts brought about by the construction of transport infrastructures. The calculated *CV* values of time-based accessibility and fare-based accessibility in two two scenarios are used as major indicators for the evaluation of the degree of spatial disparity variability changes. The results are shown in Table 3. The increased (13.95%) *CV* value of the time-based accessibility from scenario 1 to scenario 2 implies that the whole city has departed from the relief of disparity from the travel-time reduction after the development of rail transit. The inner city, middle city, and outer city areas witnessed a significant increase in the spatial disparity of travel time after the construction of the rail transit. The rail transit provides citywide intensification of the spatial inequality of transit fare, as reflected by the *CV* values from scenario 1 (0.27) to scenario 2 (0.28). Clearly, in the bus and rail transit scenario, the equity benefits of travel-fare expenses were received by the inner city and middle city, with the disparity in the transit fare within the outer city intensified.

The *CV* of the time-based accessibility of communities in the bus and rail transit scenario is higher than that in the bus-only scenario, but its average and standard deviation are lower, indicating that the rail transit has greatly reduced the average travel time of communities, and the gaps of services acquired from the main commercial complexes among communities are widening. The standard deviation, the average, and the *CV* of the fare-based accessibility of communities in the bus and rail transit scenario are higher than those in the bus-only scenario, which means that the operation of rail transit increased the disparity of public transit service prices among communities, improving fairness in distribution. By comparing the *CV* values of time-based accessibility and fare-based accessibility, we found that the *CV* value of time-based accessibility is higher than that of the fare-based accessibility in the two scenarios, indicating that disparities in travel time among communities are greater than those in transit fare and the inequity degrees of travel times are higher than those of the transit fares.

Table 3. Coefficient of variation (*CV*) of time-based accessibility and fare-based accessibility.

		Number of Communities	Bus-Only Scenario			Bus and Rail Transit Scenario			Rate of Change	
			Standard Deviation	Mean Value	Coefficient of Variation	Standard Deviation	Mean Value	Coefficient of Variation	Mean Value	Coefficient of Variation
Time-based accessibility	All communities	2658	74.26	172.26	0.43	53.61	107.35	0.49	38.28%	0.28
	Inner Guangzhou	902	13.48	116.73	0.12	10.57	67.35	0.16	42.16%	0.17
	Middle Guangzhou	802	28.88	156.01	0.19	23.24	95.57	0.24	38.61%	0.25
	Outer Guangzhou	954	83.51	238.44	0.35	59.49	155.08	0.38	34.33%	0.38
Fare-based accessibility	All communities	2658	1.61	5.87	0.27	3.19	11.27	0.28	93.05%	0.23
	Inner Guangzhou	902	0.49	4.33	0.11	0.83	8.24	0.10	90.92%	0.18
	Middle Guangzhou	802	1.21	5.71	0.21	1.77	10.85	0.16	93.08%	0.26
	Outer Guangzhou	954	1.01	7.46	0.14	2.44	14.48	0.16	95.03%	0.26

Figure 7a,b shows the spatial distribution of Z-scores of the relative improvement and absolute improvement of the travel time in urban communities. The green areas indicate that their accessibility gains are above average and raise the overall equity, whereas red areas illustrate that their accessibility improvements are below average and reduce the overall equity. The communities with percentages of improvement above average occur in Panyu District and Conghua District. Comparing the two accessibility maps, we find that communities in the center of Panyu District, Baiyun District, and Conghua District have Z-scores greater than 0.5 for absolute and relative improvements. Most communities in Liwan District, Yuexiu District, Haizhu District, and Tianhe District have above-average relative improvements in time-based accessibility, and their absolute improvements are below average. This circumstance occurs because these communities have high time-based accessibility levels in the bus-only scenario due to good bus transit connections. Communities in Zengcheng District and western Baiyun District are disadvantaged in the bus-only scenario; both their absolute improvements and improvement rates are at significantly low levels as well. Therefore, the accessibility gap between these communities and the rest is even greater, and the inequity related to rail transit is increased.

Tables 4 and 5 illustrate how accessibility benefits are distributed across the population of each district, and z-values of the accessibility benefits are used to show the percentage f population within each district that falls into the range of half standard deviation increments. Table 5 shows the distribution of the normalized values of absolute improvements across the population of each district and reveals that the majority of the population in districts of the inner city is lower than the mean of the entire city, whereas the districts of the outer suburbs suggest the opposite. To explore the distribution of relative improvements in relation to the population of each district in Table 5, the analysis results can be compared with those in Table 4, which shows the results of the absolute improvements. Overall, the largest upward shifts toward the mean are experienced by districts in the inner city, in which a large percentage of the total population shifts from below the mean of absolute improvements of the city to above the mean of relative improvements.

Communities in Conghua District and Panyu District have a disadvantaged position in the bus-only scenario. Due to an initially poor position, rail transit lines bring high absolute improvements but low relative improvements to most residents in these two districts. The majority of the population in Conghua District has higher z-value scores for absolute improvements and relative improvements than the average value in the bus and rail transit scenario. The position of the population in Baiyun District relative to the mean is worse, with low absolute improvements and low relative improvements. Notably, 25.79% of Conghua District's population has relative improvement z-values greater than 1.5.

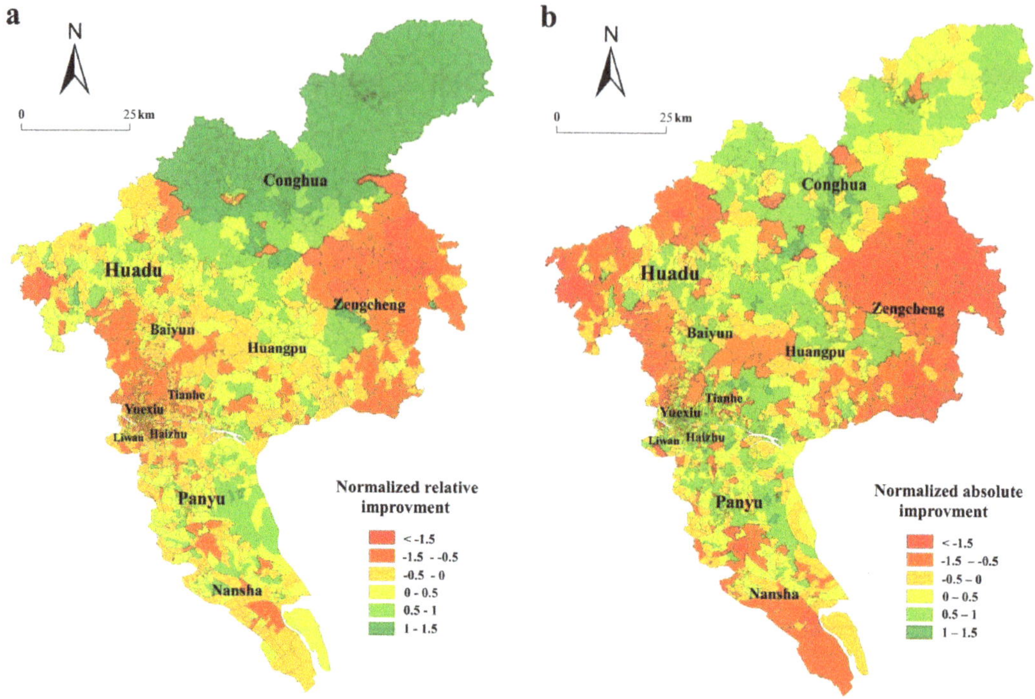

Figure 7. Normalized values of relative (**a**) and absolute (**b**) improvements in time-based accessibility.

Table 4. The absolute improvements in time-based accessibility by population according to districts after rail transit lines were established.

z-Value Scores		<−1.5	−1.5–−0.5	−0.5–0	0–0.5	0.5–1.5	1.5<
Inner Guangzhou	Yuexiu district	0.00%	2.54%	3.67%	0.23%	0.00%	0.00%
	Haizhu district	0.00%	1.68%	6.32%	0.75%	0.09%	0.36%
	Liwan district	0.03%	1.42%	3.38%	0.57%	0.00%	0.00%
	Tianhe district	0.03%	2.70%	5.35%	0.86%	0.48%	0.54%
Middle Guangzhou	Baiyun district	0.19%	5.57%	6.15%	2.32%	0.25%	0.00%
	Huangpu district	0.00%	0.55%	2.40%	2.65%	0.77%	0.00%
	Panyu district	0.00%	0.58%	4.12%	5.30%	2.14%	0.08%
Outer Guangzhou	Huadu district	0.02%	0.25%	2.02%	2.72%	1.21%	0.02%
	Conghua district	0.00%	0.02%	0.01%	0.09%	1.20%	1.24%
	Zengcheng district	0.35%	2.50%	3.51%	1.27%	0.71%	0.02%
	Nansha district	0.00%	0.07%	1.15%	0.53%	0.29%	0.00%

According to Figure 8, the increasingly larger differences are mainly located in Zengcheng District, southern Nansha District, and Huadu District, which are on the edge of Guangzhou. The time-based accessibility of these communities is lower than the average in the bus-only scenario, and the rate of change is also lower than the average in the bus and rail transit scenario, which widens the gap between these communities and other areas with superior time-based accessibility. On the contrary, central urban areas have high time-based accessibility and a high rate of change, so rail transit also widens the gap between communities in central urban areas and other communities. Increasingly smaller differences are crucially located in Conghua District, northern and southern Baiyun District, Huadu District, Nansha District, and southern Panyu District, among which Conghua

District, Northern Baiyun District, Huadu District, and southern Panyu District have low time-based accessibility and a high rate of change. Correspondingly, western and southern Baiyun District and other places have high time-based accessibility and low rate of change. The rail transit reduces the differences in the time-based accessibility between communities of this type and other communities. The spatial equity of time-based accessibility is widely observed in these communities.

Table 5. The relative improvements in time-based accessibility by population according to districts after rail transit lines were established.

z-Value Scores		<−1.5	−1.5–−0.5	−0.5–0	0–0.5	0.5–1.5	1.5<
Inner Guangzhou	Yuexiu district	0.09%	0.97%	1.48%	2.36%	4.03%	0.01%
	Haizhu district	0.03%	0.47%	1.77%	5.02%	4.51%	0.36%
	Liwan district	0.18%	0.70%	1.56%	2.42%	2.13%	0.04%
	Tianhe district	0.18%	1.47%	2.17%	3.19%	4.20%	0.54%
Middle Guangzhou	Baiyun district	1.44%	5.40%	3.31%	3.83%	3.64%	0.20%
	Huangpu district	0.05%	0.86%	0.78%	2.35%	2.36%	0.30%
	Panyu district	0.32%	1.20%	2.99%	3.92%	4.03%	0.31%
Outer Guangzhou	Huadu district	0.37%	1.22%	1.32%	2.01%	1.60%	0.00%
	Conghua district	0.06%	0.14%	0.36%	0.81%	2.58%	0.87%
	Zengcheng district	2.45%	3.14%	1.78%	1.21%	0.77%	0.02%
	Nansha district	0.06%	0.70%	0.81%	0.27%	0.29%	0.00%

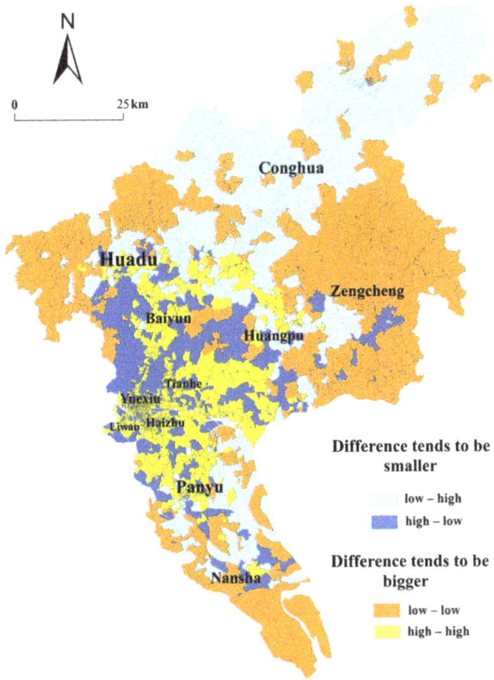

Figure 8. The distribution of the classification results of spatial equity by communities.

5. Conclusions

As investments in citywide rail transit projects increase, often accompanied by lofty goals for invigorating economic development, this paper serves to highlight the importance of designing a relatively equitable public transit system throughout the urban area. With the use of a travel data matrix obtained from the Baidu API, this paper examined the influence of rail transit on the public transit accessibility by using travel time and transit fare for all communities of Guangzhou. On this basis, we analyzed the spatial equity impacts of the rail transit and the potential for improving accessibility for the population.

The main conclusions are as follows. (1) In the bus-only scenario, the time circles of the CBD are concentric. Eighty percent of the population can reach the nearest commercial complexes within 68 min, with a fare cost lower than 2 RMB. The travel time and transit fares of communities and their needs are closely correlated. In the bus and rail transit scenario, the time circles of the CBD extend along rail transit lines similar to "fingers". Eighty percent of the population can reach the nearest commercial complexes within 56 min, a with fare cost lower than 4 RMB. Travel time and transit fares in the community have less of an association. The rail transit significantly lowers the correlation between the transit fare and travel time. (2) The rail transit increases the internal disparities of time-based accessibility that generally increases from the inner city via the middle city to the outer city and widens the gaps in locational advantage and the level of accessibility and public transit service prices among communities and exaggerates unfairness. The degree of inequity in travel time is higher than the degree associated with the transit fare in the two scenarios. (3) The Zengcheng District and western Baiyun District are in the disadvantaged position in the bus-only scenario, where absolute improvements and relative improvements in travel time in the bus and rail transit scenario are lower than the average. Therefore, rail transit widens the gap between communities in these areas and the rest, leading to an increase in inequity. (4) The rail transit affects most residents in Conghua District and Panyu District, whose absolute improvements are high and relative improvements are low. Most of the population in the inner city is below the mean of absolute improvements and above the mean of relative improvement. The position of the population in the Baiyun District relative to the mean is worse, with low absolute improvements and low relative improvements. The rail transit reduces the time-based accessibility differences between communities in Conghua District, the peripheral area of Baiyun District, the center of Huadu District and southern Panyu District, and other communities, where the spatial equity of time-based accessibility is widespread in these communities.

Researching the impact of rail transit on transit accessibility is an efficient approach to assess the spatial imbalance of the given effect. However, accessibility for each individual and the public transit needs of each community are different. Our limitation of this accessibility approach results from its aggregate nature, which reflects general differences in the community level and does not combine the residents' travel demand. In the future, the coordination of the rail transit network and bus transit network could be analyzed further based on the spatial distribution of residents' travel demand and urban economic activities.

This study helps us to understand the influence of the distribution of rail transit on the travel behavior of public transit and more clearly outlines the structure of the public transit system of Guangzhou. The difference in the travel structure hides transport unfairness. The difference in accessibility in different travel modes, to a certain extent, determines the difference in transport equity.

The rail transit supply in Guangzhou is concentrated in the central urban area, which will help solve the problem of urban transport congestion and CBD accessibility, but the distribution of public transit has a low density and there is only one regular bus line in many communities in peripheral areas. In the future, the government could improve the existing public transit services in communities with low-level accessibility by providing regular bus lines and small buses connecting to rail transit stations. Future policies could combine the merits of the reliability and speed of rail transit and the wide coverage of bus transit for the gradual optimization of the public transit network by region. The planning of new rail transit stations and lines should emphasize coordination with the bus transit system and could draw attention to the connection with the public bicycle network, so that the influence range of the rail transit network can be increased from 500 m on foot to 2000 m by bicycle, which encourages more people to choose rail transit as their daily travel mode choice. This paper only analyzed the impact of rail transit on conventional bus travel based on the travel cost. In the future, the coordination of the rail transit network and bus transit network could be analyzed further based on urban expansion, the spatial distribution of residents' travel demand, and urban economic activities.

Author Contributions: Conceptualization, H.C., W.Y. and T.L.; methodology, H.C., W.Y. and T.L.; software, H.C.; validation, W.Y. and T.L.; formal analysis, H.C.; investigation, H.C.; resources, W.Y.; data curation, T.L. and T.L.; writing—original draft preparation, H.C., W.Y. and T.L.; writing—review and editing, H.C., W.Y. and T.L.; visualization, H.C.; supervision, W.Y. and T.L.; project administration, H.C. All authors have read and agreed to the published version of the manuscript.

Funding: This work was supported by the National Natural Science Foundation of China (No. 41701169), Natural Science Foundation of Hunan Province (2021JJ40155), The Science and Technology Program of Guangzhou (202102021041).

Institutional Review Board Statement: Not applicable.

Informed Consent Statement: Not applicable.

Data Availability Statement: Data are available upon request.

Conflicts of Interest: The authors declare no conflict of interest.

References

1. Delmelle, E.C.; Casas, I. Evaluating the spatial equity of bus rapid transit-based accessibility patterns in a developing country: The case of Cali, Colombia. *Transp. Policy* **2012**, *20*, 36–46. [CrossRef]
2. Ermagun, A.; Tilahun, N. Equity of transit accessibility across Chicago. *Transport. Res. D.* **2020**, *86*, 102461. [CrossRef]
3. Dong, H. Rail-transit-induced gentrification and the affordability paradox of TOD. *J. Transp. Geogr.* **2017**, *63*, 1–10. [CrossRef]
4. Pucher, J.; Renne, J.L. Socioeconomics of urban travel: Evidence from the 2001NHTS. *Transp. Q.* **2003**, *5*, 49–77.
5. Benenson, I.; Martens, K.; Rofe, Y.; Kwartler, A. Public transport versus private car GIS-based estimation of accessibility applied to the Tel Aviv rail transitpolitan area. *Ann. Regional Sci.* **2011**, *47*, 499–515. [CrossRef]
6. Yang, W.; Chen, H.L.; Wang, W.L. The path and time efficiency of residents' trips of different purposes with different travel modes: An empirical study in Guangzhou, China. *J. Transp. Geogr.* **2020**, *88*, 102829. [CrossRef]
7. Cheng, Y.; Loo, B.P.Y.; Vickerman, R. High-speed rail networks, economic integration and regional specialization in China and Europe. *Travel Behav. Soc.* **2015**, *2*, 1–14. [CrossRef]
8. Monzón, A.; Ortega, E.; López, E. Efficiency and spatial equity impacts of high-speed rail extensions in urban areas. *Cities* **2013**, *30*, 18–30. [CrossRef]
9. Kim, H.; Sultana, S. The impacts of high-speed rail extensions on accessibility and spatial equity changes in South Korea from 2004 to 2018. *J. Transp. Geogr.* **2015**, *45*, 48–61. [CrossRef]
10. Jin, F.; Jiao, J.; Qi, Y.; Yang, Y. Evolution and geographic effects of high-speed rail in East Asia: An accessibility approach. *J. Geogr. Sci.* **2017**, *27*, 515–532. [CrossRef]
11. Andersson, D.E.; Shyr, O.F.; Fu, J. Does high-speed rail accessibility influence residential property prices? Hedonic estimates from southern Taiwan. *J. Transp. Geogr.* **2010**, *18*, 166–174. [CrossRef]
12. Litman, T. *Evaluating Transportation Equity: Guidance for Incorporating Distributional Impacts in Transportation Planning*; Victoria Transport Policy Institute: Victoria, Australia, 2007.
13. Chen, C.S.; Tu, C.H.; Chen, S.J.; Chen, S.J.; Chen, C.C. Simulation of Groundwater Contaminant Transport at a Decommissioned Landfill Site—A Case Study, Tainan City, Taiwan. *Int. J. Env. Res. Pub. He.* **2016**, *13*, 467. [CrossRef]
14. Yen, B.T.H.; Mulley, C.; Zhang, M. Equity in financing public transport infrastructure: Evaluating funding options. *Transp. Policy* **2020**, *95*, 68–77. [CrossRef]
15. Hansen, W.G. How accessibility shapes land use. *J. Am. Inst. Plan.* **1959**, *25*, 73–76. [CrossRef]
16. Gao, A.; Zhu, F. Study on Influence of High-speed Railways on Accessibility and Economic Ties in Southwest Area. *Railw. Transp. Econ.* **2014**, *36*, 1–5.
17. Ben-Elia, E.; Benenson, I. A spatially-explicit method for analyzing the equity of transit commuters' accessibility. *Transport. Res. A* **2019**, *120*, 31–42. [CrossRef]
18. Zhang, Y.; Huang, J.; Wen, Y. Impact of high-speed railway on spatial pattern of Chinese cities' accessibility. *Sci. Geogr. Sin.* **2015**, *35*, 387–395.
19. Gleason, J.M. A set covering approach to bus stop location. *Omega* **1975**, *3*, 605–608. [CrossRef]
20. Gutiérrez, J.; García-Palomares, J.C. Distance-measure impacts on the calculation of transport service areas using GIS. *Environ. Plan. B* **2008**, *35*, 480–503. [CrossRef]
21. Mavoa, S.; Witten, K.; McCreanor, T.; O'sullivan, D. GIS based destination accessibility via public transit and walking in Auckland, New Zealand. *J. Transp. Geogr.* **2012**, *20*, 15–22. [CrossRef]

22. Fransen, K.; Neutens, T.; Farber, S.; De Maeyer, P.; Deruyter, G.; Witlox, F. Identifying public transport gaps using time-dependent accessibility levels. *J. Transp. Geogr.* **2015**, *48*, 176–187. [CrossRef]
23. Lei, T.L.; Church, R.L. Mapping transit-based access: Integrating GIS, routes and schedules. *Int. J. Geogr. Inf. Sci.* **2010**, *24*, 283–304. [CrossRef]
24. Farber, S.; Morang, M.Z.; Widener, M.J. Temporal variability in transit-based accessibility to supermarkets. *Appl. Geogr.* **2014**, *53*, 149–159. [CrossRef]
25. Owen, A.; Levinson, D.M. Modeling the commute mode share of transit using continuous accessibility to jobs. *Transport. Res. A* **2015**, *74*, 110–122. [CrossRef]
26. Bills, T.S.; Twumasi-Boakye, R.; Broaddus, A.; Fishelson, J. Towards transit equity in Detroit: An assessment of microtransit and its impact on employment accessibility. *Transport. Res. D.* **2022**, *109*, 103341. [CrossRef]
27. Claire, P.; Dupont-Kieffer, A.; Palmier, P. Potential accessibility to the workplace by public transit and its social distribution in Lille, France: A scenario-based equity appraisal. *Transp. Policy* **2022**, *125*, 256–266. [CrossRef]
28. Jiao, J.; Wang, F. Shared mobility and transit-dependent population: A new equity opportunity or issue? *Int. J. Sustain. Transp.* **2021**, *15*, 294–305. [CrossRef]
29. Boisjoly, G.; El-Geneidy, A. daily fluctuations in transit and job availability: A comparative assessment of time-sensitive accessibility measures. *J. Transp. Geogr.* **2016**, *52*, 73–81. [CrossRef]
30. Páez, A.; Scott, D.; Potoglou, D.; Kanaroglou, P.; Newbold, K.B. Elderly mobility: Demographic and spatial analysis of trip making in the Hamilton CMA, Canada. *Urban Stud.* **2007**, *44*, 123–146. [CrossRef]
31. Ford, A.C.; Barr, S.L.; Dawson, R.J.; James, P. Transport accessibility analysis using GIS: Assessing sustainable transport in London. *ISPRS Int. J. Geoinf.* **2015**, *4*, 124–149. [CrossRef]
32. Currie, G. Gap analysis of public transport needs: Measuring spatial distribution of public transport needs and identifying gaps in the quality of public transport provision. *Transp. Res. Rec.* **2004**, *1895*, 137–146. [CrossRef]
33. El-Geneidy, A.; Levinson, D.; Diab, E.; Boisjoly, G.; Verbich, D.; Loong, C. The cost of equity: Assessing transit accessibility and social disparity using total travel cost. *Transport. Res. A.* **2016**, *91*, 302–316. [CrossRef]
34. Yang, Y.; Zhang, P.; Ni, S. Assessment of the Impacts of urban rail transit on rail transitpolitan regions using system dynamics model. *Transp. Res. Procedia* **2014**, *4*, 521–534. [CrossRef]
35. Jiang, H.; Zhang, W.; Qi, Y.; Wu, L. The land accessibility influenced by China's high-speed rail network and travel cost. *Geogr. Res.* **2015**, *34*, 1015–1028.
36. Chien, S.; Schonfeld, P. Joint optimization of a rail transit line and its feeder bus system. *J. Adv. Transport.* **1998**, *32*, 253–284. [CrossRef]
37. Kuan, S.N.; Ong, H.L.; Ng, K.M. Solving the feeder bus network design problem by genetic algorithms and ant colony optimization. *Adv. Eng. Softw.* **2006**, *37*, 351–359. [CrossRef]
38. Catherine, S. Travel behavior change in older travelers: Understanding critical reactions to incidents encountered in public transport. *Int. J. Environ. Res. Public Health* **2015**, *12*, 14741–14763.
39. Kong, Q.J.; Zhao, Q.; Wei, C.; Liu, Y. Efficient traffic state estimation for large-scale urban road networks. *IEEE Trans. Intell. Transp. Syst.* **2013**, *14*, 398–407. [CrossRef]
40. Chen, B.Y.; Yuan, H.; Li, Q.; Lam, W.H.; Shaw, S.L.; Yan, K. Map-matching algorithm for large-scale low-frequency floating car data. *Int. J. Geogr. Inf. Sci.* **2014**, *28*, 22–38. [CrossRef]
41. Yang, W.; Chen, B.Y.; Cao, X.; Li, T.; Li, P. The spatial characteristics and influencing factors of modal accessibility gaps: A case study for Guangzhou, China. *J. Transp. Geogr.* **2017**, *60*, 21–32. [CrossRef]

International Journal of
Environmental Research
and Public Health

Article

Spatial and Temporal Evolution and Driving Factors of Urban Ecological Well-Being Performance in China

Jing Bian [1,2], Feng Lan [1,2,*], Yulin Zhou [1,2], Zhenzhen Peng [1,2] and Mingfang Dong [1]

1 School of Management, Xi'an University of Architecture and Technology, Xi' an 710055, China
2 Research Center of Green Development and Mechanism Innovation of Real Estate Industry in Shaanxi Province, Xi'an University of Architecture and Technology, Xi'an 710055, China
* Correspondence: lanfeng@xauat.edu.cn

Abstract: Extensive development leads to the decline of ecological well-being, and it is necessary to improve the urban ecological well-being performance (EWP). This paper adopted the Super-slack-based measure (Super-SBM) model to evaluate the EWP of 285 Chinese prefecture level cities from 2011 to 2017. The exploratory spatial data analysis method (ESDA) was used to explore the spatial and temporal evolution characteristics of the EWP, and then the spatial Durbin model (SDM) was adopted to analyze the driving factors of the EWP. The results show that the trend of the overall average EWP has experienced a stage evolution process of "upward → downward → upward". The urban EWPs have significant spatial agglomeration and path dependence. The economic development level and technological progress had the positive impacts on the EWP, and the urbanization level, economic extroversion and industrial structure had the negative impacts on the EWP. The result reveals that there was a "U-shaped" relationship existing between urbanization level and the EWP. The negative spatial spillover effect of urbanization level on the EWP was significant. The corresponding policy implications were put forward. This study will provide strategic guidance for policy makers to optimize and enhance the urban EWP.

Keywords: ecological well-being performance; super-SBM model; spatial Durbin model; sustainable development

Citation: Bian, J.; Lan, F.; Zhou, Y.; Peng, Z.; Dong, M. Spatial and Temporal Evolution and Driving Factors of Urban Ecological Well-Being Performance in China. *Int. J. Environ. Res. Public Health* **2022**, *19*, 9996. https://doi.org/10.3390/ijerph19169996

Academic Editors: Hongxiao Liu, Tong Wu and Yuan Li

Received: 5 July 2022
Accepted: 11 August 2022
Published: 13 August 2022

1. Introduction

The rapid urbanization development in China has made significant progress, but it has also faced many challenges. Environmental deterioration, a lack of resources, a shortage of resources and other ecological problems have seriously damaged the ecosystem, which has hindered the promotion of people's well-being and the ecological civilization. The proportion of population who are urban residents will raise to 68% in 2050 [1]. With the rapid increase of global population, the problem of resource consumption has become increasingly serious [2]. As shown in Figure 1, the total energy consumption per capita in China was 3488 kgce in 2019, nearly 5.68 times those of the total consumption in 1980 [3]. Meanwhile, the problems of "urban diseases" appear, for example, as a result of insufficient water supply, urban traffic congestion, low land use efficiency, insufficient public infrastructure construction and so on, which bring severe challenges to urban development. Climate change, ecological environmental damage and other issues are intertwined, affecting people's health, economic and other well-being levels in different ways [4,5].

China has actively been promoting a new type of urbanization for solving serious ecological environment problems and improve people's well-being in recent years. China's urban construction has changed from extensive development to a new type of urbanization. The 14th Five-Year Plan (2021–2025) clearly proposed the goal of achieving new progress in the building of a Beautiful China and a new level of people's livelihood. The economic system is subordinate to the ecological system, rather than an independent and unrestricted system [6]. Ecological well-being is the expansion of the connotation of social well-being, which has raised sustainable development and quality of life to a more important position [7]. Ecological well-being performance (EWP) refers to the efficiency of transforming natural consumption into human well-being [8]. Therefore, it is an inevitable tendency for promoting the EWP and high-quality development under the constraints of ecological resources.

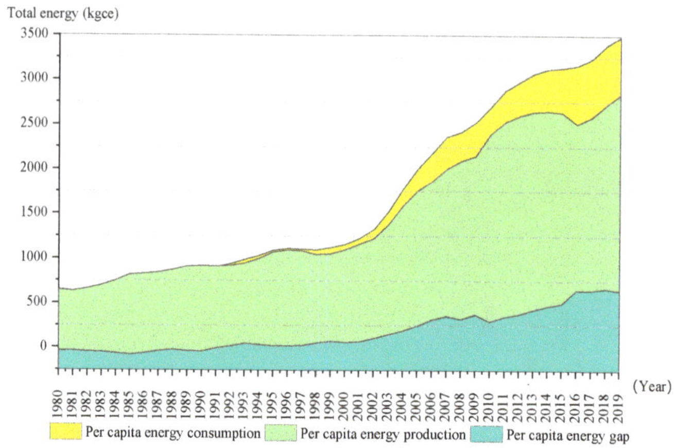

Figure 1. China's per capita energy gap from 1980 to 2019.

The research on the EWP is developing in the continuous global search for a healthy and benign economic development model, and has highly concerned scholars in recent years [9]. Some of the existing literature has carried out comprehensive analysis of the EWP evaluation, space distribution of the EWP and its driving factors. The evaluation methods of the EWP mainly include two types [10]. First is the single-factor evaluation method, and the proportion of well-being level to natural consumption is used to estimate the EWP [11]. Well-being level is usually measured by the Human Development Index (HDI) or average life expectancy, and ecological consumption is usually measured by ecological footprint (EF) [12]. For example, Daly initially used the ratio of natural consumption (service) to well-being level (ecological resource throughput) for calculating the sustainable development level of a national region [13,14]. Abdallah et al. [15] put forward the Happy Planet Index (HPI), which measured the longevity and happiness of life. The other is comprehensive evaluation method, such as Data Envelopment Analysis (DEA) and Stochastic Frontier Analysis (SFA) [16]. For example, Hu et al. used the network DEA model to estimate the EWP, and analyzed the evolution and impact of the urban EWP in the Yangtze River Delta [17]. Xiao et al. adopted the improved SFA model to estimate the EWP of 79 Chinese cities in the Yellow River Basin [18]. Bianchi et al. used the DEA model to assess the comparable development of eco-efficiency and the geographical heterogeneity in the European region between 2006 and 2014 [19]. Moutinho et al. used a two-stage DEA method to measure the eco-efficiency scores of European Union countries from 2008 to 2018 [20].

There are great differences in economic development, resource endowments and energy consumption in different Chinese regions, and different regions have spatial differences in terms of the EWP. For example, Wang et al. explored the spatial and temporal differentiation characteristics of the EWP of 30 provinces in China [21]. Li et al. selected the Yellow River Delta as the research area, and adopted the coupling coordination degree model to research spatiotemporal evolution, and used the geographic weighted regression model to analyze the influencing mechanism of the EWP [22]. Deng et al. used the kernel density estimation method and Markov chain to reveal the spatial disequilibrium and dynamic evolution characteristics of the EWP [23]. Zheng et al. used the exploratory spatial data analysis (ESDA) method to explore the spatial-temporal distribution pattern of the EWP in China [24]. Kounetas et al. used a nonparametric model to estimate patterns of convergence or divergence of environmental performance in U.S. states from 1990 to 2017 [25]. It has become an important direction to carry out the spatial and temporal evolution research of the EWP and simulate the change trend of the EWP.

Many studies have been conducted to explore the driving factors of the EWP. The traditional quantitative methods such as Tobit model, Logarithmic Mean Divisia Index (LMDI) method and other econometric model have been extensively employed to explore driving factors of the EWP. For example, Dietz et al. [26] studied the relationship between human well-being environment intensity and economic growth in 58 countries, and they argued that the result was opposite to Kuznets curve. Jorgenson, et al. [27] analyzed the dynamic relationship between human well-being energy intensity and economic growth in 12 European countries. Behjat and Tarazkar [11] employed the autoregressive distributed lag model to reveal influencing factors of the EWP in Iran from 1994 to 2014, finding that the population growth has a negative effect on the EWP. Silva et al. selected Brazil as a case study, and adopted spatial regression models to explore the impacts of the affluence and income inequality on the environmental intensity of well-being [28]. Ahmed [29] analyzed the driving factors of natural resources abundance, human capital, and urbanization on the ecological footprint in China. Xiao and Zhang [30] used Tobit model to explore the influence mechanism of the EWP, and the study showed that green technology innovation efficiency had a positive impact on the EWP.

By summarizing the existing literatures, the research of the EWP evaluation, space distribution and driving factors are in the stable increase period, and there are relatively rich research results and considerable research foundation. However, there are still some deficiencies in relevant literatures. The EWP study mainly focuses on the national and provincial scales, but pays little attention to the urban EWP. The data of these scales have some limitations in reflecting the internal heterogeneity of regional development, so it is essential to carry out the research on urban EWP. Furthermore, the spatial and temporal evolution characteristics of the EWP has not been sufficiently explored, and the spatial effect of driving factors on the EWP needs to be further considered. The traditional econometric analysis usually ignores the existence of spatial correlation and does not consider whether the EWP has the effects of spatial spillovers, affecting the accuracy of analysis results. Consequently, it is necessary to study the spatial and temporal evolution characteristics of the EWP and analyze the spatial effects of driving factors on the EWP.

Therefore, the aims of this study were: (1) to use the Super-SBM model for estimating the EWP of 285 prefecture-level cities in China, and to propose the ESDA method for exploring the spatial and temporal evolution characteristics of the EWP; (2) to adopt spatial econometric model for analyzing driving factors of the EWP; (3) to design strategies for enhancing the EWP with urban characteristics. The major novelty of this study is summed up as follows. (1) Existing literatures usually ignores the spatial effects of the urban EWP. The ESDA method is employed in this study, which can reveal spatial agglomeration characteristics of the EWP. (2) On the basis of considering temporal heterogeneity and spatial heterogeneity, the spatial panel Durbin model is used to test the spatial spillover effect of the EWP, which further enriches the theoretical research in this field. (3) The EWP presents a new study perspective for urban sustainable construction, and the analyzed results and policy implications are conducive to accelerate the implementation of urban management decisions and promote regional sustainable development.

2. Method and Data

2.1. The Super-SBM Model with Undesirable Outputs

The Super-SBM model is used to evaluate the EWP, which can also break through the shortcomings of traditional DEA. It consists of the following aspects [31,32]:

(1) Assuming that there are n cities (decision-making unit, DMUs) with the X input and Y output matrices. It can be expressed as below:

$$X = (x_{ij}) \in R^{m\times n} > 0,\ Y = (y_{ij}) \in R^{s\times n} > 0$$

Then, the EWP production possibility set (P) is expressed as:

$$P = \{(x,y)|x \geq X\lambda, y \leq Y\lambda, \lambda \geq 0\}$$

where λ is the non-negative vector in R^n.

The DUM (x_0, y_0) can be described as

$$x_0 = X\lambda + s^-$$

$$y_0 = Y\lambda - s^+$$

With $\lambda \geq 0$, $s^- \geq 0$ and $s^+ \geq 0$. The vectors $s^- \in R^m$ and $s^+ \in R^s$ are the input redundancy and insufficient output. Applying s^- and s^+, the EWP ρ is as below:

$$\min\rho = \frac{1 - \frac{1}{m}\sum_{i=1}^{m} s_i^- / x_{io}}{1 + \frac{1}{s}\sum_{i=1}^{s} s_i^+ / y_{io}} \tag{1}$$

$$\text{Subject to} \begin{cases} x_0 = X\lambda + s^- \\ y_0 = Y\lambda - s^+ \\ \lambda \geq 0, s^- \geq 0, s^+ \geq 0 \end{cases} \tag{2}$$

where ρ is the EWP value, $0 < \rho \leq 1$; and λ is the linear coefficient of a DMU; x and y are input and output variables respectively; m and s are the number of input and output indicators respectively.

(2) The Super-SBM model is shown as follows:

$$\delta^* = \min\delta = \frac{\frac{1}{m}\sum_{i=1}^{m} \overline{x_i} / x_{io}}{\frac{1}{s}\sum_{r=1}^{s} \overline{y_r} / y_{ro}} \tag{3}$$

$$\text{Subject to} \begin{cases} \overline{x} \geq \sum\limits_{j=1,\neq 0}^{n} \lambda_j x_j, \\ \overline{y} \leq \sum\limits_{j=1,\neq 0}^{n} \lambda_j y_j, \\ \overline{x} \geq x_o, \overline{y} \leq y_0 \\ \overline{y} \geq 0, \lambda \geq 0 \end{cases} \tag{4}$$

δ^* is the EWP value. The evaluative DMU is relatively effective when $\delta^* \geq 1$. The higher the δ^* is, the higher the urban EWP is. The evaluative DMU is relatively ineffective when $\delta^* < 1$.

2.2. Spatial Autocorrelation Analysis

There are two important steps to build a spatial econometric model. The first step is to test the spatial correlation, and the second step is to select and establish an appropriate spatial econometric model. It is necessary to check whether the selected variable data has spatial dependence. The exploratory spatial data analysis (ESDA) method was used to verify whether the selected samples have spatial autocorrelation. The global Moran's I index is given as follows [33,34]:

$$I = \frac{\sum\limits_{i=1}^{n}\sum\limits_{j=1}^{n} \omega_{ij}(Y_i - \overline{Y})(Y_j - \overline{Y})}{S^2 \sum\limits_{i=1}^{n}\sum\limits_{j=1}^{n} \omega_{ij}} \tag{5}$$

A Moran scatter plot reflects the correlation of the internal structure of global space and the correlation of local space [35]. The local Moran' I index is expressed as:

$$I_i = \frac{(Y_i - \overline{Y})}{S^2} \sum_{j=1}^{n} \omega_{ij}(Y_j - \overline{Y}) \quad (6)$$

where $S^2 = \frac{1}{n}\sum_{i=1}^{n}(Y_i - \overline{Y})^2$; $\overline{Y} = \frac{1}{n}\sum_{i=1}^{n} Y_i$, n is the number of cities; Y_j is the EWP value of city j; ω_{ij} is the spatial weight matrix.

The rules are as follows: the scope of Moran's I statistic is [−1,1]. If Moran's I > 0, it indicates that there is a positive spatial correlation between the research variables, and there is a certain agglomeration phenomenon. It is shown as agglomeration effect of high EWP value among the local city and its adjacent cities ("High-High"), or it is shown as agglomeration effect of low EWP value ("Low-Low"). If Moran's I < 0, it indicates that it is a spatial negative correlation. The spatial characteristics of the local city and its adjacent cities have spatial disparity, and it shows "High-Low" or "Low-High" spatial agglomeration effect [36,37]. If Moran's I is zero, it indicates that the research variables are independent and randomly distributed. When $|Z| > 1.96$, $P < 0.05$, the spatial correlation is significant [38].

2.3. Spatial Econometric Model

Spatial econometric models mainly include spatial the autoregressive model (SAR), spatial error model (SEM) and spatial Durbin model (SDM). When the spatial lag term of explanatory variables affects the explanatory variables, it is necessary to consider the establishment of SDM [39]. The basic modality of spatial econometric models can be described as follows:

$$y = \lambda Wy + X\beta + WX\theta + \mu_i + \delta_i + \varepsilon_{it} \quad (7)$$

$$\varepsilon_{it} = \rho W\varepsilon_{it} + \varphi_{it}, \varphi \sim N(0, \sigma_{it}^2 I_n) \quad (8)$$

where W is the spatial weight matrix, β is the coefficient of the explanatory variable, ρ is the spatial effect coefficient, θ is the impact parameter of the explanatory variable in the adjacent region, μ_i, δ_t are the spatial and temporal effects respectively. ε_{it} is the random disturbance term.

The spatial panel model covers a variety of commonly used models.

(1) $\theta = 0$, $\rho = 0$ and $\lambda \neq 0$, it is simplified as SAR model;
(2) $\theta = 0$, $\lambda = 0$ and $\rho \neq 0$, it is simplified as SEM model;
(3) $\lambda \neq 0$ and $\theta = 0$, it is simplified as SDM model.

Lesage and pace [40] decomposed the impact of explanatory variables in spatial econometric model on explained variables into direct effect, indirect effect and total effect. The direct effect means the effect of explanatory variables on the local cities, and the indirect effect is the effect of explanatory variables on its adjacent cities, namely spatial spillover effect [41,42]. The total effect is the joint effect of the direct effect and indirect effect [43].

2.4. The Integrity Research Method

The integrity research method for this paper is given in Figure 2. First, the index system and driving factors were established through a literature review. Second, the empirical analysis of 285 Chinese prefecture level cities from 2011–2017 was performed, and the Super-SBM model was adopted to evaluate the EWP. Third, the ESDA method was used to explore the spatial temporal evolution characteristics of the EWP. Finally, the SDM was constructed to study the driving factors and spatial spillover effects of the EWP.

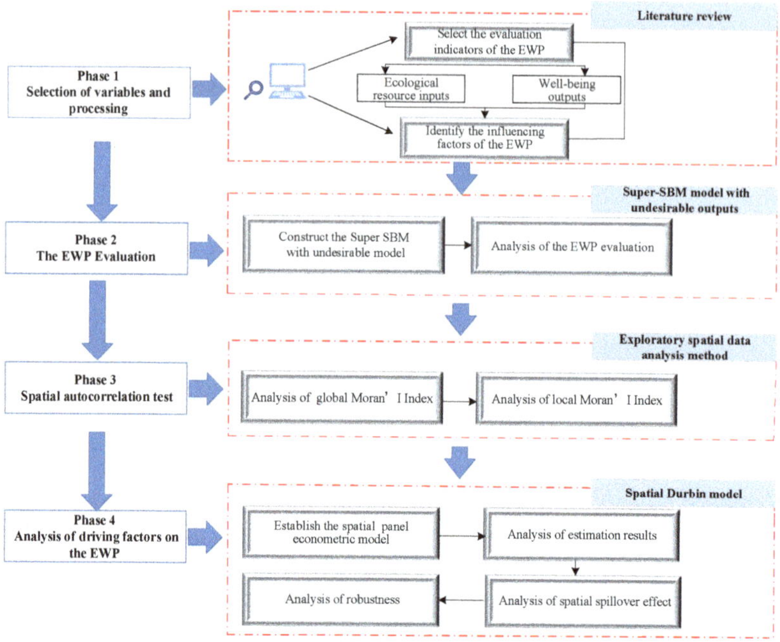

Figure 2. The flowchart of the research methods used in this study.

2.5. Data Collection

According to the data of the National Bureau of statistics, there were data of 298 Chinese prefecture level cities in 2017. Considering availability and accuracy of data, 285 prefecture level cities in China were selected as the research area. Some statistical standards of EWP indicators data changed from 2018 onwards; therefore, the research period of this paper is 2011–2017. The seven years, China has gone through a number of major reform measures, which has a certain theoretical research value. The data are gathered from "China city statistical yearbook 2012–2018" and "China statistical yearbook 2012–2018". Based on the classification of China's economic zones by the National Bureau of statistics, 285 cities are divided into the cities of eastern region, middle region and western region.

3. Selection of Variables and Processing

3.1. The Selection of the EWP Indicators

This paper constructs the EWP index system structure from the perspective of the two aspects of ecological input and well-being output. Ecological footprint is an appropriate indicator to measure natural consumption. However, it is difficult to obtain data of ecological footprint at urban scale. This paper chooses comprehensive input indicators to measure natural consumption, such as energy, water consumption and land consumption. Objective well-being usually reflects people's basic economic and environmental needs [44]. The HDI has become widely recognized, which includes education, medical care and economics dimensions. Therefore, referring to that dimension of HDI, this paper proposed the evaluation indicator structure of the EWP, as shown in Figure 3.

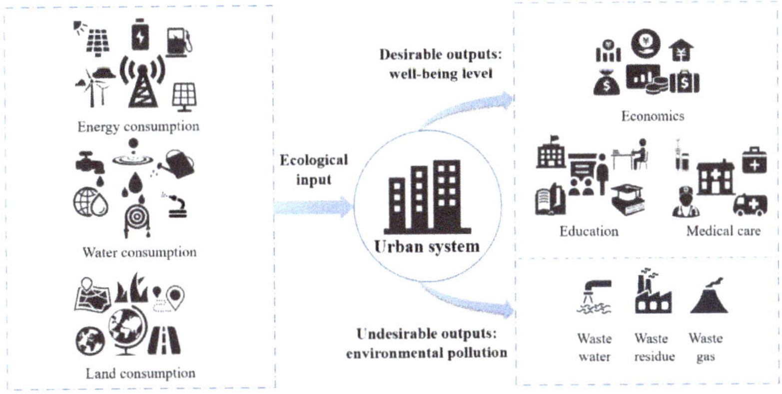

Figure 3. The evaluation indicator structure of the EWP.

3.1.1. The Selection of Input Indicators

Energy consumption. Due to the data of coal, oil and other energy sources at the city level not being available, referring the experience of Zhang et al. [45], the urban electricity consumption (x_1) is selected to measure energy consumption in this paper.

Water consumption. Water resources are not only the basic element of maintaining human life and health, but also the essential resource in production activities. The water consumption of residential use (x_2) is adopted to reflect water consumption [46].

Land consumption. Though the spatial expansion of urbanization needs to use a great deal of land, and the total amount of land resources is limited. The built-up area (x_3) [45] is represented to reflect land consumption in this paper, as shown in Table 1.

Table 1. The indicator descriptions of the EWP evaluation.

Dimension	Criteria	Indicators	Unit
Input indicators	Ecological input	Per capita urban electricity consumption (x_1)	Kwh
		Per capita water consumption of residential use (x_2)	Ton
		Per capita built-up area (x_3)	m^2
Output indicators	Well-being	Per capita GDP (y_1^d)	Yuan
		The number of students enrolled per 10^4 persons (y_2^d)	Person
		The number of doctors per 10^4 persons (y_3^d)	Year
	Environment pollutants	Per capita wastewater discharge (y_1^{ud})	Ton
		Per capita SO_2 (y_2^{ud})	kg
		Per capita soot /dust (y_3^{ud})	kg

Note: In order to negate the influence of inflation, the economic data of the GDP were normalized to that of 2011.

3.1.2. The Selection of Output Indicators

(1) Desirable outputs.

The level of economic development. This is often regarded as an important indicator of the economic situation of a country (region); GDP (y_1^d) [42] is selected to reflect urban economic development level in this study.

The level of education development. The number of students enrolled is widely used to measure educational development level. For instance, Shen and Zhou [47] adopted it to characterize the education development level. Consequently, the number of students enrolled per 10^4 persons (y_2^d) is selected to reflect a city's educational development level.

The level of medical care. Due to the continuous data on per capita life expectancy in Chinese prefecture level cities is difficult to obtain. The number of doctors per 10^4 persons (y_3^d) is used to reflect the medical care level [38].

(2) Undesirable outputs.

Considering the availability of data, wastewater, waste gas and solid waste are included in the EWP index system as undesirable outputs. Wastewater discharge (y_1^{ud}), SO_2 (y_2^{ud}) and soot/dust emission (y_3^{ud}) are selected as undesirable outputs to reflect environmental pollutant indicators [48]. The indicators of this empirical study are the per capita consumption of each sample unit, that is, the total amount of indicators divided by the permanent population.

3.2. *The Selection of the EWP Driving Factors*

3.2.1. Economic Development Level

Economic factors are particularly important for the urban EWP, which reflect the scale or degree of economy development in different periods. For example, Grossman and Krueger analyzed on the relevance of environmental quality and economic development, finding that the relationship between them is an inverted U-shaped curve [49]. Per capita GDP is selected for eliminating the influence of population size [50]. In addition, the square of per capita GDP is introduced for exploring the relationship between the EWP and economic development level.

3.2.2. Urbanization Level

Urbanization level affects the EWP through multiple mechanisms. The main view is that urbanization level promotes urban economic development and improves the energy and resource utilization efficiency, which has a positive effect on the EWP. However, some scholars argued that the low quality of urbanization causes large energy and resource consumption, and increases the pressures of urban environment, which has a negative effect on the EWP [51]. The proportion of urban population to total urban population at city level is selected to measure urbanization level.

3.2.3. Urban Compactness

Moderate urban scale can optimize the comprehensive benefits of a city. It is believed that urban compactness not only means providing more urban space with the least land, but also reflects the pursuit of a better quality of life in the city. In contrast, any scholars argued that strict planning of urban boundaries may be ineffective, and the establishment of the compact city is not an effective policy [52]. Referring the research of Long et al. [53], the population density is chosen to measure urban compactness in this paper.

3.2.4. Industrial Structure

Industrialization has facilitated economic growth. However, it has aggravated environmental pollution due to high pollution and consumption, affecting people's quality of life and health, and hindering the improvement of regional well-being. For example, manufacturing, power, gas and construction cause serious environmental pollution. Inspired by Zhong et al. [54], the proportion of added value of secondary industry to GDP at city level is used for measuring the industrial structure.

3.2.5. Economic Extroversion

The "pollution paradise hypothesis" assumes that developed countries prefer to choose countries with relatively loose environmental regulation requirements to set up pollution-intensive enterprises [55]. In contrast, there is another hypothesis that economic extroversion prompts multinational enterprises to bring original capital, strengthen cooperation and connection of international enterprises through the spillover effect and competition effect. The economic extroversion is estimated by the proportion of foreign direct investment to GDP at city level in this paper.

3.2.6. Urban Greening Level

There are many literatures have studied on urban greening. For example, Jin, et al. [56] explored the relevance between well-being and greening levels in South Korea, finding that greening level has a positive correlation on the well-being. Conversely, some scholars have analyzed the impact of urban greening on air quality, finding that the impact of the greening coverage rate on urban air is not significant [57]. Referring to the experience of Luo, et al. [58], the urban green area is selected for measuring the urban greening level.

3.2.7. Technological Progress

The introduction of advanced technology, the innovation of management levels and improvement of labor quality can generate technology and knowledge spillover, and reduce the emission of environmental pollutants. However, some enterprises with low level of environmental pollution control and relatively low production capacity are not sensitive to technology input, and the cost of technology input forms a "crowding out effect" on enterprise production [59]. The technological progress is defined by the proportion of science and technology investment to GDP at city level in this paper. The driving factors of the EWP are shown as in Table 2.

Table 2. The definition of variables.

Explanatory Variables	Definitions of Variables	Variable Abbreviation	Key References
Economic development level	Per capita GDP (10^4 yuan/person)	PGDP	[60]
	(Per capita GDP)^2	PGS	
Urbanization level	The proportion of urban population to total urban population (%)	UR	[61,62]
Urban compactness	The population density(Person/square kilometer)	PD	[53]
Industrial structure	The proportion of added value of secondary industry to GDP (%)	IS	[54]
Economic extroversion	The proportion of foreign direct investment on GDP (%)	FDI	[63]
Urban greening level	Per capita urban green area (m^2/person)	UG	[64]
Technological progress	The proportion of science and technology investment on GDP (%)	TP	[65]

4. Results and Discussion

4.1. Spatial and Temporal Evolution Characteristics on the EWP

4.1.1. Analysis of the EWP Evaluation

This paper used MaxDEA 8 Ultra software to measure the EWP. The average EWP value was 0.6229, which was relatively low. Among 285 cities, Sanya, Shenzhen and Haikou ranked in the top three in terms of the EWP, and the average values were 1.870, 1.257 and 1.228, respectively. However, the EWP values of Guigang, Yichun and Baishan were 0.126, 0.124 and 0.106, respectively. This may be related to these top EWP performers benefiting from the rapid development of regional economy and the advance of technology.

From Figure 4, it can be seen that the EWP fluctuates every year in China. The average EWPs of eastern, central and western cities were 0.6297, 0.6217 and 0.6158. From the perspective of regional differences, the EWPs show the decreasing pattern of "east-middle-west". Relying on the physical and geographical advantages of coastal areas, the eastern region cities assimilate foreign investment. Most of them are exposed to high-end technology, forming a cumulative effect of economic, educational, medical and other advantages [66]. In contrast, the EWP of western cities is lower than that of other areas. A lot of cities in the western region are geographically faraway, and the transportation and education systems are underdeveloped, so this type of urban construction needs to promote high-quality economic development.

In this study, Arcgis software was applied to study the EWP space distribution of 285 prefecture-level cities in China. Figure 5a–d shows the spatial distribution of urban EWP in 2011, 2014, 2017 and 2011-2017. Figure 5a presents the cities such as Shenzhen, Guangzhou and other eastern cities, mainly distributed throughout the excellent group (EWP ≥ 1) in 2011. Compared with 2011, the number of cities of eastern region at the excellent group decreased in 2014. The number of central and western cities in the worst group (EWP < 0.2) decreased in 2017, indicating that China has made remarkable achievements in new urbanization development since 2014. It can be seen from Figure 5d that from 2011 to 2017, the average EWP values of central and western cities retain at relatively low level. The economic level of these cities is relatively dropped behind, and the limited environmental governance investment, the low levels of education and medical treatment are also the inducing factors. As a result of the unreasonable industrial structure and the relatively backward technical level, the ecological input and well-being output are not proportional.

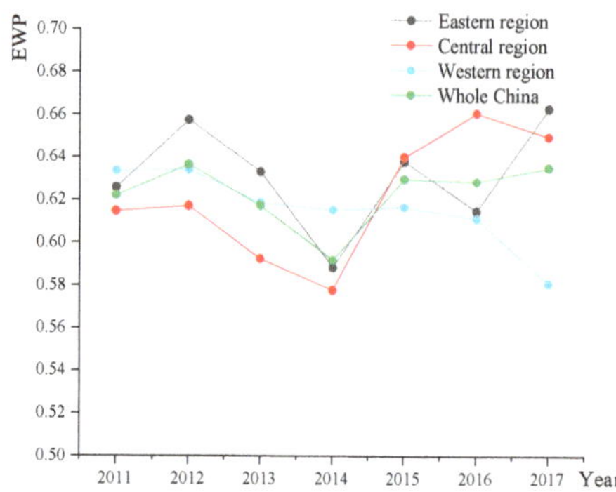

Figure 4. The EWP in Chinese eastern, central, western region and whole China from 2011–2017.

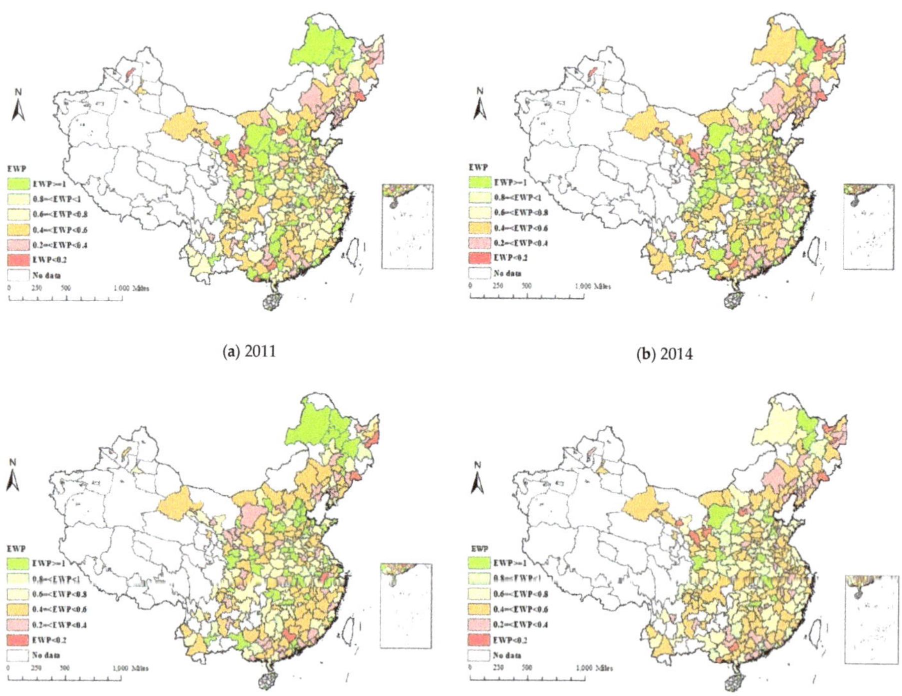

(**a**) 2011 (**b**) 2014

(**c**) 2017 (**d**) Average value of the EWP from 2011–2017

Figure 5. Spatial pattern of the urban EWP in 2011, 2014, 2017 and 2011–2017.

4.1.2. Spatial Autocorrelation Analysis

Spatial autocorrelation analysis was applied to explore the spatial distribution of the EWP in 285 Chinese cities from 2011 to 2017. Geoda software was adopted to analyze Moran's I index of the EWP. The whole of Moran's I values were greater than 0. Table 3 shows that the autocorrelation tendency of the EWP from 2011 to 2017 was roughly "up-down-up".

Table 3. Moran's I statistical values for the EWP of 285 Chinese cities from 2011 to 2017.

Year	Moran's I	*p* Value	Z Value
2011	0.0869	0.0070	2.3557
2012	0.1419	0.0010	3.5734
2013	0.1032	0.0030	2.7583
2014	0.0885	0.0100	2.3945
2015	0.0210	0.2530	0.6357
2016	0.0178	0.2840	0.6055
2017	0.1290	0.0010	3.4757
2011–2017	0.0900	0.0060	2.4520

(1) In 2011–2012, the Moran's I value raised from 0.0869 to 0.1419, and was significant at a 1% level. The Moran's I value was 0.1419, which implied that the positive spatial correlation of the EWP was very significant in 2012.
(2) The spatial correlation of the EWP has experienced a continuous decline from 2013–2016. The agglomeration degree of the EWP began to increase in 2017, indicating that the implementation achievements of new urbanization and relative policies have been gradually effective.

Since there existed spatial agglomeration of the EWP in Chinese prefecture cities, this paper further conducted Moran scatter plots from 2011 and 2017 to analyze the local spatial correlation of the EWP. Figure 6 shows that 151 cities (52.98%) have positive spatial dependence. Among them, 80 cities (28.07%) were located in quadrant I (HH type), 71 cities (24.91%) were located in quadrant III (LL type). The spatial correlation of the remaining 134 cities was not similar, and they accounted for 47.02%. The results show that most cities and their adjacent cities in China have similar spatial cluster characteristics to the EWP. Compared with 2011, the spatial agglomeration distribution characteristics of the EWP in 2014 were more significant. It shows that the green development achievements have been gradually obvious. The results show that 168 cities (58.95% of total number) have positive spatial correlation on the EWP. Among them, 68 cities (23.86%) and 100 cities (35.09%) were HH type and LL type in 2017, and 119 cities (41.05%) were LH type and HL type. It can be seen that Moran scatter plots have the following characteristics:

(1) In 2011, 2014 and 2017, the agglomeration types of the EWP in China was mainly HH type and L-L type. This indicated that the spatial dependence characteristics of the EWP was significant, and most cities and their neighboring cities show similar agglomeration characteristics.
(2) During the study period, HH type mainly concentrates in the eastern region. These cities have advantages in geographical location and resources. The HH types have high spatial agglomeration, and become the growth pole of driving the EWP in surrounding cities.
(3) LL type mainly appears in China's heavy industry, energy base and western remote regions. Some cities in LL types have backward economy and fragile ecological environment, leading to the low value agglomeration of the urban EWP.

The Moran's I index analysis implies that the clustering phenomenon of the urban EWP is significant. The traditional regression model usually ignores the spatial correlation and spatial heterogeneity, and the estimation results are not accurate enough. Therefore, the spatial econometric method was conducted to further test the spatial effects of driving factors on the EWP.

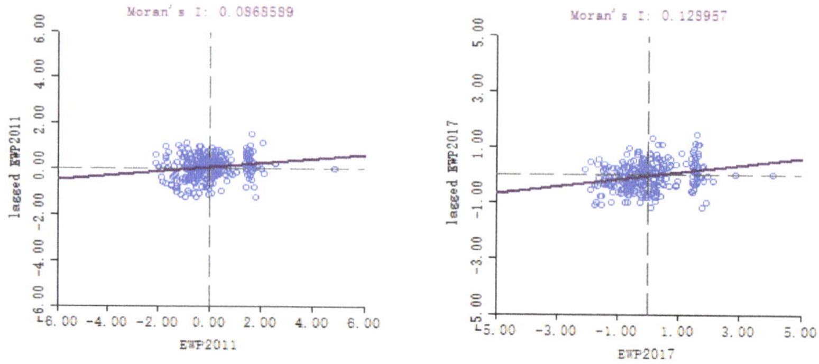

Figure 6. Moran scatter plots of the EWP in 2011 (**left**) and 2017 (**right**).

4.2. Analysis of Driving Factors on the EWP

4.2.1. Selection and Estimation of Spatial Panel Model

First, whether the model has spatial correlation was tested, and the OLS panel model with non-spatial effects was estimated (Table 4). From the result analysis of ordinary panel regression, LM Test and robust LM lag test rejected the original hypothesis of no spatial lag and no spatial error, and both of them were significant at 1% level. Consequently, only using the common OLS regression may cause the imprecise results. The spatial effect panel model was used to make spatial regression analysis, which is expressed as follows:

$$EWP_{it} = \alpha + \beta(PGDP_{it}, PGS_{it}, UR_{it}, \ln PD_{it}, IS_{it}, FDI_{it}, UG_{it}, TP_{it}) \\ +W_{ij}\theta(PGDP_{it}, PGS_{it}, UR_{it}, \ln PD_{it}, IS_{it}, FDI_{it}, UG_{it}, TP_{it}) + \rho W_{ij}EWP_{it} + c_i + \mu_i + \varepsilon_{it} \quad (9)$$

where, EWP_{it} represents the EWP of the i city of the t time, and β is the regression parameters of the driving factors; $W_{i,j}$ is the adjacent space weight matrix, and the first-order spatial weight matrix of rook is selected in this study; ρ is the interaction parameter of the EWP in adjacent cities, θ is the influence parameter of driving factors of adjacent cities on the EWP, and c_i, μ_t are spatial and temporal effects respectively, which are random disturbance terms.

Table 4. The non-spatial panel model test results of the EWP.

Statistic	Mixed Effect	Spatial Fixed	Time Fixed	Double-Fixed
Adjusted R^2	0.1740	0.0508	0.1830	0.0576
σ^2	0.0595	0.0172	0.0586	0.0168
loglikfe	−11.4752	24.7052	2.5676	25.1069
LM Lag	71.1922 ***	53.3217 ***	72.0546 ***	40.2838 ***
Robust LM Lag	39.6311 ***	11.1342 ***	31.5794 ***	6.5575 ***
LM Error	131.0704 ***	65.6392 ***	126.7096 ***	49.2843 ***
Robust LM Error	99.5093 ***	23.4518 ***	86.2344 ***	15.5580 ***

Note: *** indicates significant level of 1%.

The results of SDM, SAR and SEM spatial panel models are as shown in Table 5. LR test and Wald tests were adopted to determine if the SDM could be simplified as SAR or SEM. The Wald test and LR test both rejected the original hypothesis. This implies that the SDM could not be simplified as a SEM and SAR. Hausman test statistics rejected the original hypothesis at 1% significance level test, indicating that the fixed effects should be chosen. The results reveal that the SDM model is more appropriate than SAR and SEM. Therefore, the SDM with double-fixed effects is considered in this paper.

Table 5. The spatial effect panel regression results of the EWP.

Variables	SDM		SAR		SEM	
	Coefficients	**T**	**Coefficients**	**T**	**Coefficients**	**T**
PGDP	0.0174 ***	3.6777	0.0084 **	2.1429	0.0106 **	2.5289
PGS	0.0222 ***	3.6157	0.0249 ***	4.6640	0.0254 ***	4.4866
UR	−0.0041 ***	−4.9031	−0.0043 ***	−5.0739	−0.0042 ***	−4.9428
ln PD	0.0460	0.7145	0.0231	0.3682	0.0343	0.5376
IS	−0.0021 **	−2.0816	−0.0021 **	−2.3939	−0.0022 **	−2.3868
FDI	−0.0028 *	−1.9392	−0.0026 *	−1.8408	−0.0027 *	−1.9149
UG	−0.0005	−2.0442	−0.0005	−2.0198	−0.0006	−2.2200
TP	0.0597 ***	4.9796	0.0564 ***	4.8698	0.0585 ***	4.94709
W× PGDP	−0.0257 ***	−3.110				
W× PGS	−0.0029	−0.2805				
W× UR	−0.0015	−0.8384				
W× ln PD	−0.2228	−1.4155				
W× IS	0.0020	1.1574				
W× FDI	0.0032	1.2116				
W× UG	0.0011 **	2.0626				
W× TP	0.0290	1.3504				
σ^2	0.0188		0.0191		0.0190	
R-squared	0.7764		0.7724		0.7660	
Log-likehood	1282.9770		1265.5207		1269.5204	
Statistical Tests	**Z-Value**	***p*-Value**				
Wald_spatial_lag	30.2098 **	0.0355				
LR_spatial_lag	34.9126 **	0.0187				
Wald_spatial_error	22.9044 ***	0.0035				
LR_spatial_error	26.9133 **	0.0134				
Hausman test	−68.0646 ***	0.0000				

Note: * indicates significant level of 10%; ** indicates significant level of 5%; *** indicates significant level of 1%.

(1) The coefficient of economic development level had a positive impact on the EWP. It implies that economic development still takes a significant part in improving the EWP, and urban economic development is the premise and basis to improve the EWP. Economic development can bring human social benefits and increase the well-being of urban residents.

The coefficient of the square of per capita GDP was positive significantly at a 1% level. There was the U-shaped line between the economic development level and the EWP, proving that the EWP decreases with the increase of economic activity at the initial stage of development, and then the EWP gradually improves with the improvement of economic development.

(2) The coefficient of the urbanization level was −0.0041, and had significantly negative impacts on the EWP. On one hand, it is mainly attributed to the urban crowding caused by rapid transfer of population from rural areas to cities. Moreover, the extensive urbanization not only promotes the economic development, but also causes great pressure on the ecological resources and natural environment. The mismatches between the rapid expansion of urbanization and urban infrastructure, education and medical resources reduce the well-being of residents.

(3) The influence coefficient of population density was 0.046, and it was not significant. It demonstrated that the population density had a positive effect on the EWP, and this observation reflected that of Song, et al. [67]. The compact city can make efficient use of land resources and can provide urban residents with higher quality public infrastructure and municipal services.

(4) The industrial structure had a negative impact on the EWP, and it was significantly at 5% level, which was consistent with the literature of Fan et al. [68]. China's urbanization development is mainly based on industrialization, and the industrial structure is detrimental to the EWP optimization. It is crucial to expedite the transformation and upgrade of industrial structure to improve the EWP.

(5) The coefficient of economic extroversion was significant at 10% level. It shows that China's economic extroversion has an inhibitory effect on the urban EWP, indicating that China's policies for introducing and utilizing foreign capital are relatively loose. Under the

loose policy conditions, enterprises discharge a large number of pollutants, resulting in the reduction of the EWP.

(6) The coefficient of urban greening level was not significant. It indicated that the degree of urban greening had not taken a positive role in ameliorating the EWP, which is inconsistent with expectations. Generally speaking, the higher the urban greening level is, the higher the air quality might be. However, the regression result of urban greening level is not significant, which indicates that urban greening level is not balanced. For example, the amount of greenery in many urban centers and old districts is insufficient, and the overall urban greening level in the western region is lagging behind.

(7) A significant positive-correlation influence was found between the technological progress level and the EWP at 1% level. It implies that promoting the technological progress level is conducive to promote the EWP. China has increased the input of innovation resources, and the ability of independent innovation has been significantly enhanced in the late years. The high level of technological progress produces agglomeration and spillover effects, and improves the urban EWP.

4.2.2. Analysis of Spatial Spillover Effect

The Matlab 2019a software was used to obtain the spillover effect of each variable (Table 6). The direct and indirect effects of per capita GDP were significant at 1% level. This implied that economic development had a positive spillover impact on the local EWP and a negative spillover effect on its adjacent cities. The total effect of per capita GDP was negative correlation and did not pass the significance level test. To some extent, it reflects that the city with higher economic development level may not have a positive effect on the EWP of adjacent cities.

Table 6. The spatial spillover effect of the EWP.

Variables	Direct Effect		Indirect Effect		Total Effect	
	Coefficient	T	Coefficient	T	Coefficient	T
PGDP	0.0167 ***	3.6381	−0.0274 ***	−2.8400	−0.0108	−1.2248
PGS	0.0220 ***	3.7379	0.0021	0.1711	0.0241 *	1.9455
UR	−0.0042 ***	−5.0343	−0.0030	−1.3533	−0.0072 ***	−2.9001
ln PD	0.0381	0.6219	−0.2616	−1.3815	−0.2235	−1.1584
IS	−0.0020 **	−2.0481	0.0020	0.9272	−0.00004	−0.0197
FDI	−0.0030 *	−1.8749	0.0032	0.9610	0.0006	0.1698
UG	−0.0005 *	−1.9544	0.0012 *	1.8744	0.0007	0.9750
TP	0.0203	−5.0581	−0.0595 *	0.7982	−0.0391	−1.4223

Note: * indicates significant level of 10%; ** indicates significant level of 5%; *** indicates significant level of 1%.

The direct effect and total effect of urbanization development were significant at 1% level. It indicates that the low-quality urbanization development level has a negative impact on the local EWP and its adjacent cities. The low-quality urbanization mainly contains excessive of natural consumption and over-expanded urban construction. Urbanization development is a vital factor of the urban EWP.

The direct effect of industrial structure was significant at 5% level, indicating that the industrial structure had a negative effect on the local EWP. Although the increase of the secondary industry promotes the development of a local economy, it also brings environmental pollution, ecological deterioration and other environmental problems, which reduces the resource utilization efficiency and social well-being level. The coefficients of indirect effect and total effect were not significant.

The direct effect of economic extroversion was significant at 10% level, which implied that economic extroversion had a negative spillover effect on the local EWP. The indirect effect and total effect were not significant. It shows that foreign direct investment has caused environmental pollution, but the new technology and management experience have formed a diffusion effect in the adjacent cities.

The direct effect of technological progress was not significant. Its indirect effect coefficient had negative relation to the EWP. It shows that although technological progress promotes the EWP of the local region, it does not promote the EWP of surrounding cities. It has not played a positive role in driving the adjacent cities in the study period. This may be related to the adaptability of independent research and development achievements of some cities in China. The total effect of urban greening level was not significant. The direct effect and indirect spatial spatial of urban compactness on the EWP were not significant, and the multifaceted role of urban compactness was not fully reflected.

4.2.3. Analysis of Robustness

The robustness of SDM is tested by changing spatial weight, and the adjacent space–weight matrix is replaced by the distance–space–weight matrix. The coefficient significances of driving factors on the EWP are basically consistent with Table 7, which further confirms that the results of this study are basically credible. The analysis results of robustness test prove that after changing the spatial weight, the EWP regression result of SDM with the adjacent space weight matrix is robust.

Table 7. The analysis results of robustness test.

Variables	SDM		SLM		SEM	
	Coefficients	T	Coefficients	T	Coefficients	T
PGDP	0.0164 ***	3.3192	0.0074 *	1.8925	0.0095 **	2.2372
PGS	0.0239 ***	3.9582	0.0257 ***	4.7986	0.0263 ***	4.6925
UR	−0.0041 ***	−4.8689	−0.0043 ***	−5.1489	−0.004 ***	−5.0713
ln PD	0.0380	0.5962	0.0298	0.4722	0.0372	0.5850
IS	−0.0015 **	−1.5868	−0.0020 *	−2.2145	−0.0019 **	−2.1001
FDI	−0.0026 *	−1.7957	−0.0026 *	−1.8719	−0.0027 *	−1.8616
UG	−0.0004	−1.6258	−0.0005	−2.0074	−0.0005	−2.0798
TP	0.0577 ***	4.9013	0.0565 ***	4.8537	0.0585 ***	4.9841
W× PGDP	−0.0164 ***	−2.7416				
W× PGS	−0.0030	−0.3334				
W× UR	−0.0009	−0.5931				
W× ln PD	−0.1382	−1.1232				
W× IS	0.0004	−0.2447				
W× FDI	0.0024	0.7403				
W× UG	0.0008 **	1.9851				
W× TP	0.0276	1.5969				
σ^2	0.0190		0.0193		0.0192	
R-squared	0.7738		0.7706		0.7660	
Log-likehood	1274.9153		1260.6148		1263.5959	
Statistical Tests	**Z-Value**	***p*-Value**				
Wald_spatial_lag	24.6256 ***	0.0018				
LR_spatial_lag	28.6011 ***	0.0004				
Wald_spatial_error	19.7626 **	0.0113				
LR_spatial_lag	22.6389 ***	0.0039				
Hausman_ test	−35.2411 ***	0.0058				

Note: * indicates significant level of 10%; ** indicates significant level of 5%; *** indicates significant level of 1%.

5. Conclusions and Policy Implications

5.1. Conclusions

The spatial and temporal evolution characteristics of 285 Chinese prefecture level cities were analyzed from 2011–2017. The SDM was built to reveal the driving factors of the EWP with the spatial panel data. The main conclusions showed that:

First, from the evaluation of the EWP, the average EWP value of 285 Chinese cities was 0.6229, and the efficiency value was relatively low. The trend of overall EWP has experienced a stage evolution process of "upward → downward → upward". The cities of average EWP in eastern and central region rated first and second, and cities in western region ranked third according to the EWPs. Due to the coastal geographical advantages and priority support of policies, the levels of economy, education and medical care in the eastern region are evidently superior.

Second, from the spatial autocorrelation analysis, it is reasonable to conclude that Moran's I index of the EWP during the study period was greater than 0. There was spatial agglomeration of the EWP in 285 Chinese cities, and the spatial effect was significantly positive. The spatial agglomeration types of the EWP were mainly HH type and LL type. The urban EWPs have significant spatial agglomeration and path dependence.

Third, the economic development level and technological progress had significant roles in promoting the EWP. The urbanization level, industrial structure and economic extroversion had negative effects on the EWP. The impacts of urban compactness and urban greening level on the EWP were not significant. The negative spatial spillover effect of urbanization level on the EWP was significant.

Nonetheless this study has some deficiencies, and the topic needs further research. With regard to aspects of the EWP evaluation, there are some limitations in data collection of city scale. When these data can be developed in the future, the evaluation indicators need to be enriched, such as the education development level, medical development level and subjective well-being indicators. The time span of the EWP established in this paper only includes 7 years. If the follow-up research can establish the EWP time series with a time span of 10–20 years, it can further study the spatial and temporal evolution pattern of the EWP scientifically. The driving factors of the EWP only analyze social and economic factors such as economic development level, urbanization level and so on, but not include natural environmental factors such as climate and altitude.

5.2. Policy Implications

First, it is found that there are positive spatial autocorrelation and path dependence of the EWP. The cities in the eastern region have higher economic, educational and medical level, and superior environmental supervision. The EWPs of 285 Chinese prefecture-level cities have obvious spatial correlations. Urban agglomerations in the eastern region should continue to maintain the spatial agglomeration mode of high EWP value, and promote the coupling development of ecology and economy. The regions with low EWP should avoid the diffusion of LL type clusters and improve the ability of independent innovation.

Second, the range of ecological environment and resource-carrying capacity should be promoted. It is therefore recommended that the government solve the problem of agricultural transfer population, improve medical and educational conditions, and enhance the construction of urban infrastructure.

Finally, the technological innovation of ecological environment should be improved. The incentive policies for technological innovation need to be improved, such as tax incentives, innovation subsidies and so on. In addition, the transformation achievements of science and technology in the ecological environment should be strengthened. The scientific and technological support systems for ecological environments need to be built.

Author Contributions: J.B. performed the research framework and wrote the manuscript; F.L. designed the research framework; Y.Z. and Z.P. assisted in analysis data. M.D. contributed the methodology. All authors have read and agreed to the published version of the manuscript.

Funding: This work has been supported by the Natural Science Basic Research Program of Shaanxi (Program No. 2022JQ-733), the National Natural Science Foundation of China (Program No. 71874136), the Humanity and Social Science Youth Foundation of Ministry of Education of China (Program No. 20YJC630239).

Institutional Review Board Statement: Not applicable.

Informed Consent Statement: Informed consent was obtained from all the subjects involved in the study.

Data Availability Statement: Not applicable.

Conflicts of Interest: The authors declare no conflict of interest.

References

1. DESA World Urbanization Prospects—The 2018 Revision. Available online: https://www.un.org/development/desa/publications/2018-revision-of-world-urbanization-prospects.html (accessed on 16 May 2018).
2. Zhang, D.; Xu, J.; Zhang, Y.; Wang, J.; He, S.; Zhou, X. Study on sustainable urbanization literature based on Web of Science, scopus, and China national knowledge infrastructure: A scientometric analysis in CiteSpace. *J. Clean. Prod.* **2020**, *264*, 121537. [CrossRef]
3. National Bureau of Statistics. *China Energy Statistical Yearbook*; China Statistics Press: Beijing, China, 2018.
4. Fu, L.; Ren, Y.; Lu, L.; Chen, H. Relationship between ecosystem services and rural residential well-being in the Xin'an river Basin, China. *Ecol. Indic.* **2022**, *140*, 108997. [CrossRef]
5. Bian, J.; Ren, H.; Liu, P. Evaluation of urban ecological well-being performance in China: A case study of 30 provincial capital cities. *J. Clean. Prod.* **2020**, *254*, 120109. [CrossRef]
6. Daly, H. Economics in a full world. *Eng. Manag. Rev. IEEE* **2005**, *293*, 100–107.
7. Luo, B.L.; Hong, W.J.; Geng, P.P. Empowering people, strengthening capacity and ensuring inclusiveness: Enhancing farmers' subjective well-being in reducing relative poverty. *Manag. World* **2021**, *37*, 18.
8. Zhang, S.; Zhu, D.; Shi, Q.; Cheng, M. Which countries are more ecologically efficient in improving human well-being? An application of the Index of Ecological Well-being Performance. *Resour. Conserv. Recycl.* **2018**, *129*, 112–119. [CrossRef]
9. Liu, L.; Wu, J. Ecosystem services-human wellbeing relationships vary with spatial scales and indicators: The case of China. *Resour. Conserv. Recycl.* **2021**, *172*, 105662. [CrossRef]
10. Hou, J.; Ruan, X.; Lv, J.; Guo, H. Two-Stage Super-Efficiency Slacks-Based Model to Assess China's Ecological Wellbeing. *Int. J. Environ. Res. Public Health* **2020**, *17*, 7045. [CrossRef] [PubMed]
11. Behjat, A.; Tarazkar, M.H. Investigating the factors affecting the ecological well-being performance in Iran from 1994 to 2014. *Environ. Dev. Sustain.* **2021**, *23*, 13871–13889. [CrossRef]
12. Caillon, S.; Cullman, G.; Verschuuren, B.; Sterling, E.J. Moving beyond the human-nature dichotomy through biocultural approaches: Including ecological well-being in resilience indicators. *Ecol. Soc.* **2017**, *22*, 27. [CrossRef]
13. Daly, H. The economics of the steady state. *Am. Econ. Rev.* **1974**, *64*, 15–21.
14. Daly, H. A further critique of growth economics. *Ecol. Econ.* **2013**, *88*, 20–24. [CrossRef]
15. Abdallah, S.; Thompson, S.; Marks, N. Estimating worldwide life satisfaction. *Ecol. Econ.* **2008**, *65*, 35–47. [CrossRef]
16. Zhang, Y.; Mao, Y.; Jiao, L.; Shuai, C.; Zhang, H. Eco-efficiency, eco-technology innovation and eco-well-being performance to improve global sustainable development. *Environ. Impact Assess. Rev.* **2021**, *89*, 106580. [CrossRef]
17. Hu, M.; Sarwar, S.; Li, Z. Spatio-Temporal Differentiation Mode and Threshold Effect of Yangtze River Delta Urban Ecological Well-Being Performance Based on Network DEA. *Sustainability* **2021**, *13*, 4550. [CrossRef]
18. Xiao, L.; Xiao, Q. Study on the Differentiation and Spatial Convergence of Urban Ecological Welfare Performance Patterns in the Yellow River Basin. *Soft Science* **2021**, *35*, 46–53. [CrossRef]
19. Bianchi, M.; del Valle, I.; Tapia, C. Measuring eco-efficiency in European regions: Evidence from a territorial perspective. *J. Clean. Prod.* **2020**, *276*, 123246. [CrossRef]
20. Moutinho, V.; Madaleno, M. A Two-Stage DEA Model to Evaluate the Technical Eco-Efficiency Indicator in the EU Countries. *Int. J. Environ. Res. Public Health* **2021**, *18*, 3038. [CrossRef] [PubMed]
21. Wang, S.; Zhang, Y.; Yao, X. Research on Spatial Unbalance and Influencing Factors of Ecological Well-Being Performance in China. *Int. J. Environ. Res. Public Health* **2021**, *18*, 9299. [CrossRef]
22. Li, J.; Gong, Y.; Jiang, C. Spatio-temporal differentiation and policy optimization of ecological well-being in the Yellow River Delta high-efficiency eco-economic zone. *J. Clean. Prod.* **2022**, *339*, 130717. [CrossRef]
23. Deng, Y.J.; Yang, X.; Chen, G.J. Spatial unbalanced and dynamic evolution of China's ecological welfare performance level. *J. China Univ. Geosci.* **2020**, *20*, 115–127.
24. Zheng, D.F.; Wang, Y.Y.; Liu, X.X.; Jiang, J. Temporal-spatial pattern and potential analysis of China's ecological well-being zone based on ecosystem services. *J. Ecol. Rural Environ.* **2020**, *36*, 645–653.
25. Kounetas, K.E.; Polemis, M.L.; Tzeremes, N.G. Measurement of eco-efficiency and convergence: Evidence from a non-parametric frontier analysis. *Eur. J. Oper. Res.* **2021**, *291*, 365–378. [CrossRef]
26. Dietz, T.; Rosa, E.A.; York, R. Environmentally Efficient Well-Being: Rethinking Sustainability as the Relationship between Human Well-being and Environmental Impacts. *Hum. Ecol. Rev.* **2009**, *16*, 114–123.
27. Jorgenson, A.K.; Alekseyko, A.; Giedraitis, V. Energy consumption, human well-being and economic development in central and eastern European nations: A cautionary tale of sustainability. *Energy Policy* **2014**, *66*, 419–427. [CrossRef]
28. da Silva, J.M.C.; Li, H.; Barbosa, L.C.F. The ecological intensity of human well-being at the local level. *Environ. Sustain. Indic.* **2020**, *8*, 100061. [CrossRef]
29. Ahmed, Z.; Asghar, M.M.; Malik, M.N.; Nawaz, K. Moving towards a sustainable environment: The dynamic linkage between natural resources, human capital, urbanization, economic growth, and ecological footprint in China. *Resour. Policy* **2020**, *67*, 101677. [CrossRef]
30. Xiao, L.; Zhang, X. Spatio-temporal characteristics of coupling coordination between green innovation efficiency and ecological welfare performance under the concept of strong sustainability. *J. Nat. Resour.* **2019**, *34*, 312–324. [CrossRef]
31. Tone, K. A slacks-based measure of efficiency in data envelopment analysis. *Eur. J. Oper. Res.* **2001**, *130*, 498–509. [CrossRef]

32. Tone, K. A slacks-based measure of super-efficiency in data envelopment analysis. *Eur. J. Oper. Res.* **2002**, *143*, 32–41. [CrossRef]
33. Anselin, L. Spatial Econometrics: Methods and Models. *J. Am. Stat. Assoc.* **1990**, *85*, 160.
34. Anselin, L. *Spatial Econometrics: Methods and Models*; Springer: Berlin/Heidelberg, Germany, 1988.
35. Anselin, L. Local Indicators of Spatial Association—LISA. *Geogr. Anal.* **2010**, *27*, 93–115. [CrossRef]
36. Li, L.; Hong, X. Spatial Effects of Energy-Related Carbon Emissions and Environmental Pollution—STIRPAT Durbin Model Based on Energy Intensity and Technology Progress. *J. Ind. Technol. Econ.* **2017**, *36*, 65–72.
37. Zhang, Z.; Li, Y. Coupling coordination and spatiotemporal dynamic evolution between urbanization and geological hazards–A case study from China. *Sci. Total Environ.* **2020**, *728*, 138825. [CrossRef] [PubMed]
38. Li, J.; Li, S. Energy investment, economic growth and carbon emissions in China—Empirical analysis based on spatial Durbin model. *Energy Policy* **2020**, *140*, 111425. [CrossRef]
39. Peng, W.; Yin, Y.; Kuang, C.; Wen, Z.; Kuang, J. Spatial spillover effect of green innovation on economic development quality in China: Evidence from a panel data of 270 prefecture-level and above cities. *Sustain. Cities Soc.* **2021**, *69*, 102863. [CrossRef]
40. Lesage, J.P.; Pace, R.K. *Introduction to Spatial Econometrics*; Chapman and Hall/CRC: Boca Raton, FL, USA, 2009.
41. Wang, H.; Cui, H.; Zhao, Q. Effect of green technology innovation on green total factor productivity in China: Evidence from spatial durbin model analysis. *J. Clean. Prod.* **2021**, *288*, 125624. [CrossRef]
42. Tan, J.; Wang, R. Research on evaluation and influencing factors of regional ecological efficiency from the perspective of carbon neutrality. *J. Environ. Manag.* **2021**, *294*, 113030. [CrossRef]
43. Elhorst, J.P. Matlab Software for Spatial Panels. *Int. Reg. Sci. Rev.* **2014**, *37*, 389–405. [CrossRef]
44. Armitage, D.; Béné, C.; Charles, A.T.; Johnson, D.; Allison, E.H. The Interplay of Well-being and Resilience in Applying a Social-Ecological Perspective. *Ecol. Soc.* **2012**, *17*, 15. [CrossRef]
45. Zhang, Y.; Shen, L.; Shuai, C.; Bian, J.; Zhu, M.; Tan, Y.; Ye, G. How is the environmental efficiency in the process of dramatic economic development in the Chinese cities? *Ecol. Indic.* **2019**, *98*, 349–362. [CrossRef]
46. Wang, R.; Feng, Y. Research on China's Ecological Welfare Performance Evaluation and Improvement Path from the Perspective of High-Quality Development. *Math. Probl. Eng.* **2020**, *2020*, 5476089. [CrossRef]
47. Shen, L.; Zhou, J. Examining the effectiveness of indicators for guiding sustainable urbanization in China. *Habitat Int.* **2014**, *44*, 111–120. [CrossRef]
48. Bian, J.; Zhang, Y.; Shuai, C.; Shen, L.; Ren, H.; Wang, Y. Have cities effectively improved ecological well-being performance? Empirical analysis of 278 Chinese cities. *J. Clean. Prod.* **2019**, *245*, 118913. [CrossRef]
49. Grossman, G.M.; Krueger, A.B. *Environmental Impacts of a North American Free Trade Agreement*; National Bureau of Economic Research: Cambridge, MA, USA, 1992; Volume 8, pp. 223–250.
50. Feng, Y.; Zhong, S.; Li, Q.; Zhao, X.; Dong, X. Ecological well-being performance growth in China (1994–2014): From perspectives of industrial structure green adjustment and green total factor productivity. *J. Clean. Prod.* **2019**, *236*, 117556. [CrossRef]
51. Bai, Y.; Deng, X.; Jiang, S.; Zhang, Q.; Wang, Z. Exploring the relationship between urbanization and urban eco-efficiency: Evidence from prefecture-level cities in China. *J. Clean. Prod.* **2018**, *195*, 1487–1496. [CrossRef]
52. Anas, A.; Rhee, H.J. When are urban growth boundaries not second-best policies to congestion tolls? *J. Urban Econ.* **2007**, *61*, 263–286. [CrossRef]
53. Long, L.; Wang, X.; Guo, B. Evaluation of urban ecological well-being performance based on revised DEA Model—A case study of 35 major cities in China. *J. Nat. Resour.* **2017**, *32*, 595–605.
54. Zhong, Z.; Peng, B.; Xu, L.; Andrews, A.; Elahi, E. Analysis of regional energy economic efficiency and its influencing factors: A case study of Yangtze river urban agglomeration. *Sustain. Energy Technol. Assess.* **2020**, *41*, 100784. [CrossRef]
55. Copeland, B.R.; Taylor, M.S. North-South Trade and the Environment. *Q. J. Econ.* **1994**, *109*, 755–787. [CrossRef]
56. Jin, H.Y.; Kwon, Y.; Yoo, S.; Yim, D.H.; Han, S. Can urban greening using abandoned places promote citizens' wellbeing? Case in Daegu City, South Korea. *Urban For. Urban Green.* **2021**, *57*, 126956. [CrossRef]
57. Yang, S.; Ma, S. Influences on air quality by urban economic growth: Based on the panal data from provincial cities. *Urban Probl.* **2015**, *12*, 4–11. [CrossRef]
58. Luo, Z.; Zhang, J.; Liu, Y.; Zhu, L. Spatiotemporal evolution of urban greening in China and the affecting factors between 2000 and 2017. *Arid Land Geogr.* **2020**, *43*, 481–490.
59. Fang, S.; Xiao, Q. Research on regional ecological well-being performance and spatial effect in China. *China Popul. Resour. Environ.* **2019**, *29*, 1–10.
60. Zhang, W.; Wang, M.Y. Spatial-temporal characteristics and determinants of land urbanization quality in China: Evidence from 285 prefecture-level cities. *Sustain. Cities Soc.* **2018**, *38*, 70–79. [CrossRef]
61. Wang, Y.; Zhao, T.; Wang, J.; Guo, F.; Kan, X.; Yuan, R. Spatial analysis on carbon emission abatement capacity at provincial level in China from 1997 to 2014: An empirical study based on SDM model. *Atmos. Pollut. Res.* **2019**, *10*, 97–104. [CrossRef]
62. Zhang, Y.; Fu, Y.; Kong, X.; Zhang, F. Prefecture-level city shrinkage on the regional dimension in China: Spatiotemporal change and internal relations. *Sustain. Cities Soc.* **2019**, *47*, 101490. [CrossRef]
63. Zhao, P.-J.; Zeng, L.-E.; Lu, H.-Y.; Zhou, Y.; Hu, H.-Y.; Wei, X.-Y. Green economic efficiency and its influencing factors in China from 2008 to 2017: Based on the super-SBM model with undesirable outputs and spatial Dubin model. *Sci. Total Environ.* **2020**, *741*, 140026. [CrossRef]

64. Wu, Z.; Xing, G.; Huang, D. Performance evaluation and influencing factors of urban ecological welfare in Huaihe economic belt based on SBM-Tobit model. *J. Liaoning Univ. Technol.* **2020**, *40*, 173–176.
65. Zhou, Y.; Kong, Y.; Sha, J.; Wang, H. The role of industrial structure upgrades in eco-efficiency evolution: Spatial correlation and spillover effects. *Sci. Total Environ.* **2019**, *687*, 1327–1336. [CrossRef]
66. Zhang, J.; Zeng, W.; Wang, J.; Yang, F.; Jiang, H. Regional low-carbon economy efficiency in China: Analysis based on the Super-SBM model with CO_2 emissions. *J. Clean. Prod.* **2017**, *163*, 202–211. [CrossRef]
67. Song, M.; Song, Y.; An, Q.; Yu, H. Review of environmental efficiency and its influencing factors in China: 1998–2009. *Renew. Sustain. Energy Rev.* **2013**, *20*, 8–14. [CrossRef]
68. Fan, Y.; Bai, B.; Qiao, Q.; Kang, P.; Zhang, Y.; Guo, J. Study on eco-efficiency of industrial parks in China based on data envelopment analysis. *J. Environ. Manag.* **2017**, *192*, 107–115. [CrossRef] [PubMed]

MDPI
St. Alban-Anlage 66
4052 Basel
Switzerland
Tel. +41 61 683 77 34
Fax +41 61 302 89 18
www.mdpi.com

International Journal of Environmental Research and Public Health Editorial Office
E-mail: ijerph@mdpi.com
www.mdpi.com/journal/ijerph

www.ingramcontent.com/pod-product-compliance
Lightning Source LLC
Chambersburg PA
CBHW061844210926
51850CB00006B/968
9783036583518